LET THERE

The Story of Light
from Atoms to Galaxies

LET THERE BE
Light

The Story of Light
from Atoms to Galaxies

Alex Montwill & Ann Breslin

University College Dublin, Ireland

Imperial College Press

ICP

Published by

Imperial College Press
57 Shelton Street
Covent Garden
London WC2H 9HE

Distributed by

World Scientific Publishing Co. Pte. Ltd.
5 Toh Tuck Link, Singapore 596224
USA office: 27 Warren Street, Suite 401-402, Hackensack, NJ 07601
UK office: 57 Shelton Street, Covent Garden, London WC2H 9HE

British Library Cataloguing-in-Publication Data
A catalogue record for this book is available from the British Library.

Cover image: Apollo Sunrise.
Credit: Apollo 12 Crew, NASA.
Source: http://apod.nasa.gov/apod/ap960620.html

LET THERE BE LIGHT
The Story of Light from Atoms to Galaxies

ISBN-13 978-1-86094-850-3
ISBN-10 1-86094-850-2
ISBN-13 978-1-84816-328-7 (pbk)
ISBN-10 1-84816-328-2 (pbk)

Typeset by Stallion Press
Email: enquiries@stallionpress.com

Printed by Fuisland Offset Printing (S) Pte Ltd. Singapore

*This book is dedicated to
Ann and Liam,
our life partners.*

Preface

Physics is a complex subject. Students meet many concepts and phenomena; mass and energy, electric charge and magnetism, light and heat, atoms and molecules, stars and galaxies, to name just a few. Much time and effort is devoted to learning experimental methods and new mathematical techniques. There is often little opportunity to stand back and take an over-all view.

As the overall picture begins to unfold and the various parts of the jigsaw come together, a deeper understanding develops. Apparently unrelated phenomena are seen to be different aspects of the same thing. It emerges that the many laws and formulae are derived from a small number of basic fundamental principles. The steps are logical and comprehensible, but often very subtle, and it is in following these steps that we discover the beauty and fascination of the subject.

As is implied by the title, the book focuses on light, or more generally, electromagnetic radiation. Many of the properties of light derive from the most fundamental principles and laws of physics. Fermat's principle of least time leads to the laws of reflection and refraction. Maxwell's logic leads to the propagation of electromagnetic radiation. Einstein derived his equation $E = mc^2$ by a logical path from the hypothesis of symmetry of empty space and the constancy of the speed of light in vacuum. Planck's discovery of light quanta leads to the apparent paradoxes of quantum mechanics! Nature, it would seem, is at the same time both comprehensible and incomprehensible.

The book is written in textbook style, at a level somewhere between the rigorous and the popular. We feel that it is suitable as background reading for third level students and may also be enjoyed by readers who have an interest in science and are comfortable with basic mathematics. Wherever possible, mathematical derivations are given in appendices rather than inserted into the main body of the text. These appendices may be consulted by readers who wish to delve more deeply into the subject.

Many of the topics in the book formed part of a number of series of science slots on Irish National radio (RTE1). Each series consisted of about 20 ten-minute slots on a common theme under titles such as *'From Greeks to Quarks'*, *'Forces at Work'*, *The Mind Laboratory'*, *'Portraits in Physics'*, *'Street Science'* and *'Letters from the Past'*. These programs continued over a period of almost 10 years attests to the fact that there were people out there listening!

To put the physics into a historical context and to show the human side of some of the 'main players' in the story, most chapters conclude with a biographical sketch. These sketches are, in the main, anecdotal and light hearted.

Chapter 1 Introducing Light.   We give a historical synopsis of the perception of light through the ages, and the discovery of some extraordinary properties in the last two centuries. Sometimes light behaves as a wave, and sometimes as a stream of particles. When Max Planck, Niels Bohr and others looked more deeply into the puzzle, they opened a Pandora's box full of secrets of Nature.

Chapter 2 Geometrical Optics — Reflection.   The path taken by light is always the one which takes the shortest time. This law (Fermat' principle), is applied to the change of direction of light when it is reflected. *Historical figure: Pierre de Fermat.*

Chapter 3 Geometrical Optics — Refraction. When light comes to the boundary between two media it has two choices; it can be reflected at the boundary or transmitted into the second medium. Fermat's law still applies, but such 'freedom of choice' gives a hint of non-predictability in the laws of Nature. We look at practical applications of refraction in lenses and treat the human eye as an optical instrument.

Chapter 4 Light from Afar — Astronomy. Starlight reaches us after a long, journey across the universe. The ancient astronomers showed remarkable ingenuity in their deductions on the size and distance from the earth of the moon and sun. Equally impressive were the measurements and calculations made in the Middle Ages, considering that they were carried out by observers on the earth which is both rotating and in orbital motion. *Historical figure: Galileo Galilei.*

Chapter 5 Light from the Past — Astrophysics. The laws of physics that apply on earth are seen to apply throughout the universe and do not appear to have changed with time. We can retrace the steps back to the birth of the universe. *Historical figure: Isaac Newton.*

Chapter 6 Introducing Waves. We look at the characteristics of waves in general. *Historical figure: Jean Baptiste Fourier.*

Chapter 7 Sound Waves. To most of us, the properties of sound are more familiar than those of light. Sound can be reflected, can bend around corners, can interfere constructively, and can resonate. This chapter draws on the wealth of human experience with sound to highlight the properties of waves. *Historical Interlude — The 'Sound Barrier'.*

Chapter 8 Light as a Wave. Armed with the knowledge of what we can expect from a wave we examine the wave properties of light, a very special wave which can exist in empty space independent of a material medium. Under certain conditions, light + light = darkness! *Historical figure: Thomas Young.*

Chapter 9 Making Images.   From photographs to holograms.

Chapter 10 There was Electricity, There was Magnetism, and Then There was Light.   Electric charges mysteriously attract and repel one another at a distance. New forces appear when these charges are in motion. It took the genius of Maxwell to combine the experimental laws of electromagnetism to predict the propagation of electromagnetic waves. *Historical figure: James Clerk Maxwell.*

Chapter 11 'Atoms of Light' — The Birth of Quantum Theory. Just when it seems that all is understood, a small problem with the spectrum of light emitted by hot surfaces turns out to be a symptom of something big. Max Planck put his head on the block by suggesting nature is not continuous... *Historical figure: Max Planck.*

Chapter 12 The Development of Quantum Mechanics.   The consequences of Planck's hypothesis are studied from different angles. Surprisingly completely different mathematical approaches lead to the same final conclusions. Quantum mechanics leads to a new philosophy and physics changes beyond all recognition. *Historical figure: Niels Bohr.*

Chapter 13 Atoms of Light Acting as Particles.   The evidence that light is like a stream of particles is conclusive. These 'Atoms of Light' are called *photons. Historical figure: Robert A. Millikan.*

Chapter 14 Atoms of Light Behaving as Waves.   Nature seems to be playing a trick. Just when we have accepted photons as particles, and looked at individual photons one by one, they behave not as particles, but as waves! *Historical figure: Richard Feynman.*

Chapter 15   Relativity. Part 1: How It Began. We begin with the hypothesis of symmetry of space and time. Empty space has no preferred places, there is no universal clock. The speed of light is the same for everybody. Following logical steps of Einstein and others, we arrive at the startling result concerning the nature of time, contrary to natural instinct. *Historical figure: Hendrik A. Lorentz.*

Chapter 16 Relativity. Part 2: Verifiable Predictions. Continuing in the logical steps of the previous chapter, we arrive at a prediction with great practical consequences. Matter and energy are equivalent and related by the equation $E = mc^2$. From the hypotheses of symmetry of space and constancy of the speed of light comes the prediction that matter can be converted into energy and energy into matter. *Historical figure: Albert Einstein.*

Chapter 17 Epilogue. The creation of matter out of energy revealed a whole new world of fundamental particles, including the theoretically predicted or *'heavy atom of light'*, the W particle.

Acknowledgements

We wish to thank our colleagues in the UCD School of Physics for their help and support. Khalil Hajim, Sé O'Connor and Eoin O'Mongáin gave us valuable advice on various parts of the manuscript. Gerry O'Sullivan provided us with the facilities of the School while David Fegan, Alison Hackett and John Quinn helped in many ways.

Special thanks are due to John White for introducing us to Adobe Illustrator; without his help it would not have been possible to produce the diagrams in this book. John drew many of the diagrams and undertook the numerical calculation of graphs and field diagrams.

We are grateful to Tim Lehane who painstakingly proof read the entire text. James Ellis provided expert technical support including the production of the holograms and a number of other images. Bairbre Fox, Marian Hanson and Catherine Handley were always there to help us when needed.

We received valuable assistance from *Magicosoftware.com* and from Vincent Hoban of the Audio-Visual centre at UCD. Eugene O'Sullivan of John J Mcdonald and Co. gave us free legal advice.

We would like to thank Lizzie Bennett of Imperial College Press who introduced us to the world of publishing and Chee-Hok Lim and Alex Quek of World Scientific with whom we enjoyed a very fruitful working relationship.

We are indebted to all the people who gave us experimental data, high resolution images and copyright permissions. These are acknowledged individually in the text.

Contents

Chapter 1

Introducing Light

Light plays a central role in our lives. It is the universal messenger which enables us to be aware of the objects around us and of the rest of the universe. Without light we would not receive the life-giving energy from the sun. Much more than that, light or *electromagnetic radiation* is at the centre of the physical laws. Without it the universe as we have come to know it would simply not exist!

Visible light forms only a tiny part of the *electromagnetic spectrum*. Our eyes are sensitive to a certain range of *wavelengths* of that spectrum, but not to gamma rays, X-rays, radio waves, and infrared and ultraviolet radiation.

Light travels at a speed which is almost beyond our imagination. In this chapter we describe the early methods of measuring that speed. We also discuss the wonderful process of vision, how our eyes can distinguish colour and our brains can reconstruct an image.

The remainder of this chapter gives a preview of the rest of the book. The story is an exciting one, full of the unexpected, teaching us that we must accept Nature as it is, not as we think it should be.

A major surprise came in the year 1900, when Max Planck proposed that light can only have certain *quantised* values of energy — a precursor of the extraordinary property of the *duality* of light. This means that it has apparently contradictory attributes: sometimes it behaves as a *particle*, and at other times it behaves as a *wave*. This property of light was the first clue to the very basic *quantum* laws of Nature, which were revealed when Niels Bohr and his collaborators probed into the 'world of the very small'.

1

1.1 The perception of light through the ages

Philosophers throughout the ages have struggled to explain exactly what light is and why it behaves as it does. It was not always realised that we see luminous objects, such as candles and the sun, because they emit light and the eye receives that light. We also see many other objects, such as the moon, trees, and each other, simply because the light from a luminous object like the sun is reflected from them. We can gaze into each other's eyes not because they are luminous, but because they reflect light which originally came from the sun and perhaps, on the way, had been reflected by the moon!

1.1.1 *The ancient Greeks*

The Greek philosophers from as early as Pythagoras (*c.* 582 BC–*c.* 497 BC) believed that light came from 'visible' things and that our eyes received the tiny *particles* of light. The philosopher and statesman Empedocles (5th century BC), originator of the idea of four elements — earth, air, fire and water (and two moving forces, love and strife) — also made a number of assertions about light. He believed that light came from luminous objects but that light rays also came out from the eyes. In addition, he proposed that light travels at a finite speed.

The Greek mathematician Euclid (*c.* 325 BC–*c.* 265 BC), perhaps better known for his works on geometry, is also believed to have thought that the eyes send out rays of light and that this gives the sensation of vision. Euclid studied mirrors, and the law of reflection is stated in a book entitled *Catoptrics*, thought to have been written by him in the 3rd century BC.

1.1.2 *The Middle Ages*

Ibn Al-Haitham (965–1040) did not accept the theory that objects are seen by rays emanating from the eyes and maintained that light rays originate at the objects of vision. He studied the

Ibn Al-Haitham. *Courtesy of The Pakistan Academy of Science.*

passage of light through various media and carried out experiments on the refraction of light as it crossed the boundary between two media. He became known as 'the father of modern optics' and was the author of many books — one of the best-known, *Kitab Al-Manathr*, was translated into Latin in the Middle Ages. It speculated on the physical nature of light, described accurately the various parts of the eye, and was the first to give a scientific explanation of the process of vision. This was a monumental work, based on experiment rather than dogmatism.

Rene Descartes (1596–1650) considered light as a sort of pressure transmitted through a mysterious elastic medium called the ether, which filled all space. The remarkable diversity of colours was attributed to rotary motions of the ether.

Galileo Galilei (1564–1642) developed the experimental method and prepared the way for a proper investigation of the properties of light. The transmission of light had been thought to be instantaneous but Galileo tried to measure the speed of light by putting two people on hills separated by about a mile. One opened a lantern and the other raised his hand when he saw the light. No time difference was detected, which is not surprising since the time interval, based on the currently accepted speed of light, would have been about five microseconds. (There are one million microseconds in one second.)

The *law of reflection* was known to the ancient Greeks. To put it simply, it says that light is reflected from a surface at an angle which is symmetrically opposite to the angle at which it came in.

The *law of refraction* was discovered experimentally in 1621 by the Dutch mathematician Willebrord Snell (1580–1626). It deals with what happens when light goes from one medium into another. Snell died in 1626, without publishing his result. The first mention of it appeared in the *Dioptrique* by Rene Descartes, without reference to Snell, but it is generally believed that Descartes had in fact seen Snell's unpublished manuscript.

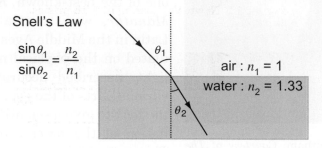

Snell's Law

$$\frac{\sin\theta_1}{\sin\theta_2} = \frac{n_2}{n_1}$$

air : $n_1 = 1$

water : $n_2 = 1.33$

n_1 and n_2 are called *refractive indices* and are properties of the respective media, while the angles θ_1 and θ_2 are as indicated in the diagram. Note the bending of the light as it travels from one medium to another.

The laws of reflection and refraction are the basis of the whole of *geometrical optics* and form the subject matter of Chapters 2 and 3. Both these laws can, in turn, be derived from an even more fundamental law — discovered by the French mathematician Pierre de Fermat (1601–1665) — formulated as the *principle of least time*. (For a biographical note on Fermat see 'A Historical Interlude' at the end of the next chapter.)

1.2 Colours

1.2.1 *The visible spectrum*

In 1666, Isaac Newton showed that white light is made up of a continuous spectrum of colours, from red to orange, yellow, green and finally to blue, indigo and violet. He passed a beam of sunlight through a prism, and saw it fan out into its constituent colours. By putting a piece of paper on the far side of

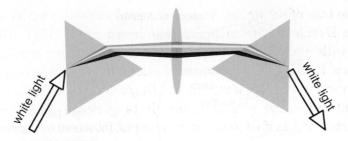

Figure 1.1 Newton's experiment with prisms.

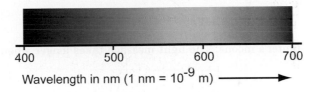

Wavelength in nm (1 nm = 10^{-9} m) ⟶

Visible part of electromagnetic spectum.

the prism, he was able to look at 'individual' colours. He was able to recreate white light by bringing the colours together again using a second prism.

Figure 1.1 is a schematic representation, in that one would not normally see the spectral colours by looking at the beam from the side. In addition, principally owing to the finite width of the incoming beam, it is not possible to recombine the colours completely. In practice the final image is white in the centre with a combination of colours on each side.

1.3 Measuring the speed of light

1.3.1 *The astronomical method*

In 1676, the Danish mathematician Olaus Römer (1644–1710) found that eclipses of Jupiter's moons do not occur at the times predicted by Newtonian mechanics. They are about 11 minutes too early when Jupiter is closest to the earth and about 11 minutes too late when it is furthest away. Römer concluded that the

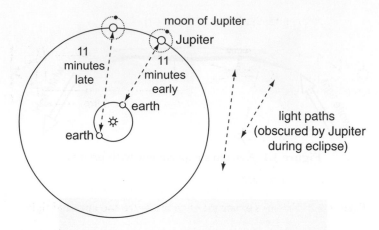

Figure 1.2 Jupiter's moons. The light message takes longer when Jupiter is further away.

discrepancy occurs because light takes longer to travel the larger distance (as indicated in Figure 1.2), and on the basis of the measured time difference of about 22 minutes, he calculated the speed of light to be $2.14 \times 10^8 \, \text{ms}^{-1}$. Although not a particularly good estimate in modern times, this value is certainly of the right order of magnitude and a remarkable achievement at the time.

1.3.2 *Terrestrial measurement*

In 1849, the French physicist Hyppolyte Fizeau (1819–1896) made the first terrestrial measurement of the speed of light, in a simple but ingenious way. A beam of light was passed through one of 720 notches around the edge of a rotating wheel, was reflected by a mirror and retraced its path, as shown in Figure 1.3. When the returning light passed through a notch, an observer could detect it; if it hit the disc between notches, the light was eclipsed. The 'round-trip' distance from the open

Hyppolyte Fizeau

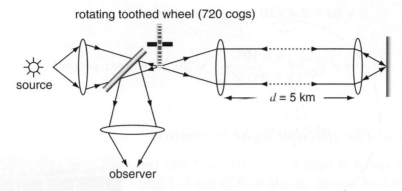

rotating toothed wheel (720 cogs)

source

observer

$d = 5$ km

Figure 1.3 Fizeau's experiment to measure the speed of light.

notch to the mirror and back to the edge of the disc was measured. Fizeau timed the eclipses and measured the rotational speeds of the disc at the time of the eclipses. With this information he calculated the speed of light in air and obtained a value only about 4% different from the currently accepted value of 299,792,458 ms^{-1}.

Fizeau demonstrated considerable craftsmanship when he constructed the 720-cog wheel with the light focused accurately to pass through the gaps!

We can estimate the speed at which Fizeau needed to rotate his wheel as follows:

This wheel has only 70 cogs

Suppose that the distance $d = 5$ km. How fast must the wheel rotate so that a tooth has replaced a neighbouring gap by the time light which has passed through the gap has returned from its 10 km back-and-forth journey? (Assume that the speed of light $c = 3 \times 10^8$ ms^{-1}.)

Remembering that there are 720 teeth and 720 gaps, light must cover the back-and-forth journey 1440 times during each revolution of the wheel. This has to be repeated n times every second.

$$c = 2d \times 2 \times 720 \times n$$

$$\Rightarrow n = \frac{3 \times 10^8}{4 \times 5000 \times 72} = 20.2 \text{ revs per second}$$

1.3.3 *The speed of light in context*

The speed of light c is 3×10^8 ms^{-1} and the speed of sound in air is 330 ms^{-1}. *Light travels about 1 million times faster than sound!*

The reaction time of an athlete to the start signal is about 0.3 seconds. In that time *sound* will travel a distance of about 100 metres, which means that it will have just about reached the finishing line of the 100 m sprint.

By comparison, *light* will have gone 100,000 km, or 2.5 times around the world!

Jesse Owens in Berlin 1986 Olympic Games.

In the everyday world the speed of light can be considered almost infinite. As we look at a landscape, light reaches us from different objects practically instantaneously. There is no appreciable delay between the light reaching us from a tree in the garden, and from the top of a mountain on the horizon. Radio waves and telephone messages reach us within a fraction of a second from the most distant places in the world. Light from stars and distant galaxies, however, may take billions of years to reach us.

The constant c is one of the fundamental constants of our universe, and nobody can tell why it has the value it has. It is interesting to speculate how different the laws of physics would be if the constant c had another value, particularly if it were much smaller — say, of the order of the speed of sound. Obviously the speed of communication would be reduced; it would take hours for news to reach us from other parts of the world. Travel by jet plane would not be safe, because we would not know what was ahead of us! These changes, however, are insignificant compared to the fundamental differences in space and time which would exist in such a 'slow light' universe. The features of *Einstein's theory of relativity* would now be part of the everyday world.

1.4 The process of vision

1.4.1 *'Look and see'*

We open the curtains and see the landscape. Grass, trees, mountains and perhaps some people. An 'ordinary' occurrence, unremarkable for those fortunate enough to be able to see. But what exactly happens in the process of vision? How are we made aware of distant objects? A picture is formed in our minds, apparently instantaneously, enabling us to visualize a whole scene. Somehow thousands of distant physical objects send us a series of messages, which our brain is able to unscramble and interpret.

Light is the messenger which brings us the information. Reflected from every leaf and every blade of grass, light is scattered in all directions. A tiny fraction of it happens to hit the eye. In an instant the information is coordinated to reconstruct the whole panorama!

1.4.2 *The journey of a photon*

In the example of our perception of the countryside, the story begins about 8 minutes earlier, when light is born on the surface of the sun. It does not matter, for the moment, what we mean by saying 'light is born'. We will just picture light as a tiny particle (a *photon*), which suddenly materialises at the surface of the sun.

Countless photons leave the sun every second and go out into the universe in all directions. They travel through space with a speed of $3 \times 10^8 \, \text{ms}^{-1}$ for billions of years, unchanged and unhindered. A very small fraction 'just happen' to go in the direction of the earth, and reach us, having traversed about 100 million miles

Lough Conn, Co. Mayo, Ireland.

of empty space. In turn, a small fraction of these might hit a leaf on a tree and be absorbed, giving energy to help the tree to grow. Others get reflected, and some of these, an even tinier fraction, reach our eyes and are focused on the retina. There are still enough photons remaining to activate the 'photosensitive' cells in the retina. (These photosensitive cells are closely packed in three layers and interconnected by tiny fibres.)

Having arrived safely after their long journey, the photons have done their job. They give up their energy to electrons, which flow through nerve fibres to the brain, as electrical currents.

The sun not only emits light which enables us to see, it is also our main source of energy.

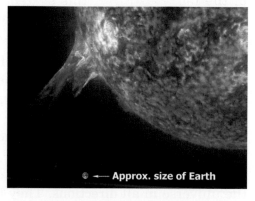

Approx. size of Earth

Solar flare. *Courtesy of NASA / ESA.*

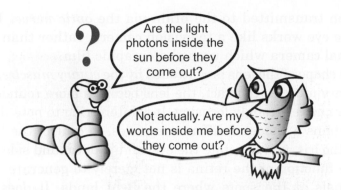

About 600 million tonnes of hydrogen are 'burned' on the sun every day (equivalent to about 10^{25} J of energy). The earth receives just about $5 \times 10^{-8}\,\%$ of this, which amounts to 2×10^{16} J a day, more than ample for our needs. The sun is 'captured' in this image from the Solar and Heliocentric Observatory, as its surface erupts in a large 'prominence'. An image of the earth is shown here to illustrate the scale of the eruption.

1.4.3 *The eye is like a digital camera*

The front of the eye forms a complex optical system which focuses the light on the *retina*. To ensure that the image is of the highest quality this system is capable of rapid adjustments to control the viewing direction, focusing distance, and the intensity of the light admitted. The retina consists of millions of *photosensitive* cells which send out an electrical signal when struck by light. (This process is called the *photoelectric effect*, about which we shall have much more to say in Chapter 13.) These signals

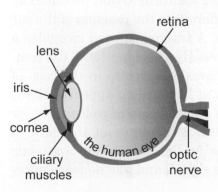

The eye.

are then transmitted to the brain via the *optic nerve*. In this way the eye works like a television camera, rather than a conventional camera which uses photographic film.

The shape of the lens is controlled by the *ciliary muscles*. When the ciliary muscles contract, the lens becomes more rounded and therefore more strongly focusing. It is interesting to note that the eye contains a range of different muscles, to control the lens, to adjust the iris, and to rotate the eyeball up, down and sideways!

The function of the retina is not merely to generate electrical signals at the spots where the light lands. It does much more, and acts as a kind of microcomputer, pre-analysing the information before it is transmitted to the brain. In particular it is responsible for the sensation of *colour*. The human eye is sensitive only to a small range of the *electromagnetic spectrum*. Different colours are characterised by different *wavelengths* of the *electromagnetic wave*. Information from the two eyes is compared and coordinated and tiny differences between the two images are used to measure perspective and distance and even to estimate the speed of approaching objects.

Two computers — the back of the eye and the brain

It is interesting to study the reaction of the eye to mixtures of photons of different wavelengths. Thus, for example, a mixture of 'red' and 'green' light produces a sensation identical to that produced by 'yellow' light alone. This is quite different to the reaction of the ear to sounds of different wavelengths. A trained ear can recognise a chord of music as a mixture of notes. By 'listening hard' the components of the chord can be distinguished. Not so in the case of light. No matter how hard we look at light which appears yellow, it is impossible to tell whether it is a pure beam of yellow light, or a mixture of red and green. The microcomputer at the back of the eye has sent the signal 'yellow' to the brain. The brain, our 'mainframe computer', accepts this signal, and has no more information on the original input data.

1.4.4 *Reconstructing the object*

Consider photons which are reflected from some point on a tree — say, from a leaf at the very top. Focusing by the lens at the front of the eye means that every photon which reaches the lens from that particular point on the tree is directed towards the same spot on the retina. Similarly, photons from every other leaf on the tree are brought together at certain corresponding points on the retina. An 'image' of the tree is formed at the back of the eye, which is recognised by the brain as 'a tree'. Recognition of any object is a job for the brain rather than the retina. As babies grow, their brains develop the ability to reconstruct objects such as 'my fingers', 'mother' and 'teddy'. They also learn to allow for the fact that the image in the retina is upside down with respect to the object!

So far, we have represented light by 'rays', and have not been concerned with either particle or wave properties of light. In fact, we have freely interchanged concepts of 'photon', and 'wave' in the preceding paragraphs, depending on context. Later we will formalize the various aspects from which the subject may be approached.

1.4.5 *Why is the grass green?*

Colour is not an absolute physical property of a surface but a function of the kind of light reflected by the surface. At dusk everything looks grey, but when the yellow street lights come on, the colours of things change quite significantly.

Living plants absorb light energy in the process of *photosynthesis*, enabling them to grow and bloom. It so happens that red and violet light are the most effective, and are absorbed, whilst green light is mostly reflected by leaves and grass. The green light we see is the light *not used* in photosynthesis.

Chlorophyl in plants reflects green light.

1.4.6 *Seeing in the dark*

'Darkness' means that there is not enough light present in the visible spectrum to activate the eye. There may still be electro-magnetic radiation of longer or shorter wavelengths. Surfaces at moderate temperatures emit radiation mainly in the *infrared* region. Even though we cannot 'see' this radiation, it can be registered on special photographic film. On right is an example of an infrared photograph taken from the air, showing such heat-emitting surfaces.

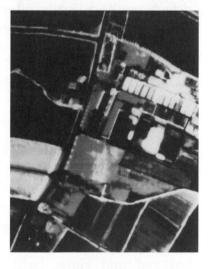

Thermal images, winter at dawn.
Courtesy of Eón Ó'Mongáin, UCD School of Physics.

This aerial view of part of the Malahide Road in Dublin was taken in the infrared. Roofs of buildings are clearly seen to be at a temperature higher than that of the surroundings. The picture was taken in winter, just before dawn, at which time road surfaces and bare soil fields are cooler than vegetation. Evergreen hedges along the borders of the fields are relatively warm. Notice the line of trees along the right hand side of the road.

There is no hiding place in the darkness of the night for criminals trying to escape from justice. The intruders in the images below had no idea that the radiation they were emitting was being recorded by an infrared camera!

Thermal images: intruders at night.

Courtesy of Sierra Pacific Innovation, www.imaging1.com.

An added advantage of infrared camerawork is that such radiation is transmitted through clouds. Not only does it work by night, but it is also unhampered by cloud cover.

1.4.7 *The branches of optics*

We can approach the study of light from any of its aspects:

Geometrical optics: the path of light is represented by a ray without reference to either waves or particles.

Physical optics: emphasis on the wave nature of light.

1.5 The nature of light

1.5.1 *Contradictory evidence*

Newton firmly believed that light was carried by particles (corpuscles). He published his theory of light in the book *Optiks* (1704). It is ironic that Newton, the firm devotee of the corpuscular theory of light, was the first person known to have observed Newton's rings, which are caused by the interference of light, a wave phenomenon.

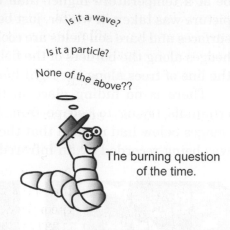

Is it a wave?

Is it a particle?

None of the above??

The burning question of the time.

At the beginning of the 20th century, a number of experiments appeared to confirm the corpuscular theory. Light had the properties of a particle. Other experiments indicated the opposite. Light behaved as a wave. Reconciling such contradictory evidence appeared to be the main outstanding problem in *Natural Philosophy*.

1.5.2 *Light as a wave*

Waves can interfere with one another, sometimes reinforcing, sometimes cancelling out. They can also bend around corners. As far back as 1802, Thomas Young had shown that light beams from two thin slits coming together can apparently 'mutually destruct' at certain points. Seen as two waves, the two beams *interfere destructively*, giving darkness. This can be explained as the crest of one wave superimposing on the trough of another and the two 'cancelling out'. In other places, where two crests or two troughs come together, they *interfere constructively* and we get an increased intensity — not as dramatic, and easier to accept intuitively. In order to observe the phenomenon, we must arrange that at certain points in space there is always constructive interference (waves *in phase*), and at others destructive interference (waves *out of phase*). The effect is analogous to *nodes* and *antinodes* in a vibrating string. When we discuss Young's experiment in more detail (Chapter 8), we will see that the experiment is relatively easy to perform. Suffice it to say that it demonstrates a remarkable effect associated with waves, namely:

$$\text{light} + \text{light} = \text{darkness}$$

1.5.3 *Maxwell's electromagnetic waves*

James Clerk Maxwell

James Clerk Maxwell

Another piece of evidence in support of the wave nature of light came from the theoretical work of James Clerk Maxwell (1831–1879). Maxwell put together laws of electricity and magnetism which had been discovered by Karl Gauss (1777–1855), Andre Ampère (1775–1836) and Michael Faraday (1791–1867). These laws were, in Maxwell's time, well established but were considered as separate and independent. Maxwell's achievement was to unify phenomena in *electrostatics*, *magnetism* and *current electricity* by expressing the laws mathematically, in the form of four simultaneous differential equations.

The experimental evidence which Maxwell synthesised:

1. *Coulomb's law*, which describes the force exerted on one another by electric charges *at rest*. It can be expressed mathematically in another form called Gauss's theorem.
2. *Gauss's theorem*, applied to magnetism, expresses the fact that magnetic monopoles do not exist.
3. *Ampère's* discovery that an electric charge *in motion* produces a magnetic field.
4. *Faraday's* discovery that a *changing* magnetic field produces an electric field. Maxwell extended the symmetry

to a *changing* electric field in turn producing a magnetic field.

The equations describing these laws must all be true at the same time (they are *simultaneous* equations). When Maxwell put the four equations together, and solved them, he established the consequences of the four laws being true at the same time. The result was a prediction that by accelerating an electric charge, one would create a signal which would propagate through space: *An oscillating charge would give rise to an electromagnetic wave, travelling through space at a fixed speed.* Maxwell was able to calculate this speed, and obtained an answer practically identical to the measured speed of light. This could hardly be a coincidence. Light *must* be an *electromagnetic wave*.

1.5.4 *Light as a particle*

In 1900, Max Planck (1858–1947) made a discovery which appeared to be incompatible with the wave theory of light. In interpreting the spectrum of electromagnetic radiation emitted by a *black surface* at high temperature ('blackbody radiation'), he found that a theoretical model which gave very good agreement with the entire spectrum had to be based on a new and unexpected assumption. He proposed that the oscillating electric charges which give rise to light emission can only have discrete energies. These energies come in units (*quanta*) of hf, where h is a universal constant and f is the frequency of the oscillations. An oscillator can have energy nhf, where n is a whole number, but amounts of energy in between just do not exist; for some unknown reason they are forbidden in nature.

Such a model was very difficult to accept, as it proposed the idea of *quantisation*, at that time quite foreign to 'natural philosophy'. Classical physics assumed that physical quantities have a continuous range of values. It was taken as self-evident that there are no restrictions on the energy of a physical system, or for that matter, on any observable physical quantity.

In 1905, Albert Einstein extended the idea to light itself. The energy transmitted by light also comes in quanta. Each photon carries a quantum of energy hf, and to all intents and purposes behaves like a particle. It can knock an electron out of a metallic surface, something which a wave, with its energy spread out, rather than concentrated at a point, could never do.

Not only does a photon behave as a particle with energy, but it also has momentum and can collide and bounce off an electron. The result is very much like a collision between two billiard balls (see Chapter 13 — 'The Compton effect').

1.5.5 *An illustration of duality?*

It is not possible to give a good analogy with wave–particle duality in the 'household world'. Certainly we can observe different properties of the same thing, depending on what we look for. A person may be at the same time a doctor, a parent, an athlete and a democrat, each aspect independent of the others. A certain machine can be a word processor, a computer, an electronic-mail communicator and a game station, all at the same time. These analogies are not very good, because being an athlete and being a democrat are not mutually exclusive, whereas in the household world particles never look like waves, or vice versa!

We can try another illustration, an optical illusion in which the appearance of an object depends on the observer's point of view. We reconstruct a concrete object in our mind when we interpret a picture. What that object is may differ from time to time quite dramatically. The illustration below is entirely *symbolic* and certainly should be taken as such. It has no direct connection with light, with photons, or with quantum theory.

The picture is an example of pictographic ambiguity, where more than one 'image' is contained in a single drawing. A similar picture was published in 1915 by the cartoonist W.E. Hill and is called 'My wife and my mother-in-law'. At first glance one sees immediately one image, but not the other. What is the chin of the young woman from one perspective becomes the nose of the old lady from another point of view.*

Again, the analogy with wave–particle duality is not close. The object which one sees is an abstraction in the imagination of the beholder. The physical reality is the material of the canvas, the painting oil, the frame of the picture, without ambiguity! It is not surprising that it is not possible to find a good illustration of wave–particle duality in the classical world. It is a *quantum phenomenon* in 'the world of the very small'.

Two images in one.

1.6 The birth of quantum mechanics

1.6.1 *Particles have wave properties*

In 1924, a dramatic idea was advanced which put a new slant on our view of waves and particles. In his doctoral thesis, submitted to the University of Paris, Louis de Broglie (1892–1987) proposed that not only light but all matter has properties of both particles and waves. The examiners were not convinced. The idea seemed quite absurd, and de Broglie had no experimental evidence to support his conjecture. To confirm their view they consulted Einstein, who, probably to everyone's surprise, recommended acceptance of the thesis. Specifically, de Broglie proposed that the relation between p, the momentum of a particle, and its associated wavelength, λ, involves Planck's constant,

---

* An older version of the picture appears in an advertisement for the Ohio Buggy Company, captioned 'Here is my wife, but where is my mother-in law?'.

and is the same as that between the momentum and the wavelength of a photon, namely:

$$\lambda = \frac{h}{p} \qquad \text{(The de Broglie equation)}$$

At the time de Broglie was not aware that there was already some experimental evidence to support his assertion. Within a year came what is now recognised as the official confirmation, in a paper by C.J. Davidson and L. Germer. When the two scientists sent a beam of electrons through a crystal, the electrons were scattered to form a pattern exactly like the diffraction pattern of light waves.

1.6.2 *The Copenhagen interpretation*

In 1921, Niels Bohr (1885–1963) founded the Institute of Theoretical Physics in Copenhagen to develop the mechanics for dealing with the world of the ultimate constituents of matter — atoms, atomic nuclei and photons of light. The most eminent physicist from all over the world attended Bohr's institute at one time or another, and the results of their deliberations became known as the *Copenhagen interpretation of quantum mechanics*. It soon became clear that the wave–particle duality of light is a symptom of a much deeper principle at the core of the laws of Nature. In the atomic world physical entities exist in *superposition of states*. For example, an atom may exist as a mixture of states of different energies, and acquires a given energy state only when energy is measured. A photon is neither a particle nor a wave but acquires one or other identity when it is observed. We have to revise our understanding of *reality*. In the world of atoms, molecules and light quanta, physical objects do not have an independent existence.

It is somewhat ironic that Albert Einstein, who had been the first to approve de Broglie's particle–wave hypothesis and had later made his vital contribution to the quantum theory of the photoelectric effect, became one of the greatest critics of the Copenhagen interpretation of quantum mechanics. He could not come to terms with the suggestion that physical attributes of particles and systems depend on what measurements are made, or whether or not measurements are made at all. How could the act of observation change physical reality? In one of his many letters to Bohr he expressed what was to him the absurdity of it all: "How could a mouse change the universe by looking at it?" Einstein's own theory of relativity was derived by logical steps from initial logical assumptions about space and time and the constancy of the speed of light. To him 'the most incomprehensible thing about Nature was that it is comprehensible'. The philosophy of quantum mechanics appeared to be neither logical nor comprehensible!

Einstein developed the theory of relativity on the basis that the speed of light is a universal constant. In empty space, there are no 'milestones' and therefore no reference points to define absolute speed. This was the starting point of Einstein's logical train of thought which led to his famous equation $E = mc^2$.

1.6.3 *The universal messenger*

Light is the principal actor on the stage of the universe. It brings information from distant stars and galaxies. It tells us about the distant past. It plays a key role in our understanding of the most basic laws of Nature. This book will attempt to tell the exciting story of light!

Chapter 2

Geometrical Optics: Reflection

The quickest route

In geometrical optics we are not particularly interested in the nature of light, but we want to know *where it goes and what route it chooses* when there are many possible routes.

Everybody knows that, all else being equal, light travels from A to B by the most direct route, i.e. in a straight line. But sometimes light may undergo reflection at one or more points along its journey, or it may pass through different media, such as glass or water. Depending on the angle at which it enters a new medium, it will be bent (*refracted*). The light no longer travels in a straight line from A to B, but chooses *the quickest possible route*. This is known as *Fermat's principle of least time*. All rules of geometrical optics on angles of reflection and refraction follow from that principle.

I don't care about its nature, I just want to know its path.

In this chapter we deal with *reflection*. The phenomenon of *refraction*, which is the change of direction when light enters a different medium, will be considered in Chapter 3.

23

2.1 Fermat's law

2.1.1 *Light takes the quickest route*

Let us represent the path of the light by a 'ray' as a line with an arrow tip indicating the direction. A point source of light will in general give rise to rays spreading out from the point, uniformly in all directions. Some, or all, of these rays may then be reflected from certain surfaces, or bent (*refracted*) by a transparent medium. They can be guided by a lens or mirror and focused to pass through another point. In a complex optical instrument, many lenses and mirrors may repeat the procedure of refraction and reflection, to produce a final image.

Pierre de Fermat (1601–1665), proposed a fundamental law which applies to all light rays:

Fermat's law of least time
When light goes from point A to point B, it will always take the quickest possible path.

Fermat's law is beautiful in its simplicity, and generality. Imagine a light ray travelling by a complicated route through different media, or through an optical instrument. No matter how complicated the path between any two points, it will always choose the quickest possible route! The entire subject of geometrical optics is governed by this one law.

2.1.2 *The path in empty space*

The simplest application of the law of least time occurs in empty space, where light follows a straight line between any two points. Of all the rays which set out from a source, the only one to reach a particular point will be the ray which originally sets out in the right direction, i.e. straight towards the point. That particular ray will have taken the *quickest* path and, in this case, *the quickest path is also the shortest path.* In empty space a straight line is the quickest route.

This applies anywhere where the speed of light is constant and there is an unobstructed path between the points.

2.1.3 *The quickest path via a reflection*

Suppose that the direct route from A to B is blocked by a screen, but that there is a possibility of getting around the screen via reflection at a mirror, as shown in Figure 2.1. Which is the shortest and quickest remaining route from A to B?

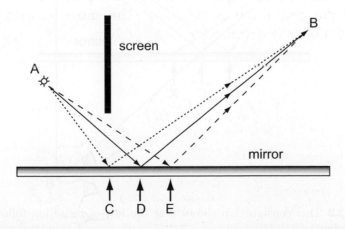

Figure 2.1 The direct route is blocked, but a mirror provides another way.

Some reasonable suggestions:

1. The point of reflection is C, directly under the screen.
2. The point of reflection is D, such that the angle between the ray and the mirror is the same for the incident and reflected rays.
3. The point of reflection is E, halfway horizontally between A and B.

2.1.4 *The law of reflection*

Construct a point B*, under the mirror, such that BC = B*C, as seen in Figure 2.2. Let the ray be reflected at some point, X, at the mirror.

No matter which point X we choose, XB = XB* and the journey from A→X→B is the same length as that from A→X→B*, no

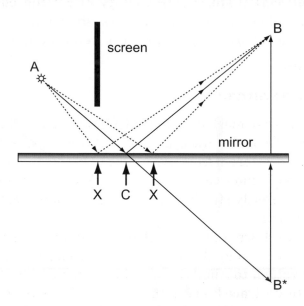

Figure 2.2 This construction shows that the law of reflection follows from Fermat's law.

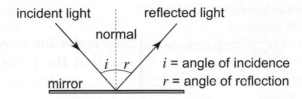

Figure 2.3 Law of reflection $i = r$.

matter where the ray reflects at the mirror. But the straight line (solid path) is the shortest distance from A to B* and therefore also the shortest path from A to B. For that path, and *for that path only*, the angle between the ray and the mirror is the same for the incident and reflected rays.

It is usual to define angles of incidence and reflection with respect to the *normal* (the line perpendicular to the surface at the point of reflection), as shown in Figure 2.3.

Law of reflection: *The angle of incidence is equal to the angle of reflection*. This is the first practical consequence of *Fermat's law*.

2.2 Mirrors

2.2.1 *A plane mirror*

The simplest and most common mirror of all is, of course, a plane mirror. Consider Alice, standing in a well-lit room. Some of the light scattered from Alice goes to the mirror and is reflected. We all know that the result is an image of Alice, looking back, apparently standing at an equal distance on the other side of the mirror.

In Figure 2.4, we trace the path of the light rays as they obey the law of reflection at the mirror. Let us concentrate on those rays which return to Alice's eyes and are seen by her. Only two rays need to be traced: one from the top of her head and the other from her toe. Both rays return directly to her eyes, as illustrated in the diagram. There is no point in extending the

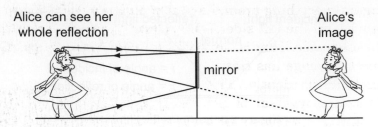

Figure 2.4 A full-length plane mirror.

mirror as any light rays from Alice reflected from points higher or lower have no chance of reaching her eyes. They would hit the mirror at the wrong angle, and go off into space, and will not be seen — at least not by Alice!

Alice has mounted the mirror in the best position to get a full-length reflection. It is exactly half her height and extends from a point halfway between the floor and her eye level, to a point midway between eye level and the top of her head.

> What we see in a plane mirror is an illusion, reconstructed in our mind. There is of course nothing behind the mirror. The reflected rays *appear* to diverge from something. Such an image is called a *virtual image*.

2.2.2 *Reversal from left to right*

The image you see in a plane mirror is standing upright and facing in your direction. You see what appears to be a copy of yourself, "turned around", and looking back, face to face. On closer examination, you realise that what you see is not an exact copy, and is not quite how others see you.

The image in the mirror has 'turned around' in a peculiar way. Imagine that you went behind the mirror and turned around to 'face yourself'. The act of *turning* consisted of swinging everything around a central axis, your right arm now appearing on the left side of the mirror, your left arm rotating to the other side. That is not what happened to the image. Every dot on the image has turned around independently. This results in a reversal of left and right, clockwise and anti-clockwise.

The young sportsman in the photograph does not quite see himself as others see him. He is holding the ball in the *other* hand. The number 13 on his shirt is written *backwards*. Even more noticeable is the image of the clock; it is going *anti-clockwise*, and the numbers do not look right!

The world in a mirror.

2.2.3 *Reflection from a curved and uneven surface*

So far we have only considered reflections at a *plane* surface. The same *law of reflection* applies if the mirror is curved, provided that we measure the angles relative to the *normal* at the point of reflection.

When a light ray meets a rough reflecting surface, the normal at the point of reflection can have practically any direction, resulting in *diffuse* reflection at all angles, as seen in Figure 2.5.

2.2.4 *A spherical concave mirror*

A *spherical* mirror consists of a reflecting surface which is curved, so that it forms part of the surface of an imaginary

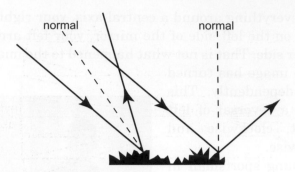

Figure 2.5 The incoming ray 'sees' only the normal at the precise point of reflection.

(large) hollow sphere. In the case of a *concave* mirror, the 'hollow' (concave) side is facing the incoming rays. Such a mirror has interesting properties which a plane mirror does not have, basically owing to the fact that the rays reflected from a concave surface tend to be brought together. For example, a parallel beam of light (provided it is not too far away from the axis) can be brought together at a point called the *focus*, and conversely a source of light placed at the focus will give rise to a reflected beam parallel to the axis. These properties can have many practical applications, some of which are illustrated in Figures 2.6 and 2.7.

A spherical mirror, however, will not bring all rays perfectly to the focus, and rays far from the optic axis will not cross that axis at quite the same point after reflection as those which came in close to the axis. To make a mirror with perfect focusing properties, the surface

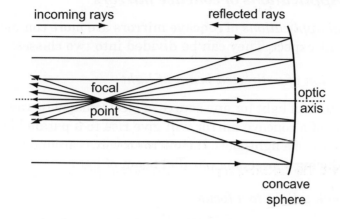

Figure 2.6 A concave mirror will reflect parallel rays through a focal point.

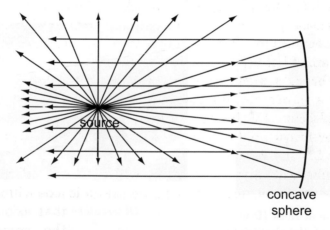

Figure 2.7 The same mirror will give a beam of parallel reflected light if the source is placed at the focus.

has to be *paraboloidal* rather than spherical. The geometrical properties of a parabola exactly fulfil the criteria for perfect focusing. In *Appendix 2.1* this property of a parabolic mirror is derived directly from Fermat's principle.

2.2.5 *Applications of concave mirrors*

Practical applications of concave mirrors are more common than one might expect. They can be divided into two classes:

1. *Creation of a directed beam of light*

    A source of light (or any source of electromagnetic rays) at the focus of a concave mirror will give rise to a parallel beam. In the case of a lighthouse the mirror revolves around the source, producing a sweeping parallel beam of light.

2. *Bringing light to a focus*

    Light shining over the entire surface area of a large concave mirror will be concentrated at the focus.

Hale Telescope. *Courtesy of Palomar Observatory and Caltech.*

Lighthouse at Hook Head. *Courtesy of the Commissioners of Irish Lights.*

VLA. *Courtesy of NRAO/AUI/NSF.*

Reflecting telescope

An almost perfect example of a parallel light beam is light from a distant star. The fact that light from a dim source can be effectively gathered over the entire surface of a large mirror and focused at a point finds an obvious application in astronomical telescopes. The Hale telescope at Mount Palomar, for example, uses a parabolic mirror, with a diameter of 5.1 m (200 inches).

Mirrors are preferable to lenses in many optical applications. We will meet lenses in the next chapter; suffice it to say here that some wavelengths of the electromagnetic spectrum are absorbed by the glass of a lens — a problem which does not arise in the case of a mirror. In addition, high precision quality mirrors are easier and cheaper to manufacture than large lenses.

Radio waves from cosmic sources are brought to the focus of each mirror in a similar way to visible light.

The Very Large Array (VLA) of radio telescopes near Soccoro, New Mexico.

2.2.6 *The 'death rays' of Archimedes*

According to legend, Archimedes (287–212 BC) set fire to an invading fleet of Roman ships at the siege of Syracuse using an arrangement of mirrors and 'burning glasses'. This event is commemorated in a painting by Guiglio Parigi which may be seen in the Galleria degli Uffizi in Florence. That the artist was unfamiliar with the laws of physics is apparent, in that the painting shows divergent (rather than convergent) light rays — and these would be quite harmless!

While the arrangement depicted by Parigi may not be scientifically correct, the idea of using mirrors to focus sunlight at a distant point may not be as absurd as it at first appears. George Louis LeClerc, Comte de Buffon (1707–1788), set out in 1747 to produce fire using Archimedes' method. He assembled 168 mirrors in an arrangement designed to focus sunlight at a distance of about 50 m. It is said that he succeeded in igniting a plank of wood instantly — showing that the system could indeed be used as a formidable weapon at that range! He also claimed to have melted 3 kg of tin at 6 m using just 44 of his mirrors. Buffon's work, while it had not been subjected to rigorous scientific scrutiny, had enough credibility to be recognised, and his portrait appears on a postage stamp issued by the French post office.

In 2004, the Discovery Channel television programme *MythBusters* tried to ignite a fire in a set-up similar to that described in legend at the battle of Syracuse, but the attempt failed. The programme concluded that the legend could not be true, and that the myth of Archimedes' death rays had been 'busted'.

Burning mirrors. *Courtesy of David Wallace and MIT Mechanical Engineering 2.009 class.*

David Wallace and his 2.009 class at the Massachusetts Institute of Technology (MIT) were not convinced. In a feasibility project they assembled a total of 127 plane mirrors to form a giant concave mirror, which they used to focus sunlight on a wooden model of a ship at a distance of about 30 m. After about 10 minutes' exposure to the death rays the hull of the ship burst into flames. The MIT team admits that the conditions at the siege of Syracuse were not quite the same as in their reconstruction. The mirrors available to Archimedes would have been made of copper, not glass, and the ships would probably have been further away than 30 m. Nonetheless, they did show that the account of Archimedes and the burning ships is credible. The project has also confirmed the results of Buffon's experiments.

A historical interlude: Plerre de Fermat (1601–1665)

Born near Montauban, France, in 1601, Pierre de Fermat was a lawyer and a government official. In 1648, he rose to the position of chief spokesman for the parliament, and president of the Chambre de l'Edit, which had jurisdiction over lawsuits between Huguenots and Catholics. Despite his nominally high office, he seems not to have been particularly devoted to matters of law and government; indeed, a confidential report by the *intendant*, sent to Jean Baptiste Colbert* in 1664, was quite critical of his performance.

PIERRE DE FERMAT 1601·1665   RF

$$x^n + y^n = z^n$$

$$x^n + y^n = z^n$$
n'a pas de solution pour des entiers n > 2

4,50 F
0,69 €

*Creation et gravure Andre Lavergne.
Courtesy of La Poste France.*

For Fermat, mathematics was a hobby. One can surmise that his 'day job' bored him and that he devoted more of his time to mathematics than was considered proper by his superiors. He published very little of his work and most of his results were found after his death, written on loose sheets of paper or on margins of textbooks. He had a mischievous streak and enjoyed teasing other mathematicians by stating results and theorems without revealing the proof. His son Samuel found what is his now-famous Last Theorem, scribbled as a marginal note in his father's copy of *Diophantus' Arithmetica*.

Fermat's Last Theorem states that the equation

$$x^n + y^n = z^n$$

---

* Chief Financial Minister of Louis XIV.

(Continued)

has no non-zero integer solutions for *x*, *y* and *z* when *n* > 2. Fermat wrote: "I have discovered a truly remarkable proof which this margin is too narrow to contain."

In 1908, the mathematician Paul Wolfskehl of Darmstadt, Germany, bequeathed a sum of 100,000 marks to the Academy of Sciences in Göttingen. It was to be awarded to the first person to publish a complete proof of Fermat's theorem for all values of *n*. Nobody succeeded in finding a proof until 1997, when Andrew Wiles, an English mathematician, following 11 years' work, published what has been generally accepted as a complete proof. It showed that Fermat was correct in his result although, given the complexity of the ultimate theorem, Fermat's own 'remarkable proof' may well have been wrong.

Fermat was also interested in the mathematics of *maxima* and *minima*, and developed methods for finding tangents to curved lines and surfaces. It is this work which led to his *principle of least time*. His mathematical methods formed the central issue of a debate with Descartes in the spring of 1638. Descartes, who had an aggressive temperament and a sharp tongue, considered Fermat a rival and disapproved of his mathematical reasoning. As a result there was little cooperation between the two men, who were arguably the greatest mathematicians of the time. The two made a formal peace when finally Descartes admitted the error of his criticisms of Fermat's methods, but even then there seems to have been little love lost between them.

It is remarkable that Fermat was able to come to his conclusions not only geometrically, but also analytically, using methods similar to calculus, the invention of which is normally attributed to Isaac Newton (1642–1727) and Gottfried Leibnitz (1646–1716) about 50 years later.

The *principle of least time* implies a fixed and therefore finite speed of light. In those days this speed was totally unknown. Opinion in Fermat's era still followed the belief of Aristotle that the speed of light

(Continued)

was infinite until the first measurement of *c* was made in 1677 by Olaf Römer (see Chapter 1).

Fermat was not sure whether or not light is transmitted instantaneously, and even less sure that it had a finite speed which differed from medium to medium. Nevertheless, he undertook a mathematical derivation of the physical law of refraction on the basis of the postulates that:

(1) *'the speed of light varies as the rarity of the medium through which it passes'* and
(2) *'nature operates by the simplest and most expeditious ways and means'.*

To Fermat's surprise (expressed in a letter to Clerselier in 1662), his mathematical analysis led to the experimental law now known as *Snell's law of refraction.*

Appendix 2.1 The Parabolic mirror

The parabolic mirror — a perfect illustration of the law of least time

A definition of a *parabola* is 'the locus of all points equidistant from a point called the focus, and a straight line called the *directrix'.* This is exactly what is needed to construct a mirror with no spherical aberration. The surface of such a mirror has the shape of a *paraboloid of revolution.*

In the case of the parabolic mirror it is easier to apply the *law of least time* directly. Consider a point source of light at an infinite distance from the mirror. Imagine that at a given instant a number of photons leave the source, and spread out in all directions. At any given time later they will occupy a growing spherical surface, which at infinite distance from the source will become a plane of photons travelling parallel to one another.

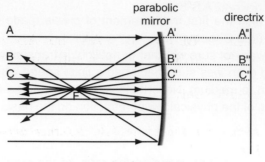

property of a parabola:
FA' = AA", FB' = BB', FC' = CC'

Figure 2.8 Parabolic mirror.

Consider three such photons arriving in a parallel beam ABC, as illustrated in Figure 2.8. What is the shortest and therefore quickest route to F, assuming that they are going to be reflected by the mirror?

The geometrical property of the parabola gives FA' = A'A", FB' = B'B" and FC' = C'C"

$$\Rightarrow \quad AF = AA'', \quad BF = BB'', \quad CF = CC''$$

Since the planes ABC and A"B"C" are parallel, AF = BF = CF.

This means that the travel times for all light rays to F from the plane ABC (and therefore also from the 'point at infinity') are equal.

To show that the equal paths are also the shortest paths:

The paths taken by the three rays in Figure 2.8 are exactly equal. They are also the shortest paths.

It is easy to see that, for example, there is no shorter path from A to F via reflection at the mirror than AA'F.

Try for instance AB'F:

$$AA'F = AA'', \quad BB'F = BB'' = AA'' \text{ but } AB' > BB'$$
$$\Rightarrow \qquad AB'F > AA'F \qquad \text{The path via B' is longer.}$$

We can conclude that:

Photons which cross the plane ABC together arrive at the focus together

Chapter 3

Geometrical Optics: Refraction

Providing more than one quickest route

When light crosses the boundary between two media, it changes direction. This phenomenon is called *refraction*. In this chapter we study the rules and applications of refraction. The basic rule is the same as always: it is *Fermat's principle of least time*. We show that the principle leads to the experimentally established *Snell's law of diffraction*.

Lenses are the most common example of the application of the laws of refraction. In making a lens, the trick is to make the shape such that all routes from a source A to a destination B on the other side of the lens take the same time, despite the fact light traverses different thicknesses of glass on different routes.

We spend the remainder of this section dealing with the geometry of the paths, and derive some simple formulae for lenses.

Making visible things we cannot see

We study the effect of various combinations of lenses which make up optical systems. One fascinating example is the optical system of the human eye. We discuss some common eye defects and how these may be corrected using suitable lenses.

Finally, we describe optical systems which enable us to look at things which are either too small or too far away to be seen with the naked eye.

3.1 Refraction

3.1.1 *The refractive index*

As we all know, from experience of city traffic, if the speeds along different routes are not the same, the shortest route is not necessarily the quickest. We have seen already that *in vacuum* the speed of light is fixed at $c = 2.99792458 \times 10^8 \, \text{ms}^{-1}$, equivalent to travelling a distance of approximately 7.5 times around the earth in *one second*. Light can also travel through certain 'transparent' media such as air, water, glass or quartz, and there the speed is less than c. The factor by which it is smaller is specified by the *refractive index n* of the medium.

The definition of the *refractive index* of medium 2 with respect to medium 1 is $_1n_2 = \dfrac{v_1}{v_2}$, where v_1 and v_2 are the speeds of light in medium 1 and medium 2, respectively.

Glass

The most common man-made substance which is transparent to visible light is, of course, glass. It is interesting to note that the Egyptians were able to make glass as far back as 1500 BC. Glass is manufactured by fusing silica (sand) with alkali; the Egyptians used soda and *natron* (sodium carbonate). About 1000 years later we find early Greek references to glass by Aristophanes (≈ 400 BC).

The following table shows the refractive indices with respect to vacuum of some materials (for light of wavelength 589 nm):

Material	n
air (1 atm, 15°C)	1.00028
water	1.33
quartz, (fused)	1.46
glass (crown)	1.52
glass (flint)	1.58

In practice it is usual to omit the subscripts when writing the refractive index $_1n_2$ if medium 1 is either vacuum or air, and simply to refer to the refractive index, n, of the denser medium.

The difference between the speed of light in vacuum (or air) and its speed in liquids or solids is quite significant. In crown glass, for example, light travels 1.52 times more slowly than in air. The practical consequence of this is the phenomenon of *refraction* — light changes direction as it crosses the boundary between one medium and another. It is not obvious why such a change of direction should occur, but perhaps not surprising.

Experience may tell us that if a car hits a muddy patch at an angle, it will tend to swerve. Similarly, waves in water tend to change direction as their speed changes in shallow water. Actually, we do not have to refer to any particular model for the

mechanism which causes the change of angle; it is required by Fermat's principle as illustrated in the lifeguard problem.

3.1.2 *The lifeguard problem*

There is an elegant way in which we can derive an exact expression for the change in angle (*Snell's law*) directly from the fundamental *principle of least time*, which we illustrate using the example of 'the path of the lifeguard'.

To find the quickest path

Consider the problem faced by a lifeguard who sees a swimmer in difficulties. He wants to get to the swimmer in the *shortest*

Figure 3.1 The lifeguard problem.

time, though not necessarily by the shortest route. His running speed is greater than his swimming speed, so the shortest route does not necessarily take the shortest time. What route should he choose?

Probably his first instinct would be to go in a straight line, but this cannot be the quickest route as the distance in the water (where he moves more slowly) is too long. Neither is it best to run to the bank opposite the swimmer and swim straight across, so that he covers the maximum possible distance on land, because the overall distance is too long. Instinct tells us that a compromise route is the quickest, and the lifeguard should dive into the water at some point in between his starting position and the point opposite the swimmer.

It turns out that the lifeguard reaches the swimmer in the *shortest time* if he runs directly to the point P in Figure 3.1, and then swims directly from there to the swimmer. The sizes of the angles i and r fix the position of P and the time is shortest when

$$\frac{\sin i}{\sin r} = \frac{\text{speed in water}}{\text{speed on land}}$$

(See Appendix 3.1)

3.1.3 *Snell's law*

Light follows the quickest path between two points. As it changes speed on crossing the boundary from a less dense to a denser medium, it is *bent towards the normal*. The ratio of the sine of the angle of incidence to the sine of the angle of refraction is *constant* for light crossing from one given medium to another. This fact was discovered by Willibrord Snell in 1621. (At that time the speed of light was unknown and indeed generally considered to be infinite.) Fermat was able to relate Snell's constant to the ratio of the speeds of light on the assumption that these are finite and fixed for different media.

We have now shown that:

$$\text{The law of least time} \Rightarrow \frac{\sin i}{\sin r} = \frac{v_1}{v_2} \quad \text{(Snell's law)}$$

The reverse journey

Assuming that the speed of light is the same in either direction, the quickest route from A to B will also be the quickest route back from B to A. If we reverse the direction of a light ray at any point, we might expect it to retrace its way back along the path by which it came. This is sometimes known as the *principle of reversibility*. Accordingly, for rays coming from a point object immersed in a denser medium such as water, the ratio $\sin i / \sin r$ is the same as before. Now, of course, the roles of the angles i and r are reversed (r represents the angle of *incidence* and i the angle of *refraction*). We will see shortly, however, that the reversed ray may 'decide' not to retrace its path back to B, but be reflected back into the water.

3.1.4 *Apparent depth*

To an angler looking down at a fish beneath the surface of water, the fish appears to be closer to the surface than it actually is.

Figure 3.2 The fish appears to be nearer the surface.

Rays from the fish are bent away from the normal as they leave the water surface. Figure 3.2 illustrates how this changes the perceived position of the fish.

The rays appear to diverge from the point O' rather than from O. The object appears to be situated at an '*apparent*' depth AO' which is related to the real depth AO:

$$\frac{AO'}{AO} = \frac{\tan r}{\tan i}$$

When the angles i and r are small, $\dfrac{AO'}{AO} \cong \dfrac{\sin r}{\sin i} = \dfrac{1}{n}$

$$\frac{\text{real depth}}{\text{apparent depth}} = n$$

This relationship holds for small angles only when the angler is more or less directly above the fish. For larger angles, the angler sees the fish at an even smaller apparent depth.

3.1.5 *The dilemma faced by light trying to leave glass*

Rays coming from the denser medium are refracted away from the normal. As we consider rays striking the surface at greater and greater angles (Figure 3.3), we will come to the point where the angle of incidence $r = \theta_c$ (the *critical* angle). At that angle $\sin i / \sin \theta_c$ is such that $\sin i = 1$. The refracted ray, at least in principle, skims along the surface at an angle of 90° to the normal. In practice, as we will see shortly, the intensity of the refracted ray decreases, becoming very small at that angle.

The *dilemma* is this: since $\sin i$ *cannot be > 1*, what is to become of a ray reaching the surface at angles $r > \theta_c$?

A ray from the denser medium striking the surface at an angle greater than the *critical angle* θ_c cannot get out, and will be *totally internally reflected*.

If the ray passes from the glass into air, the value of θ_c can be calculated from:

$$\frac{\sin 90^0}{\sin \theta_c} = n$$

$$\sin \theta_c = \frac{1}{n}$$

$$\Rightarrow \theta_c = \sin^{-1}\left(\frac{1}{n}\right)$$

Critical angles		
Substance	n	θ_c
water	1.33	48.8°
glass	1.52	41.1°

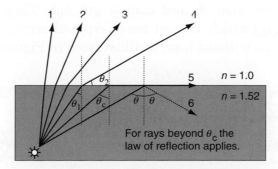

For rays beyond θ_c the law of reflection applies.

Figure 3.3 Total internal reflection.

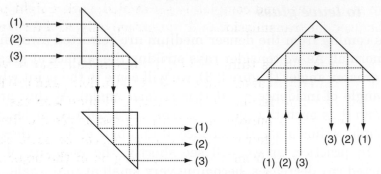

Periscope — rays emerge to give an upright image.

Figure 3.4 Totally reflecting prisms.

3.1.6 *Practical applications of total internal reflection*

Since the critical angle for glass is just under 45°, right angled prisms with the remaining angles each 45° are ideal for turning a light beam through 90° or 180°, as we see in Figure 3.4. Since no loss of intensity occurs in the process, *totally reflecting prisms* are used rather than mirrors in most optical instruments.

Light pipes and optical fibres

In a *light pipe*, the rays of light keep hitting the surface at angles greater than θ_c and cannot get out. They will travel around corners which are not too sharp and arrive at the other end of the pipe without loss, as illustrated in Figure 3.5.

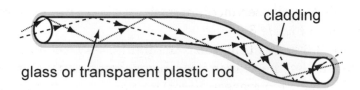

Figure 3.5 Light pipe.

Since the rays from different parts of the original source arrive at the other end completely scrambled, such a light pipe can be used for transmission of illumination only, and for creating images.

An optical fibre is thinner than a human hair, but is very strong. Optical fibre cables can transmit more data, and are less susceptible to interference than traditional metal cables. To transmit images, a bundle of very fine glass or plastic fibres, each of the order of a few microns in diameter, can be used. Each fibre transmits light from a very small region of the object, to build up a well-resolved image. Fibre-optic bundles are used widely in telecommunications.

In medicine, flexible *endoscopes* are used to look at internal organs, in a reasonably non-invasive manner. The endoscope functions as a viewing device inserted through a very small incision, while the surgical instrument is inserted through another small incision, and manipulated within the tissue. Particularly common operations include gall bladder removal, knee surgery and sinus surgery. In *laser surgery*, fibre optics can be used both to guide the laser beam and to illuminate the area.

3.1.7 *Freedom of choice when a ray meets a boundary*

The *principle of reversibility* applies only when both the starting point and the destination of the light ray are known, in situations such as the lifeguard problem (Figure 3.1). Often, when light comes to the boundary between two media it exercises 'freedom of choice' as to whether it is going to be reflected or refracted. This applies to rays going from a less dense medium such as air to a denser medium such as glass, or vice versa, as illustrated in Figure 3.6.

Figure 3.6 shows that when the light goes from glass to air, the intensity of the transmitted beam becomes fainter as the angle of incidence increases, becoming zero at the critical angle. At that point, freedom of choice is gone and the ray can only be

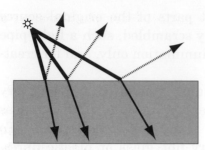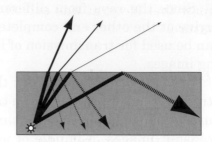

Figure 3.6 Rays crossing the boundary between glass and air are partially reflected and partially refracted.

reflected. It is difficult to understand what determines the fate of a particular ray — whether it is reflected or transmitted.

The 'shop window effect'

An example of *partial reflection* occurs when we look at a shop window. We see our own image in light *reflected* by the glass, and the goods on display, in light *transmitted* by the glass. In the first case light from outside the shop 'bounces back', and in the second light from the inside comes through. A similar 'freedom of choice' exists for light rays coming from the other direction. (In the case of the shop window this freedom is exercised twice, first at one glass surface and then at the other.)

3.1.8 *The mystery*

Partial reflection was a deep mystery to Newton. He realised that it is not determined for example by some local microscopic surface characteristic. The reason, he wrote, was *'because I can polish glass'* — the fine scratches due to polishing do not affect the light. Whether or not bumps and hollows on the microscopic or even the molecular scale can be affected by polishing is questionable. Experimentally, we find that the path of a light ray at the boundary between the two media is unpredictable, regardless of the scale at which we perform our experiments. We now know that Newton was exploring a most basic phenomenon of nature.

'The photon decides'

The phenomenon is particularly interesting from the point of view of light photons. As photons reach the surface one by one, some of them are reflected, and some continue on through the surface. Which happens is not predetermined but is purely statistical; in fact, the photon decides. We have here an example of the rules of *quantum mechanics*. The photon is governed by *basic laws of probability*. What is more, the actual probabilities may depend on seemingly absurd things such as what may or may not happen in the future, for example, whether or not it might meet further reflecting surfaces later!

Such mysterious behaviour of photons forms the starting point of *quantum electrodynamics* (QED), co-founded by Richard Feynman (1918–1988).

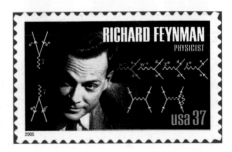

Courtesy of United States Postal Service.

3.1.9 *A practical puzzle — two-way mirrors*

In spy movies and identification parades we often see two way mirrors which behave as transparent glass windows from one side and mirrors from the other side. This seems to be impossible from physical principles, so how do they work?

Many such mirrors are partly silvered so as to reflect not all rays, but a greater fraction than reflected by normal glass. They are based on the principle of greater brightness on one side, and little light on the other.

For light hitting the glass from either side, a certain fraction of light rays are reflected, and the remainder transmitted.

Because of the difference in brightness, the vast majority of rays emerging from the glass on the bright side are reflected rays which originated on that bright side. The opposite is true on the dark side, where most of the light coming through is transmitted light. The effect may be enhanced by coating the dark side of the glass with a very thin layer of silver, turning it into a 'half-silvered mirror' which causes about 50% of the light to be transmitted and the other 50% to be reflected.

Two-way mirror in an identification parade.

The human eye can adapt to an enormous range of intensities. The eyes of the members of the identification parade are adapted to bright lights and the very small number of rays coming from the other side of the two-way mirror is not noticeable.

3.2 Lenses

3.2.1 *The function of a lens*

The basic function of a lens is to gather all the rays falling on its surface and to re-direct them by refraction in a prescribed fashion. Figures 3.7 and 3.8 illustrate how converging and diverging lenses behave.

Parallel rays converge to a point called the focus:

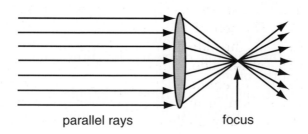

parallel rays   focus

Rays from a point source O (object) converge to a point I (image) beyond the focus:

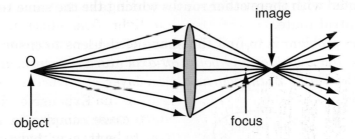

Figure 3.7 Converging lens.

A *diverging* lens causes parallel rays to diverge so that they *appear* to come from a point in front of the lens.

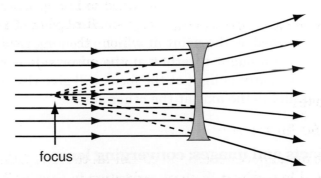

Figure 3.8 Diverging lens.

3.2.2 *Fermat's principle applied to lenses*

The *principle of least time* is most easily seen to apply directly to Figure 3.7. The job of the lens is to gather all light rays which strike it on the side facing the beam, and bring them to a focus on the other side. The lens does this by providing a number of *paths of equal length in time*. Light travels more slowly in glass than in air. The path along the straight line from O to I is actually made *longer in time* by slowing down the light as it passes through the thickest part of the lens. Rays going from O to I by one of the more roundabout routes have to traverse a smaller thickness of glass. Light rays are presented with these other routes which take the same time as the central route.

The problem is to find the shape of the lens to ensure that *the smaller width at any point off-axis exactly compensates for the extra length of the journey!*

Expensive lenses are complex in shape and may have many components, but it turns out that a single lens with spherical surfaces works quite well, particularly for rays close to the optical axis. The focal point of a converging lens is defined as the point at which incoming rays parallel to the axis are brought together at the other side of the lens. Conversely, a source of light at the focus will give rise to a beam of light parallel to the axis at the other side.

3.3 Objects and images: converging lenses

3.3.1 *Ray tracing through a thin lens*

A lens which is thicker in the middle than at the ends always acts as a converging lens, even when the curvature is convex on one side and concave on the other. It can also be shown that turning the lens around will not change its focusing properties. This means the focal length of a lens is the same on both sides, regardless of its shape, as illustrated in Figure 3.9. Accordingly, lenses are designated by a single value of *f* which applies to either side of the lens.

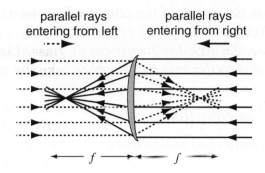

Figure 3.9 Focal points are at the same distance from both sides of a thin lens.

3.3.2 *Principal rays (thin lenses)*

We can also draw rays which pass through certain selected points for lenses, as we did for mirrors. In Figure 3.10 we have just three such rays and not four, because there is no such thing as the centre of curvature for a lens as a whole; each surface has its own centre of curvature and neither has direct significance in ray tracing for the complete lens.

Ray (1) enters parallel to the optic axis, and passes through the focus on the other side.

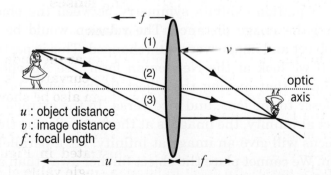

Figure 3.10 Principal rays passing through a convex lens.

*Ray (2) strikes the centre of the lens, and passes through with-
out deviation.*

*Ray (3) passes through the focus, on its way to the lens, and
exits parallel to the axis on the other.*

3.3.3 *The lens equation*

The object distance and image distances are related to the focal
length of the lens:

$$\frac{1}{u} + \frac{1}{v} = \frac{1}{f} \qquad \text{(The lens equation)}$$

$$\frac{\text{size of image}}{\text{size of object}} = \frac{v}{u}$$

These equations are derived in Appendix 3.2.

Magnification is obviously related to the above ratio and can
be defined in such a manner as to indicate the orientation of the
image. In our convention we define magnification as $m = -v/u$.
Notice that when the ratio v/u is *positive*, the image is *upright*,
and when it is *negative*, the image is *inverted*.

3.3.4 *Symmetry*

The lens equation exhibits symmetry between the object distance and the image distance. The relation would be just as
valid if object and image were interchanged. This is particularly
obvious if we look at the Newtonian form (see Appendix 3.2),
which shows that moving the object further away brings the
image closer to the lens, and vice versa. In the extreme case of
the object at infinity, the image is at the focus; placing the object
at the focus will give an image at infinity, i.e. a parallel outgoing beam. We cannot place the object further away than at infinity, therefore *the image of a real object is never closer to the lens
than the focal distance.*

3.3.5 *Breaking the symmetry*

We can now raise an interesting question relating to symmetry. The image is never closer than the focal point of the lens. We have just stated that image and object may be interchanged, and the relation remains valid. But there is nothing to stop us from *placing the object inside the focus!* Does this not break the symmetry? Let us see what happens when we do this:

The rays, having passed through the lens, appear to diverge from an image on the same side of the lens as the object. Instead of the object, we see a *virtual image* which is upright and magnified, as illustrated in Figure 3.11. The lens can be used in this way as a magnifying glass.

The magnifying glass will work provided we hold it such that the object is at or inside the focus.

We can derive relations for the magnifier directly from the lens equation:

$$\frac{1}{u} + \frac{1}{v} = \frac{1}{f} \quad \Rightarrow \quad v = \frac{uf}{u-f}$$

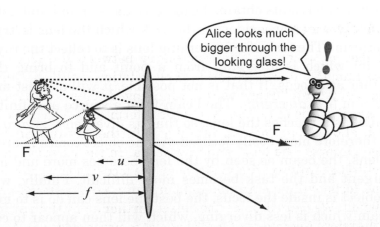

Figure 3.11 Object placed nearer to the convex lens than the focus — a simple magnifier.

The image distance, v, *is negative* if $u < f$ *and* the image is *virtual*.

Magnification $m = -\dfrac{v}{u}$ (m is positive since v is negative)

$\Rightarrow$ The image is *upright*

The closer the object is to the focus, the *smaller* is the denominator $u - f$ *and the larger the magnification.*

How about object–image symmetry? Can we somehow 'force' an image to appear inside the focus? To do this we need to have a converging beam in the first place and light always diverges from all points on a normal physical object. We need a second lens to create a converging beam which is heading towards an image when intercepted by the lens in question. This image (which the rays never reach) will act as a virtual object and lead to a final image inside the focus of the second lens.

3.3.6 *An intuitive approach — the task of a lens*

We have derived the properties of the image and its distance from the lens from the lens *equation* and also by tracing *principal rays*. The results obtained can be visualised and understood in another way by considering the task which the lens is 'trying to perform'. The task of a converging lens is to collect the rays of a beam which is *diverging* from a point and to bring them together at a focus. If that is not possible, it will at least make the beam *less diverging*. The beam from a source at infinity is parallel as it reaches the lens, so that its divergence is zero and it is therefore most easily focused. As the object comes nearer to the lens, the beam as seen by the lens becomes more and more divergent and the task becomes more difficult. Finally, when the object is inside the focus, the best the lens can do is to make a beam which is less diverging, which will then appear to come from an object which is further away, on the same side as the actual object. It will also appear larger in size because, as seen in Figure 3.11, the ray through the centre of the lens has to be projected back and upwards before it meets the other rays.

Questions

1. A lens is masked so that only half of it is in use, as seen in Figure 3.12:

   Which is correct?

   1.1. We get an image of the right side of the object only.
   1.2. We get an image of the left half of the object only.
   1.3. We still get the entire image but it is of lower intensity.

Figure 3.12 Lens mask I.

2. A lens is masked so that only the central portion is in use, as seen in Figure 3.13:

   Which is correct?

   2.1. We get an image of the central part of the object only.
   2.2. We get only a tiny image of the whole object.
   3.3. We get the entire image as before, but of lower intensity and greater clarity.

Figure 3.13 Lens mask II.

Answers

Both questions above can be answered when we recall that from any given point on the object rays go to the corresponding point on the image *via every point on the lens*. Statement No. 3 is therefore correct in both cases. We still get the entire image, but fewer rays.

In question No. 2 the rays we get are more selective in that they better fulfil the paraxial approximation. The image is therefore clearer, if less intense.

On a bright day the aperture on a camera lens is reduced as we have sufficient light to use only the central spot on the lens. The same is true for the human eye where the iris accommodates, according to lighting conditions.

3.4 Objects and images: diverging lenses

The principal rays in Figure 3.14 have been drawn according to rules similar to the previous ones:

Ray (1) enters parallel to the optic axis, diverted so that backward projection passes through the focus.
Ray (2) strikes the centre of the lens and passes through without deviation.
Ray (3) sets out from the object in the direction of the focal point on the right, and exits the lens parallel to the axis.

The rays in the diagram all come from the same point at the top of the object; similarly, rays which diverge from every other

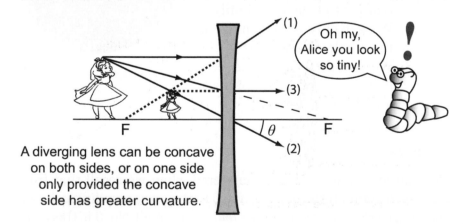

A diverging lens can be concave on both sides, or on one side only provided the concave side has greater curvature.

Figure 3.14 Principal rays passing through a convex lens.

point are made more divergent by the lens. The image is *always virtual*, on the same side as the object, and *always smaller than the object and nearer to the lens.*

3.5 Lens combinations

3.5.1 *A general method*

We can calculate the position and size of the final image produced by two lenses, in the following way. We start by finding the position and size of the image created by the first lens. This image becomes the object for the second lens. It can be a real or virtual object, depending on whether it is situated in front of or behind the second lens. The position and size of the final object are then calculated. The process may be repeated for any number of lenses.

3.5.2 *Examples — lenses in contact*

Two *thin lenses in contact* have focal lengths $+f_1$ and $+f_2$ respectively. What is the focal length (F) of the combination?

Note the emphasis that both focal lengths are positive, i.e. we have two converging lenses. A parallel beam strikes the first lens as illustrated in Figure 3.15. This will be focused at a distance f_1 by the first lens and serve as a *virtual* point

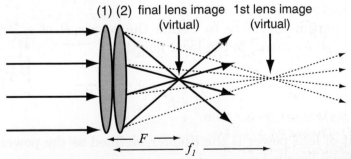

Two converging lenses together make a yet stronger combination.

Figure 3.15 Combination of two converging lenses in contact.

object for the second lens. It is a virtual object since the light rays never reached that point, but were intercepted by the second lens.

By definition, final image distance = focal length of combination.

Applying the lens equation $\dfrac{1}{u} + \dfrac{1}{v} = \dfrac{1}{f}$ to lens (2):

$u = -f_1$ (virtual object distance)
$v = F$ (final image distance)
$f = f_2$ (focal length of lens under consideration)

$$\Rightarrow -\frac{1}{f_1} + \frac{1}{F} = \frac{1}{f_2}$$

$$\Rightarrow \frac{1}{F} = \frac{1}{f_1} + \frac{1}{f_2}$$

If one of the lenses is *divergent* we insert a *negative* value for the focal length f for that lens.

3.5.3 *The power of a lens*

The shorter the focal length of a lens, the closer to itself it can focus a beam, i.e. the greater the change in the direction of light rays it produces.

It is logical therefore to define the *power P* of a lens as *inversely proportional* to the local length:

$$P \propto \frac{1}{f}$$

The unit of lens power is the *diopter*, defined as the power of a lens of focal length 1 m.

For thin lenses in contact the power of the combination is simply the sum of the powers of the individual components, remembering that the numerical value of P is *negative* if the lens is *divergent*.

Lenses and, to a lesser extent, mirrors are the central components of optical instruments. The most fundamental such instrument is the human eye, which brings the incoming light to a focus, performs a preliminary analysis of the signal, and then transmits the information on to the brain.

3.6 The eye

3.6.1 *The structure of the eye*

Light enters the eye through a thin transparent membrane called the *cornea* — the most powerful part of the *lens system*, where the main part of the bending of light occurs. Behind the *cornea*, the *crystalline lens*, controlled by the ciliary

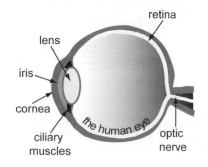

muscles, adds fine-tuning to the focusing process to produce a sharp image on the *retina*, which is a *photosensitive* membrane and acts like an array of miniature photocells. These microscopic units (there are more than 100 million of them) are called *rods* and *cones* and respond to light by generating an electrical signal. This signal is carried to the brain by the *optic nerve*. The light passes through the *pupil*, an adjustable circular opening in the centre of the *iris* which regulates the intensity of light entering the eye. The iris is beautifully pigmented and gives the eye its colour.

The ciliary muscles in the normal human eye are relaxed when one is looking at distant objects, i.e. objects 'at infinity'. The lens then focuses a parallel beam of light on the retina which is about 2 cm from the lens, as illustrated in Figure 3.16.

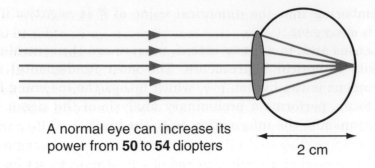

A normal eye can increase its
power from **50** to **54** diopters

2 cm

Figure 3.16 A normal relaxed eye focuses parallel light on the retina.

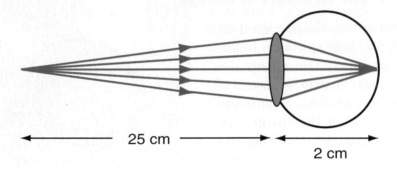

25 cm

2 cm

Figure 3.17 Least distance of distinct vision (not to scale).

When the muscles are relaxed, the power of the lens is
$P = 1/f = 1/0.02 = 50$ diopters.

When one is looking at nearby objects, the tension of each ciliary muscle is increased. As a result the curvature of the surfaces
increases and the focusing power increases. This process is called
accommodation. At maximum power, an object at a distance of
25 cm can just be kept in focus, as illustrated in Figure 3.17.

We can calculate the maximum power easily from the lens
equation:

$$P = \frac{1}{f} = \frac{1}{u} + \frac{1}{v} = \frac{1}{0.25} + \frac{1}{0.02} = 4 + 50 = 54 \text{ diopters}$$

A normal eye can increase its power from 50 to 54 diopters.

The image on the retina is *inverted* (see Section 3.3) — a fact which a baby realises when it has to reach *up* in order to touch something which must at first appear to be down below! In 1896, G.M. Stratton carried out an experiment, in which subjects had to wear special inverting glasses for an extended period. He found that, after a few days, the world which they saw through these glasses appeared to them to be quite normal. In fact, after they had taken off the glasses, they became confused and needed a further period of adjustment to get back to normal vision.

In front of the lens is the iris, within which there is a variable aperture called the pupil which opens and closes to adjust to changing light intensity. The receptors on the retina also have intensity adaptation mechanisms. As a result, the eye is adaptable to an enormous range of light intensity. For example, sun-

When light intensity doesn't matter......

light has an intensity 100,000 times greater than moonlight, which in turn is 20,000 times more intense than starlight. Yet one can easily distinguish objects at night, particularly when the moon is full.

3.6.2 *Common eye defects*

Myopia (short-sightedness)

Parallel light is focused at a point in front of the retina, as the power of the lens is too great. (Either the curvature of the lens is too large, or the eye is longer than is normal.) As a result the person cannot

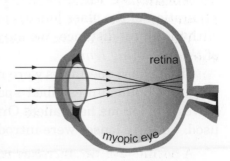

retina

myopic eye

focus clearly on points further away than a certain distance called the *far point.*

Diverging lenses will correct the defect.

Hypermetropia
(long-sightedness)

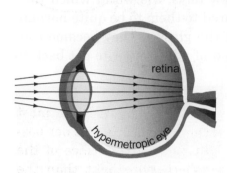

Even with maximum accommodation, the power of the lens is not sufficient to obtain an image of objects at a distance of 25 cm; the *near point* is further from the eye.

Such a person needs *reading glasses* with *converging* lenses.

Presbyopia

The inability to accommodate fully often increases with age. This defect may be corrected with bifocal lenses; the upper part of the lens is used for distant vision and the lower for close work.

Astigmatism

This defect is caused by irregularities in the lens, or elsewhere in the eye, and causes distorted vision. It can be compensated by using lenses with cylindrical or more complicated surfaces.

Whether we use contact lenses or ordinary spectacles, the separation of the glass lens from the eye is small in comparison with the object distance. We may make calculations on the basis of *thin touching lenses.*

The earliest spectacles were made in the 14th century. In 1363 Guy de Chauliac prescribed spectacles as a last remedy when salves and lotions had failed. Originally only convex lenses were used. Concave lenses were introduced later, in the 16th century.

3.7 Making visible what the eye cannot see

3.7.1 *Distant objects*

Telescopes enable us to see and examine objects which are far away; microscopes, to see and examine very small objects. The sun, the moon and the other planets in our solar system can be studied even with relatively simple telescopes. Most stars

other than the sun are so far away that, even with the most powerful telescopes, we get a *point image* and it is just not possible to observe features that might be on the surface of the stars. However, a telescope does collect more light than the naked eye, so the image is very much brighter and countless new stars become visible.

Hubble stamp. *Courtesy of An Post, Ireland.*

Betelgeuse. *Courtesy of A. Dupree (CFA), R. Gilliland (SIScI), NASA, ESA.*

In 1995, NASA published a picture (taken with the Hubble Space Telescope) of the star Betelgeuse, consisting of a central disc with the halo of an atmosphere around it. Betelgeuse is one of our nearest stars, 'only' about 430 light years away, and is very large, with a radius about 1800 times the radius of the sun.

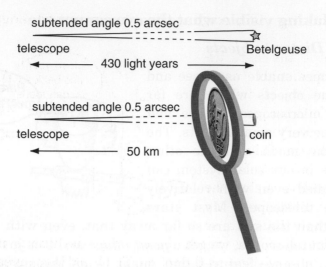

Figure 3.18 The angle subtended by one of our nearest stars.

Magnification, defined in terms of the ratio of the size of the image to the size of the object, does not of course make sense when we are viewing astronomical objects, particularly when we have a point image. However, when the image is of finite size, we can measure the *power* of a telescope in terms of the smallest *angular* size which can be resolved. This is the angle between rays from opposite sides of the object and is also a measure of the angle *subtended* by that object.

In the case of Betelgeuse, the angular size is about 0.5 *arc seconds*, equivalent to the angular size of a 50-cent coin at a distance of 50 km, as illustrated in Figure 3.18.

3.7.2 *Nearer but not clearer*

The situation is quite different in the case of an object at a finite distance from us. We can obviously increase the angular size of the object by standing closer to it. As we come nearer and nearer, we can distinguish more detail, but there is a limit. When we get inside the least distance of distinct vision, the clarity diminishes, and the image becomes blurred. Any advantage gained by increased angular size is lost.

'Inside' the least distance of distinct vision, the images of nearby points overlap, and the letters on the worm's chart merge. Things are nearer, but not clearer. The problem is that the rays of light from any point on the object are so

divergent when they reach the eye that the lens is not strong enough to focus them on the retina. The 'task' is too great and the strain on the ciliary muscles will most probably lead to a headache!

We need a *real* image on the retina to transmit to the brain; rays from every point on the object should be sharply focused at a corresponding point on the retina if we are to see a 'true' image of the object.

As we get very close to an object, its angular size increases, but it becomes impossible to focus rays from any given point on the object.

Figure 3.19 shows two sets of rays, one emerging from the top of an object and the other from the bottom. The lens is not strong enough to focus either set of rays to a single point on the retina. The result is a jungle of overlapping rays reaching the retina and confused messages reaching the brain!

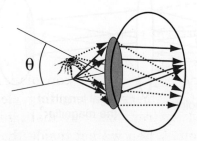

Figure 3.19 Too close for comfort.

How can we increase the angular size of the object without making the rays from individual points on the object too divergent?

3.7.3   *Angular magnification*

The simple magnifier

A single convergent lens may be used as a magnifying glass to provide the simplest method of increasing angular magnification without blurring. Placing the object inside the focal point of the lens gives a magnified virtual image, as shown in Figure 3.20. In fact, if we place the object *at the focus* of the magnifier, the rays from each point emerge parallel and can be viewed comfortably with the relaxed eye.

The angle subtended by the object as seen from the centre of the lens is: $\theta \cong \tan\theta = h/f$.

As far as the observer to the right of the lens is concerned, this is also the angle between the rays coming from points at the top and the bottom of the object. These rays are parallel and can be focused comfortably on the retina. The function of the magnifying glass has been effectively to bring the object close up to a distance f from the observer without blurring.

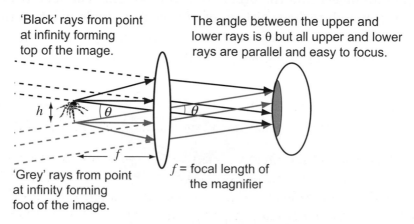

'Black' rays from point at infinity forming top of the image.

The angle between the upper and lower rays is θ but all upper and lower rays are parallel and easy to focus.

'Grey' rays from point at infinity forming foot of the image.

f = focal length of the magnifier

Figure 3.20 The simple magnifier.

Figure 3.21 Simple magnifier.

The *angular magnification* is defined as the ratio of the angular size as seen through the magnifier (θ) compared to angular size (φ) when the object is placed as close as possible to the normal eye, at the least distance of distinct vision ($d = 25\,\text{cm}$).

For small angles, angular size is equal to actual size/separation

$$\theta \approx \frac{h}{f} \quad \text{and} \quad \varphi \approx \frac{h}{d}$$

$$\Rightarrow \quad \textit{angular magnification} \quad \alpha \approx \frac{d}{f}$$

A normal good magnifying glass may have a focal length of 7.5 cm. The resultant angular magnification $\alpha \approx 25/7.5 \approx 3.3$.

This means that effectively you can observe an object as if it were 7.5 cm from your eyes instead of 25 cm. It is difficult to make large convex lenses of greater power without encountering distortion due to non-paraxial rays, although small high quality lenses with focal lengths of the order of millimetres are made for optical instruments. To achieve greater magnifying power it is normally necessary to use a combination of lenses forming a compound microscope.

Antoon van Leeuwenhoek (1632–1723) was a brilliant lens-maker who hand-crafted lenses of high quality and very short focal length. No one was able to match his skill at the time and microscopy suffered a great setback after his death until the development of compound microscopes about 100 years later.

3.8 Combinations of lenses

3.8.1 *Compound microscopes*

The simplest *compound microscope* involves two lenses. The *objective* lens has a short focal length and the object under study is placed just outside the focal point of the lens. This produces a real, inverted image, which is much larger than the object. (The magnification is typically ×50.)

This image is viewed by the *eyepiece,* which is used as a *magnifying glass*. The final image is *virtual*, *inverted* and further magnified.

Figure 3.22 illustrates the principle of the compound microscope. The objective lens, with a focal length f_1 (typically about

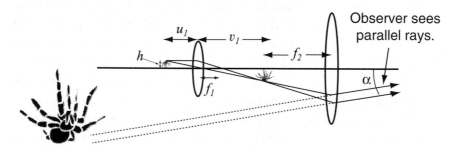

The final image is at infinity and subtends an angle α for the observer. Rays from any given point are parallel.

Figure 3.22 Compound microscope (not to scale).

3 mm), forms a real and inverted image of an object placed little more than 0.05 mm beyond its focal point, at a distance of about 15 cm. The magnification of this intermediate image is $m \approx -v/u \approx 15/0.3 = 50$. If the object has a lateral size of h, the size of the intermediate image is 50 h. This image is placed at the focus of the eyepiece and forms an object for the eyepiece. The angular magnification, $\alpha \approx d/f_2$, where f_2 is the focal length of the eyepiece and d is the least distance of distinct vision.

If $f_2 = 4$ cm,

$$\text{Angular magnification of the final image} = \frac{50 \times 25h}{4} = 312.5.$$

The notion of an image 'at infinity' suggests that such a thing might be difficult to examine. In fact, the opposite is true. The relaxed eye can comfortably focus parallel rays on the retina. The final image produced by the microscope serves as an object for the eye and can be imagined as being composed of a series of points. Rays reaching the eye from any one point are parallel. Rays from different points are not parallel. In fact, the angle between the beam of rays coming from the point at the top and the beam from the point at the bottom determines the angular magnification.

In high power microscopes, the eyepiece and objective lenses are themselves combinations of lenses. Such combinations are designed not only to increase magnification but also to minimize distortion due to lens aberrations. In addition, all but the simplest microscopes have binocular eyepieces, essential if they are to be used for an extended period of time.

When one is looking at an object close at hand, it is unnatural to focus one's eyes for infinity and even microscope operators may take time to adapt. It may also take time to adapt to binocular eyepieces so that one image is seen by both eyes, and not two. In cases where it matters, one must also get used to the fact that the image is upside down!

3.8.2 *Telescopes*

The Dutch optician Hans Lippershey (1570–1619) succeeded in 'making distant objects appear nearer but inverted' by means of a tube inside which were mounted two spectacle lenses. He applied to patent his invention, but it was refused on the basis that the instrument would be 'too easy to copy'.

The principle of the telescope is similar to that of the microscope, in that the objective lens forms an image, which serves as an object for the eyepiece and produces a final image which is inverted. The main difference between the two is that the purpose of the telescope is to look at distant objects, while the microscope is designed to look at small objects close at hand.

In the *refracting telescope* the objective is large, to gather as much light as possible, and has a relatively large focal length.

In Figure 3.23 the eyepiece is a simple convex lens. The incoming light passes through the focus of the eyepiece, which forms an inverted final image.

In the *Galilean telescope* (Figure 3.24), the objective is a convex lens and the eyepiece is a concave (diverging) lens or lens system.

A concave lens placed inside the focal point F of the objective acts in the same way as a convex lens placed at or beyond F,

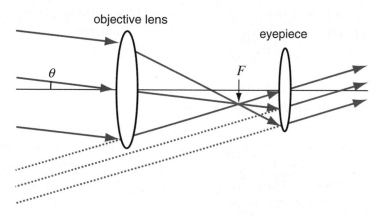

Figure 3.23 Refracting telescope.

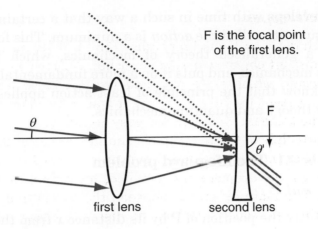

F is the focal point
of the first lens.

first lens   second lens

Figure 3.24 Galilean telescope (the image is upright).

except that the image produced by the concave lens is not inverted. This has an obvious advantage for terrestrial use.

The angular magnification is $\dfrac{\theta'}{\theta}$.

3.9 A final note on Fermat's principle

Fermat was ahead of his time in formulating a law of Nature on the basis of a minimum principle. Some years later Maupertius (1698–1759) enunciated a more general law, *'Nature is thrifty in all its actions'*, but this was a somewhat qualitative statement with much less supportive experimental evidence compared with Fermat's *principle of least time* for light.

The pioneers of generalised classical mechanics

A quantitative mathematical theory was subsequently developed over a period of a century by Leonhard Euler (1707–1783), Joseph Louis Lagrange (1736–1813) and William Rowan Hamilton (1805–1865). It was based on a statement known as the *principle of least action*. This principle applied not only to light but also to all mechanical systems, and stated that every system develops with time in such a way that a certain mathematical quantity called the *action* is a minimum. This forms the basis of a generalised theory of mechanics, which includes Newton's mechanics, and puts it on a more fundamental footing. We now know that the principle of least action applies also to relativity theory and quantum mechanics.

Appendix 3.1 The lifeguard problem

To find point P

Let us specify the position of P by its distance x from the foot of the perpendicular dropped from A, as shown in Figure 3.25.

The total time T for the journey can be expressed in terms of the fixed parameters v_1, v_2, h_1, h_2, and d as well as the only parameter over which the lifeguard has some control, namely x:

$$T = \frac{\sqrt{h_1^2 + x^2}}{v_1} + \frac{\sqrt{h_2^2 + (d-x)^2}}{v_2}$$

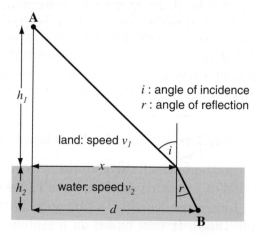

Figure 3.25 The lifeguard's path.

To calculate the minimum time, set the first derivative of time with respect to x equal to zero:

$$\frac{dT}{dx} = \left\{ \frac{x}{\sqrt{h_1^2 + x^2}} \middle/ v_1 \right\} - \left\{ \frac{d - x}{\sqrt{h_2^2 + (d - x)^2}} \middle/ v_2 \right\}$$

$$\Rightarrow \frac{v_1}{v_2} = \frac{x \big/ \mathrm{AP}}{(d - x) \big/ \mathrm{PB}}$$

$$\frac{v_1}{v_2} = \frac{\sin i}{\sin r} \quad \text{(Formula for the quickest route)}$$

The lifeguard will have to calculate the optimum position of P such that the angles i and r satisfy the above relation!

Note that in the calculation given above we expressed the time as a function of the variable x and then equated the derivative dT/dx to zero. This gave us the *stationary point* of the function which could be a *maximum, minimum or saddle point*. To determine which it is, mathematically, we could obtain the second derivative and determine whether that is negative, positive or zero at that point. Physically it is obvious that it is in fact a minimum in this example. A more precise statement of Fermat's principle refers to stationary rather than minimum values of the time.

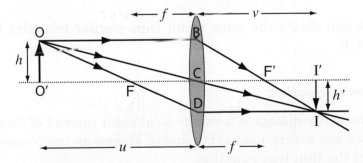

Figure 3.26 Principal rays through a convex lens.

Appendix 3.2 The lens equation

We can use Figure 3.26 to relate object and image distances to show how the size of the image is related to that of the object.

From similar triangles COO′ and CII′:

$$\frac{h'}{h} = \frac{v}{u} \tag{3.1}$$

From similar triangles F′BC and F′II′:

$$\frac{h'}{h} = \frac{v-f}{f} \quad \text{[Rays (1) and (2)]} \tag{3.2}$$

Combining equations (3.1) and (3.2)

$$\Rightarrow \frac{v}{u} = \frac{v-f}{f}$$

$$uv - uf = vf \quad \Rightarrow \quad vf + uf = uv$$

Dividing by uvf:

$$\frac{1}{u} + \frac{1}{v} = \frac{1}{f} \quad \text{(Lens equation)}$$

We can obtain the same result from similar triangles FOO′ and DCF.

Newton's equation

If we measure distances from the focal point instead of from the lens, we get a very neat relationship known as the *Newtonian form* of the thin lens equation.

Using Figure 3.26:

Object distance from focus, $x = O'F$
Image distance from focus, $x' = I'F'$

$$\frac{h}{x} = \frac{h'}{f} \Rightarrow \frac{h'}{h} = \frac{x}{f}$$

$$\frac{h'}{x'} = \frac{h}{f} \Rightarrow \frac{h'}{h} = \frac{x'}{f}$$

$$xx' = f^2 \quad \text{(Lens equation in Newtonian form)}$$

Appendix 3.3 Calculating the power of spectacles

(We can assume that the spectacle lens and the eye lens are in contact.)

A *healthy young adult* can accommodate the eye so that it focuses on objects 25 cm away (the *near point*). The power P of the fully accommodated eye can be calculated from the lens equation:

$$P = \frac{1}{f} = \frac{1}{u} + \frac{1}{v} = \frac{1}{0.25} + \frac{1}{0.02} = 4 + 50 = 54 \text{ diopters}$$

A *patient with a sight defect* can see objects clearly only when they are at a distance of between 0.8 m and 2.5 m from his eyes. What is the power of the lens which will extend his clear vision to the normal range of 0.25 m to infinity?

The problem here is *presbyopia,* described in Section 3.6. We need to prescribe separate glasses for reading and for distant vision.

Reading. The patient needs spectacles to aid accommodation and restore his near point to 0.25 m, as illustrated in Figure 3.27.

Normal eye: Fully accommodated power = 54 diopters (Example 1).

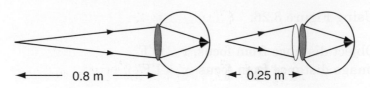

0.8 m   0.25 m

Figure 3.27 Reading glasses (not drawn to scale).

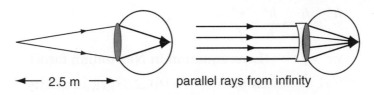

2.5 m   parallel rays from infinity

Figure 3.28 'Distance' glasses (not drawn to scale).

Patient's eye:

Fully accommodated power $= \dfrac{1}{u} + \dfrac{1}{v} = \dfrac{1}{0.8} + \dfrac{1}{0.02}$

$= 1.25 + 50 = 51.25$ diopters.

The spectacles must be fitted with convex (converging) lenses with a power of $54 - 51.25 = 2.75$ diopters to restore his near point to 0.25 m.

Distant vision. The normal eye focuses an object at infinity on the retina when fully relaxed. The defective eye focuses an object at infinity in front of the retina (it is still 'too strong') when fully relaxed.

Normal eye: 'Fully relaxed' power = 50 diopters.

Defective eye:

'Fully relaxed' power $= \dfrac{1}{u} + \dfrac{1}{v} = \dfrac{1}{2.5} + \dfrac{1}{0.02} = 50.4$ diopters.

The spectacles must be fitted with concave (diverging) lenses to restore his far point to infinity, as shown in Figure 3.28. The power of the lenses is $50 - 50.4 = -0.4$ diopters.

Chapter 4

Light from Afar — Astronomy

When light reaches us after a long, long journey

We begin with the ancient astronomers, who did remarkably well considering that they had nothing but the naked eye and a lot of ingenuity to help them! They determined that the earth is round, and were able to calculate its radius. They progressed to make calculations of the size of the moon and the size of the sun, and obtained very good estimates of the distance from each to the earth.

As measurements became more accurate, astronomers in the Middle Ages were able to make models of the paths of the planets in our solar system. The first models were very complicated, describing the paths of the planets from the point of view of the earth, which was assumed to be stationary, at the centre of the system. Nicolas Copernicus showed that this model was incorrect and put forward a new model with the sun at the centre, and the earth and the other planets in orbit around it.

Light is the messenger which brings us news from other parts of the universe. No information can travel faster than light, but the distances are so great that the news is millions and even billions of years old.

81

4.1 The earth

4.1.1 *Is the earth round?*

The earth from the spacecraft Apollo

Modern satellite pictures clearly show that the earth is round. To the ancient astronomers the shape of the earth was, however, not at all obvious, but they had both philosophical and experimental reasons for believing that it is round.

Earth. *Courtesy of NASA.*

Greek philosophers often argued that it should be possible to reconstruct Nature, and its laws, on the basis of the premise that *things exist and behave in the way that is 'most perfect'*. The case for a round earth was made by Aristotle (340 BC), on a number of such 'philosophical' grounds.

4.1.2 *Philosophical reasons why the earth should be round*

Symmetry. A sphere is the most perfect geometrical figure. Nowadays, this might sound like a rather weak argument. One must remember, however, that symmetry in Nature and in its laws is still sought as a foundation to physical theories, often with considerable success.

Gravity. Aristotle postulated that the component pieces of the earth should seek their natural home at the centre. They are attracted towards the centre and therefore tend to compress into a round form. This reasoning seems to be a combination of both philosophical and physical. About 2000 years later Isaac Newton formulated his universal law of gravity in a precise mathematical form, but the fact that things fall obviously had to be justified since time immemorial.

General intuition — Zhang-Heng (Chinese astronomer, 78–139)
'The sky is like a hen's egg, and is as round as a crossbow pellet; the Earth is like the yolk of the egg, lying in the centre.'

There seems to be no record of the reasoning behind this statement, either physical or philosophical. Nevertheless, it shows artistic expression, typical of the ancient Chinese civilization.

4.1.3 *Experimental evidence that the earth is round*

Terrestrial measurement

Eratosthenes (≈ 330 BC) knew that at noon on the first day of summer the sun was directly overhead at Syene (near today's Aswan Dam). Sunbeams shone directly down a deep vertical well,

and were reflected straight up by the water at the bottom. At exactly the same time, at his home near Alexandria, he measured the sun's rays to be about 7.5 degrees away from the vertical. Eratosthenes knew the distance from Syene to Alexandria, which in modern terms is about 500 miles, almost directly due north. Hence he calculated the circumference of the earth, as described in Figure 4.1.

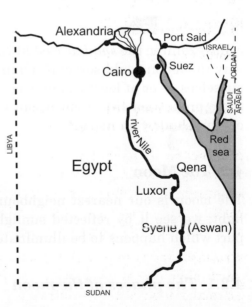

Astronomical measurement

Of course, the ancients were not in a position to travel into space in order to observe the earth 'as others see it'. However, they

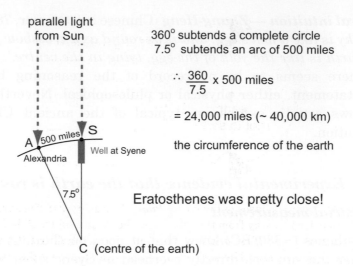

Figure 4.1 The method Eratosthenes used to show that the earth is round.

could examine the *shadow cast by the earth on the moon*. In this way the moon could serve as a 'mirror' in which we could view ourselves — or at least examine our shadow! Let us first look at some properties of the moon and then come back to study the earth's shadow on its surface.

4.2 The Moon

The moon is our nearest neighbour in space. It does not emit light; we see it by reflected sunlight, but we can only see the part which happens to be illuminated.

4.2.1 *The phases of the moon*

Ancient astronomers knew that the moon shines by reflected sunlight, and that it orbits the earth approximately once every four weeks. We can observe the changing *phases*, which depend precisely on how much of the *sunny side* of the moon can be seen from the earth at a given time, as illustrated in Figure 4.2.

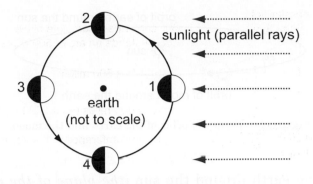

Figure 4.2 The four phases of the moon. *Phase 1 (new moon)*: The sunny side of the moon is facing away from the earth, and the moon is not visible from the earth. *Phases 2 and 4 (first and last quarters)*: We see half the illuminated side. *Phase 3 (full moon)*: The sunny side of the moon is facing the earth. We see the complete illuminated hemisphere.

4.2.2 *A lunar eclipse*

Occasionally the moon, or part of it, is in shadow for a completely different reason. The sun's rays may be obscured because the earth gets 'in the way'. This is called a *lunar eclipse*. During a total eclipse there is not quite total darkness; we see a slight glow due to sunlight scattered by the earth's dusty atmosphere.

Lunar eclipse on March 3, 2007. *Courtesy of The Irish Times, Bryan O'Brien.*

It might appear that a lunar eclipse always occurs when there is a 'full moon' (phase 3). In fact, this is usually not the case because the earth, moon and sun are rarely directly in line. Figure 4.2 is just a representation in two dimensions, whereas the real situation is one of three dimensions, as illustrated in Figure 4.3.

Here, however, it can be seen that the plane in which the moon orbits the earth is not the same plane as that of the

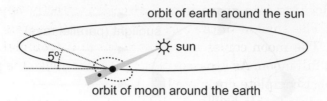

orbit of earth around the sun

☼ sun

5°

orbit of moon around the earth

Figure 4.3 Planes of orbit of the earth and the moon.

orbit of the earth around the sun (*the plane of the ecliptic*). The angle between the two planes is about 5°. A lunar eclipse occurs only when the moon is full and happens to be crossing the plane of the ecliptic when the sun, earth and moon are in a straight line.

During a lunar eclipse the moon passes across the earth. From the shape and size of the shadow we can deduce the shape and size of the earth. In Figure 4.4,

moon

shadow of the earth

Figure 4.4 Seeing our own shadow.

we are looking into space in the direction directly opposite to the sun. The shadow of the earth stretches out from us directly in front. The moon comes into view. We see its complete sunny side — a full moon. As it passes through our shadow it is plunged into almost complete darkness!

It is remarkable how well the Greeks were able to observe eclipses of the moon and to make measurements. They also realised that the size of the shadow was not the same as the size of the earth, and made the necessary corrections, which we will describe in the next section.

4.2.3 *A solar eclipse*

A solar eclipse occurs when the moon is directly in line between the earth and the sun, and casts its shadow on the earth. The situation is now reversed in the sense that the moon obscures the sun

from us rather than the earth obscuring the sun from the moon. When we are at the centre of the shadow we experience almost complete darkness; all light from the sun is obscured.

During a total eclipse, the sun is almost completely hidden by the moon, showing that the angular size ($\approx 0.5°$) of the sun and of the moon are approximately equal when viewed from the earth, as illustrated in Figure 4.5.

From similar triangles it is clear that the ratio of the diameter of the sun to the diameter of the moon is almost exactly

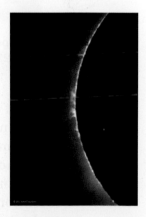

Total eclipse of the sun. *Courtesy of Steele Hill, NASA.*

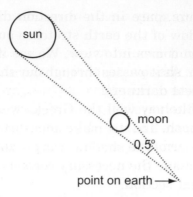

Figure 4.5a Ray diagram of a solar eclipse.

Figure 4.5b Solar eclipse on August 11, 1999. Shadow of the moon on the earth. *Courtesy of Mir 27 crew; © CNES.*

equal to the ratio of our distance from the sun to our distance to the moon. The moon's shadow on the earth is just a point, or at least very small compared to the size of the moon or the earth.

The result of a solar eclipse is clearly seen in the photograph (Figure 4.5b), taken from the *Mir* space station. It shows the shadow of the moon, with a diameter of about 100 km, as it moves across the earth. As the earth rotates, the shadow travels at about 2000 km/h. Had satellites been available to the ancient astronomers, they would, one would imagine, have been impressed!

The moon's orbit around the earth is not exactly circular, and the size of the shadow depends on the exact distance of the moon at the moment of the eclipse. Sometimes the moon may be just too far, and the rays drawn in Figure 4.7 may meet at a point above the earth. There is then no region of complete totality on the earth's surface, and a thin ring of sunlight can be seen surrounding the moon. Such an occurrence is called an *annular* eclipse.

It is interesting to point out a major difference between lunar and solar eclipses. A lunar eclipse is seen at the same time from everywhere on the earth. A solar eclipse occurs at different times along the band covered by the shadow, as the earth rotates, and the shadow travels across the earth's surface.

4.3 Sizes and distances

4.3.1 *Relative sizes of the sun and the moon*

We already have one piece of information from Figure 4.5a in the previous section, namely:

$$\frac{\text{diameter of the sun}}{\text{diameter of the earth}} = \frac{\text{distance from the earth to the sun}}{\text{distance from the earth to the moon}}$$

This ratio was derived from the fact that the moon and the sun happen to look about the same size in the sky when viewed from the earth. The equation itself does not provide any information about their actual size, which one is smaller and nearer and which one bigger and further away. Of course it is obvious that, since it is the sun which is eclipsed, the sun must be the more distant object.

Aristarchus ($\approx$ 310–230 BC) devised very clever ways of determining astronomical sizes and distances. The size of the shadow of the earth on the moon was his next clue in a remarkable chain of logic, and led to a comparison of the actual size of the earth and of the moon. As we shall see below, this second step is not quite as straightforward as the first.

4.3.2 *The shadow of the earth on the moon*

By measuring the time for the moon to go through the earth's shadow in a lunar eclipse, Aristarchus estimated that the diameter of the shadow cast by the earth on the moon is 2½ times the diameter of the moon.

First piece of information: *Size of the shadow of the earth at the moon = 2.5 × size of the moon.* How is this related to the size of the earth itself?

4.3.3 *Shrinking shadows*

The size of the shadow cast by an object depends on the direction of the light rays which illuminate it.

When light comes from an extended source like the sun, the region of the shadow can be divided into two areas: the *umbra*, which is an area not illuminated by light from any point on the sun, and the *penumbra*, which is illuminated by light from some but not all points on the sun (Figure 4.6). Usually, the umbra is quite well defined and it is that area of complete shade to which we refer in our discussion of both lunar and solar eclipses.

The same definitions apply in more common situations, such as the shade under a parasol on a beach. In most cases, owing to the brightness of the sun, the penumbra is quite bright and barely distinguishable from the sunlit area.

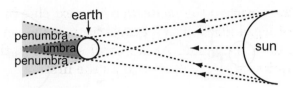

Figure 4.6 Umbra and penumbra.

If the sun were a *point* source we would expect the shadow cast by the earth on an imaginary screen situated behind the earth to spread out as the screen is moved further away from the earth. However, the sun is an *extended* source and the size of the umbra decreases as the screen is moved further away, as we may see from Figure 4.6. When the angular size of the screen becomes less than the angular size of the sun, the umbra disappears.

Common experience tells us that a parasol placed too far above the ground will not cover the whole sun and will offer little protection. In practice, scattered light will make it even less useful!

Reverting to astronomy

When the moon is obstructing the sun, the umbra of the moon's shadow has shrunk to a point by the time it reaches the earth. Similarly, when the earth is in front of the moon, the shadow of the earth gets smaller by the time it reaches the moon, but because the earth is bigger, the shadow will not be reduced to a point.

Comparing shadows

Assuming the sun is so far away that its angular size does not change over the earth–moon region, the angles of the light rays from the edge of the sun will be the same in Figure 4.7. Since the shadow of the moon tapers down to a point over the earth–moon distance, the diameter of the earth's shadow will decrease by the same amount over the same distance. Therefore

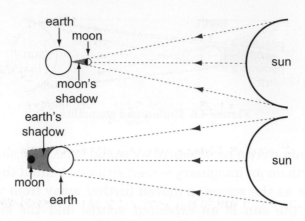

Figure 4.7 Shadows.

the shadow of the earth on the moon is narrower than the actual size of the earth by about one moon diameter.

Aristarchus was able to conclude that:

Diameter of the earth = size of the shadow of the earth on the
                moon + 1 moon diameter
          = 2.5 × diameter of the moon + 1 moon
          diameter
          = 3.5 × diameter of the moon

The circumference of the earth ≈ 40,000 km (Eratosthenes)

$$\text{The circumference of the moon} \approx \frac{40,000}{3.5} \approx 11,500 \text{ km}$$

and the diameter of the moon ≈ 3,650 km.

4.3.4 *The distance to the moon*

Aristarchus knew that the sun is not a point source. Using the information that the angular sizes of the moon and the sun as viewed from the earth are about the same, he made a correction to take into account the taper of the shadow:

Once we know the actual size of the moon, Figure 4.8 illustrates how it is easy to calculate the distance from the earth to

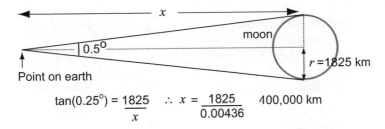

$$\tan(0.25°) = \frac{1825}{x} \quad \therefore \ x = \frac{1825}{0.00436} \quad 400{,}000 \text{ km}$$

Figure 4.8 Distance from the earth to the moon.

the moon from the angle subtended by the moon at some point on the earth.

4.3.5 *The distance to the sun*

Just one more piece of information remains to complete the jigsaw.

We have the absolute size of the moon and its distance from the earth. If we also knew the ratio of the distance from the earth to the moon and the distance from the earth to the sun, everything would fall nicely into place. Aristarchus had a very clever idea about how to obtain this information:

If we go back to Figure 4.2, which illustrates the phases of the moon, and look at it more closely, we will see that it is not completely accurate. When the moon is exactly in its first and last quarters, the sunny side in each case is tilted inwards slightly as it faces the sun. The situation is made clearer in Figure 4.9.

The line from the moon to the sun is tangential to the moon's orbital circle, and the line from the moon to the earth is a radial line. They are mutually perpendicular and the marked angles are each 90°. It is easily seen that the orbital arc from the first to the

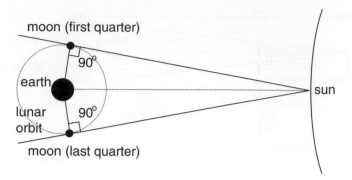

Figure 4.9 Unequal segments (not drawn to scale).

last quarter is longer than that from the last quarter back to the first. Assuming that the moon moves at a uniform speed, there will be a difference in time between the two journeys. From this time difference all the angles in the diagram can be determined, and hence the ratio of the distance between the earth and the sun and the distance between the earth and the moon. Once we know that ratio we have all the information we require.

 ### 4.3.6 *A practical problem*

The argument of Aristarchus is correct, but there was an experimental problem which he was not quite in a position to solve. As one would imagine, it is very difficult to determine the exact moment at which the moon is at a given phase. The total time taken for the moon to go through all four phases is about 30 days. Aristarchus' visual estimate was that it takes one day longer to go from the first quarter to the last quarter than from the last back to the first.

Using the data of Aristarchus, let us follow his reasoning to calculate the distance of the earth from the sun in terms of the earth–moon distance:

The total travel time $\cong$ 30 days; the long segment $\cong$ 15.5 days; the short segment $\cong$ 14.5 days.

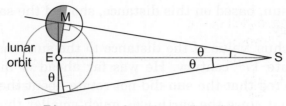

ES = distance from the earth to the sun
EM = distance from the earth to the moon

Figure 4.10 The geometry of Aristarchus' calculations.

The first quarter is 1/4 day early and the last quarter is 1/4 day late.

30 days corresponds to 360° (1 complete revolution), so 1/4 day corresponds to 3°.

This means that size of the angle θ in Figure 4.10 is 3°.

$$\frac{EM}{ES} \cong \sin 3° = 0.05$$

According to Aristarchus, the sun is about 20 times more distant than the moon.

While the method was correct, the result (which is extremely sensitive to the measurement of the time of the phases) was much too small. We now know that the distance from the earth to the sun is about 390 times larger than the distance from the earth to the moon. (The angle θ is not 3°, but 0.15°.)

Despite the fact that the final experimental answer was out by a factor of 20, one cannot but be very impressed by the reasoning of the ancient astronomers. They started with the one absolute measurement, made by Eratosthenes — that of the distance from Alexandria to Syene. There followed the determination of a series of ratios, from which they determined the sizes and distances from the earth, of the moon and the sun. All were nearly correct, except for the distance to the sun. The estimated size of the sun, based on this distance, showed the same degree of inaccuracy.

Aristarchus' value of the distance to the sun was accepted, until the late 17th century. He was far ahead of his time; he even suspected that the sun did not rotate about the earth. He reasoned that since the earth was much smaller than the sun, surely it was reasonable for it to orbit the sun rather than the other way around. This view, as we know, was not accepted for 2000 years.

4.3.7 *A summary concerning the earth, moon and sun*

The diameter of the earth is 12,750 km (a modern value, little different from the value obtained by Eratosthenes).

Relative diameters of the moon, earth and sun.

Object	Moon	Earth	Sun
What the Greeks thought	0.3	1	6
Modern values	0.27	1	109

4.3.8 *Astronomical distances*

To help obtain an intuitive feeling for these distances, let us construct a timetable for non-stop travel by a Concorde supersonic aircraft at a maximum cruising speed that is twice the speed of sound ($\approx$ 2150 km per hour or Mach 2, as in Table 4.1).

Table 4.1 Some distances and times of travel.

Journey	Distance in km	At speed of light	Concorde (one way)
Around the earth	40,000	0.13 seconds	18.5 hours
Earth–moon	3.84×10^5	1.28 seconds	7.4 days
Earth–sun	1.50×10^8	8.33 minutes	7.9 years

4.4 The planets

4.4.1 *The 'wanderers'*

The Greeks and other ancient cultures observed certain heavenly bodies which looked like very bright stars and seemed to wander across the heavens (as shown in Figure 4.11), unlike the normal fixed stars which exactly keep their positions relative to one another. The fixed stars remain in patterns we now call *constellations* while these bright objects move along paths which are certainly not straight lines or curves, and to the Greeks they were quite puzzling. They called them *planets*, from the Greek word for *'wanderer'*. The path shown in the photograph took about six months to complete. By coincidence the track of Uranus appears as the dotted line above and to the right.

4.4.2 *Ptolemy's geocentric model*

The Greeks knew of five planets: Mercury, Venus, Mars, Jupiter and Saturn. They made careful measurements of their paths, which appeared quite irregular and quite typical of a 'wanderer' moving through the skies. As illustrated in Figure 4.11, planetary paths contain loops where the planet changes direction and

Figure 4.11 The path of Mars through the constellations. *Courtesy of Tunc Trezel NASA Astronomy Picture Of the Day 16-12-2003.*

doubles back for a short period, before continuing in approximately the same direction as before.

Hipparchus of Rhodes (*c.* 180 BC–125 BC) produced a catalogue of about 850 stars which was later expanded by Claudius Ptolemy (*c.* 85–165 AD).

Both Hipparchus and Ptolemy were great astronomers and mathematicians. Considering that their measurements were made with the naked eye, looking through sighting holes aligned along wooden sticks, their accuracy was quite astounding. Some of Hipparchus' data is accurate to within 0.2° of modern values.

The two astronomers applied their mathematical skills to the data to produce a model of planetary motion. Ptolemy's work comprised 13 volumes which contained the results of many years of observation. He developed the so-called *geocentric* model, according to which the planets were in circular motion around the earth, each planet once around in its own 'year'. At the same time each planet moved in *epicycles*, or smaller circles around a point which travelled along the main orbit, as illustrated in Figure 4.12.

Ptolemy found that the simple epicycle scheme did not give predictions which were absolutely accurate, and made adjustments to his model, which made it more complicated. He had to move the centre of the planet's orbit a small distance away from the earth, and the epicycle itself was tilted from the main circle.

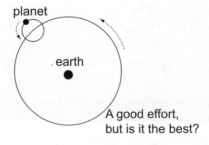

Figure 4.12 Ptolemy's geocentric model of planetary motion.

He finally produced a model which predicted the positions of all five planets with great accuracy and published his results in a book called the *Almagest*, which served astronomers and navigators for many centuries.

While the models of Hipparchus and Ptolemy predicted well the observed positions of the planets in the sky, they did not attempt to give a *reason* for planetary motion. We had to wait more than 1500 years for that!

4.5 The Copernican revolution

We now move rapidly through history to the 15th century. From the pseudo-religious point of view, the earth had to be at the centre of the universe. Natural philosophers, on the other hand, wanted to establish a model in which the earth, the other planets and the sun, and their relationship to one another, could be represented in a simpler and more logical way.

4.5.1 *Frames of reference*

Whether we realise it or not, all measurements are made with respect to a *frame of reference*. We generally to choose a frame with respect to which we get the simplest description.

Suppose that we want to describe the motion of a reflector mounted near the rim of a bicycle wheel. Viewed by an observer standing by the roadside and relating to *his* frame of reference with its origin where *he is standing*, the motion will appear as a series of cusps, rather like the epicycles of Ptolemy, as illustrated in Figure 4.13. However, such a description does not make it easy, either to visualize what is actually happening to the reflector, or to formulate the laws of physics which govern its motion.

A much simpler description is obtained by using a frame of reference with its origin at the centre of the bicycle wheel (which travels along with the wheel). As seen in that frame, the reflector is simply going round and round in a circle. There is

Figure 4.13 Path of a point on the frame of a bicycle wheel using the roadside as frame of reference.

essentially no fundamental reason for using one frame of reference rather than another.

Skipping ahead for a moment to more recent times, Albert Einstein put forward a formal postulate at the beginning of the 20th century, that all *unaccelerated* frames of reference are equivalent. This formed the first step in his theory of special relativity, to which we shall come later (Chapters 15 and 16).

4.5.2 *Copernicus and the heliocentric model*

Nicolas Copernicus (1473–1543), a Polish astronomer, mathematician and physician, and Canon at the Cathedral of Olsztyn, realised that the difficulties experienced by the Greeks in developing a model of planetary motion were largely due to the fact that they were using a most unsuitable *frame of reference.*

They were constrained by the idea that the earth is the centre of the universe, and also of the solar system. All measurements had to be made in a coordinate system in which the earth was at the origin and stationary. Everything else, including the sun, the other planets, the stars and the moon, had to revolve around the earth. Copernicus realised that it would be much simpler to adopt a *heliocentric* theory, in which the sun, and not the earth, is at the origin of our frame of reference. In this frame the sun is stationary and the planets orbit the sun. The earth is just one of these orbiting planets.

The idea was not completely original. It had been suggested more than a millennium earlier by Aristarchus, but dismissed by his peers on philosophical grounds. The Greek philosophers had found it impossible to accept that the earth was not at the centre of the universe and Aristarchus' suggestion seems to have been dismissed out of hand. Indeed, there was also considerable opposition

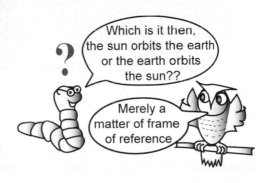

to the work of Copernicus and later of Galileo on the same 'philosophical' grounds. Copernicus' book *De Revolutionibus Orbium Celestium* was not published until the year of his death, in 1543. Galileo's confrontation with the Roman Inquisition on the basis of his 'heretical' writings will be described in the 'historical interlude' at the end of this chapter.

Once it is accepted that the sun, and not the earth, is at the centre of the planetary system, the model becomes much easier to visualise. Copernicus assumed the orbits to be circular, with Mercury and Venus nearer to the sun than the earth, and Mars, Jupiter and Saturn in the outer orbits. Using the then-current astronomical data, he was able to calculate both the radius of each orbit, and the time taken for each planet to complete a full orbit. Most of his values agreed to within a fraction of one per cent of the modern value, an outstanding achievement.

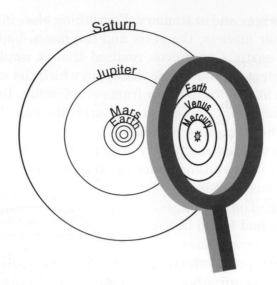

Figure 4.14 Copernican model of the solar system.

Figure 4.14 represents a simplified version of the model. The orbits are drawn approximately to scale, but are shown as being in the plane of the ecliptic. We now know that the orbital planes of the planets are inclined with respect to the ecliptic, but the inclinations are small. The greatest inclination is that of the orbit of Mercury, at 7.0° to the ecliptic.

Copernicus found that, in order to fit astronomical observations, some modifications were necessary. He had to make some of the orbits eccentric, and also make some other minor adjustments. Nevertheless the overwhelming feature of the model is its great simplicity. There is no doubt that the frame of reference built around the sun as origin gives the clearest description of the facts and, more important, gives a good framework for the formulation of the physical laws which govern the motion.

Copernicus identified two characteristic time intervals for each planet:

Sidereal period: time taken to complete one full orbit of the sun (*sidereal* — measured in relation to the stars).

Table 4.2 Vital statistics for planets known at the time of Copernicus.

Planet	Sidereal period ('planet year')	Radius of orbit (Copernican value, AU)	Average distance from sun (modern value, AU)
Mercury	88 days	0.38	0.39
Venus	225 days	0.72	0.72
Earth	1 year	1.00	1.00
Mars	1.9 years	1.52	1.52
Jupiter	11.9 years	5.22	5.20
Saturn	29.5 years	9.07	9.54

1 AU (astronomical unit) = 149.6×10^6 km = average distance of the earth from the sun.

Synodic period: time taken to reach the same configuration *as seen from the earth*. This is subjective, in the sense that it has meaning only for an observer on the earth, which itself is in orbit about the sun.

4.5.3 *Where did the epicycles come from?*

The reason why the paths of the 'wandering stars' appear to be so complicated becomes apparent when we analyse more closely the features of the Copernican model.

Copernicus chose the most suitable reference frame because that used by the people before him simply was not suitable. Take the path of Mars, illustrated in Figure 4.15 as an example. We are observing from the earth, which is travelling in an orbit 'inside' the Martian orbit, with a 'lap time' of one year. Mars, orbiting further from the sun, takes 1.9 earth years to complete a full revolution. As the earth 'overtakes' Mars on the inside, Mars appears to be moving backwards, but this is no longer true away from the

Is that how racing drivers see one another?

Guess so, when they have time to look!

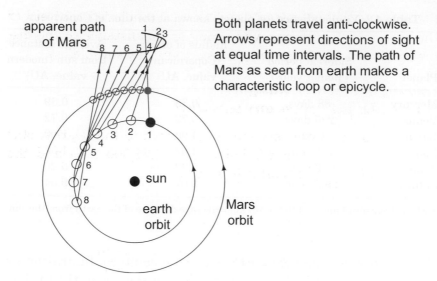

Both planets travel anti-clockwise. Arrows represent directions of sight at equal time intervals. The path of Mars as seen from earth makes a characteristic loop or epicycle.

Figure 4.15 The path of Mars, as seen from the earth.

'overtaking zone'. (Incidentally, we can see Mars in the middle of the night with the sun 'behind us', and therefore the orbit of Mars *must be outside* the orbit of the earth. Conversely, we can only see Mercury and Venus at sunrise or sunset, when they are in the direction of the sun relative to the earth.)

4.6 After Copernicus

Tycho Brahe (1546–1601), who became known for his very accurate astronomical observations, set out to test the theory of Copernicus. He reasoned that, if the earth really was in orbit, we had the advantage of observing the sky from points seperated by vast distances. The pattern of distant stars should appear different, when observed from diametrically opposite points on the orbit.

Brahe presumed that since some of the distant stars must be further away than others, the effect of *parallax* would alter their relative positions as seen from points at opposite ends of the diameter of the earth's orbit. Despite his high accuracy he could not detect any evidence of parallax, and concluded that either the theory of Copernicus was wrong or the distance to the distant stars was too large for parallax to be measurable.

Looking back with hindsight:
Why Brahe did not see any parallax?

Beyond our own solar system the distance to the nearest star (*Proxima Centauri*) is 4.3 light years (272,000 AU) and the diameter of the earth's orbit is about 2 AU.

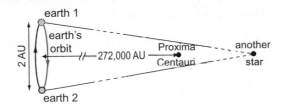

Figure 4.16 Why there is no parallax.

If Figure 4.16 were drawn to scale, the distance to *Proxima Centauri* would be about 1.5 km. What Brahe was attempting to do can be compared to trying to detect parallax between objects more than 1 km away by 'waggling' your eye back and forth through 1 cm.

Johannes Kepler (1571–1630), a former assistant to Tycho Brahe, inherited his astronomical records. In 1609 he published a book entitled *The New Astronomy*, in which he descibed his mathematical analysis of the planetary system. He showed that the reason why Copernicus had found it neccesary to make small adjustments to his model was not that there was a basic fault in the

Johannes Kepler

model, but that *the paths of the planets were not circles, but ellipses.*

Kepler's discovery

Kepler tried to find a relationship between the size of the orbits of the planets and their orbital speeds. By trial and error, he finally found a formula which gave correct results for the six planets known at that time:

Kepler's first law: The orbit of a planet about the sun is an ellipse*, with the sun at one focus.

Kepler's second law: A line joining the planet to the sun sweeps out equal areas in equal times.

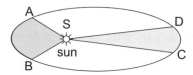

Figure 4.17 Kepler's second law.

According to Kepler's second law, the areas ASB and DSC are equal if the times for the planet to go from A to B and from C to D are equal. This means that AB must be greater than CD, and therefore the orbital speed is higher when the planet is nearer to the sun.

The eccentricity* is greatly exaggerated in Figure 4.17. As an example, the eccentricity of the earth's orbit is 0.017, which means that the major axis is just 1.00014 times longer than the minor axis.

Kepler's third law: The square of the sidereal period of a planet is proportional to the cube of the semi-major axis.

---

* For further information see Appendix 4.1, 'Mathematics of the ellipse'.

The ratio T^2/a^3 is constant for all planets.

The value of the constant of proportionality depends on the units.

If T is expressed in years and a in astronomical units, $T^2/a^3 = 1$.

Kepler was fascinated by this formula and tried to relate it to some property of *harmony* among the heavenly bodies. He even speculated that these harmonies represented *celestial music* made by the planets as they travelled in their orbits. One fact was beyond contention: the formula described the relationship between the period and the radius of orbit with great accuracy. The values of T^2/a^3 for each planet are equal to or better than 1%, as seen in Table 4.3.

Table 4.3 Kepler's third law in action.

Planet	Sidereal period T (years)	Semi-major axis a (AU)	T^2	a^3	T^2/a^3
Mercury	0.2408	0.3871	0.0580	0.0580	1.0
Venus	0.6150	0.7233	0.3782	0.3784	0.999
Earth	1.0000	1.0000	1.00	1.00	1.0
Mars	1.8809	1.5237	3.54	3.60	0.983
Jupiter	11.86	5.2028	140.66	140.83	0.999
Saturn	29.46	9.588	867.9	881.4	0.985

Galileo Galilei (1564–1642)

When Galileo heard of the construction of an optical arrangement of lenses by Hans Lippershey which made distant things look large and close, he immediately set about making

Could this be a coincidence?

Most unlikely, there must be a law behind it.

his own *telescopes*. In a letter dated August 29, 1609, about this new invention, Galileo wrote: '*I undertook to think about its fabrication, which I finally found, and so perfectly that one which I made far surpassed the Flemish one.*'

He stepped up the magnifying power of his telescope from about 6 to 30, and used it to make astronomical observations, which he described in his book *Message from the Stars*, published in 1610.

Using his telescope, Galileo was able to see mountains and craters on the moon, and was even able to estimate the heights of the mountains from their shadows; the stars appeared brighter and were much more numerous but still appeared to be point objects. (Even the most powerful modern telescopes cannot magnify the images of most stars beyond anything more than point objects.) In contrast to the stars, the planets were seen as bright discs, and Galileo was able to distinguish the 'sunny side' of Venus, and show that the planet exhibits phases, just like the moon.

This latter observation he described thus: '*I discovered that Venus sometimes has a crescent shape just as the moon does.*' This discovery may seem inconsequential, but in fact was critically inconsistent with the geocentric model of the universe. In this model Venus was always between us and the supposed orbit of the sun. Accordingly, we should never be able to see its entire illuminated surface. On the other hand, the observed phases of Venus were in good agreement with the Copernican model, in which Venus orbits the sun in the same manner as the moon orbits the earth. Galileo announced his conclusion in cryptic fashion, which literally translated states:

'*The Mother of Love (Venus) imitates the phases of Cynthia (the Moon).*'

The moons of Jupiter

Perhaps the most dramatic of Galileo's discoveries was that four moons orbit Jupiter, just as our moon orbits the earth. He was

Table 4.4 The moons of Jupiter (modern measurements).

Moon	Sidereal period T (years)	T^2	Mean dist. a from Jupiter (AU)	a^3	$\dfrac{T^2}{a^3}$
Io	4.84×10^{-3}	23.43×10^{-6}	2.818×10^{-3}	2.238×10^{-8}	1.047×10^3
Europa	9.72×10^{-3}	94.48×10^{-6}	4.485×10^{-3}	9.028×10^{-8}	1.047×10^3
Gannymede	19.59×10^{-3}	38.38×10^{-5}	7.152×10^{-3}	3.658×10^{-7}	1.049×10^3
Callisto	45.69×10^{-3}	20.87×10^{-4}	1.257×10^{-2}	1.986×10^{-6}	1.051×10^3

able to make an estimate of the relative sizes of their orbital radii and orbital times. When he communicated his results to Kepler, the latter immediately realised that his third law also applies to Jupiter's moons, although the proportionality constant is different. His mysterious law seemed to apply generally, right across the Universe! Galileo's measurements were not very accurate, but modern observations have confirmed this remarkable result, as shown in Table 4.4.

The same units of length and time have been used in Tables 4.3 and 4.4 to facilitate comparison.

Galileo was now more convinced than ever that the Copernican theory must be right. Here was another heavenly body, Jupiter, around which things were in orbit. The earth was not the only centre of a planetery system. Our moon was in orbit around us, but that was all. The sun, and the rest of the universe, did *not* revolve around us!

Checking numbers

Edward Barnard discovered the fifth moon of Jupiter in 1892. Its sidereal period is 0.498 days and its orbital radius is 181,300 km.

Let us check whether this moon also obeys Kepler's law.

$$181,300 \text{ km} = 1.2120 \times 10^{-3} \text{ AU} \qquad a^3 = 1.780 \times 10^{-9}$$
$$0.498 \text{ days} = 1.3634 \times 10^{-3} \text{ years} \qquad T^2 = 1.859 \times 10^{-6}$$

The value of T^2/a^3 is 1.044×10^3, which is in very good agreement with the values shown in Table 4.4.

A total of 13 moons are now known to orbit Jupiter. The 8 moons discovered since Galileo's time are much smaller than the Galilean moons, but every single one of them obeys Kepler's law. Io is 3600 km in diameter and orbits Jupiter every 42 hours at a mean distance of 420,000 km from the centre of the planet. A close-up picture of Jupiter's moon Io, taken from the spacecraft *Galileo*, shows Io casting its shadow on the surface of Jupiter as it orbits the massive planet.

Transit of Io across Jupiter. *Courtesy of NASA/JPL/University of Arizona.*

4.7 The solar system in perspective

In comparison to the dimensions of the earth, the dimensions of the solar system are vast. The *sun* has a mass that is 300,000 times the mass of the earth, and the furthest known planet, *Pluto,* is about 40 times further from the sun than the earth. Still, in the perspective of our entire galaxy, this is very much a 'local neighbourhood'. Our local galaxy, the *Milky Way*, contains many billions of stars, some of them much bigger than the sun. Some stars have planets but, with present techniques, we have no way of seeing them and can only infer their presence indirectly (see Chapter 7). The star *Proxima Centauri*, our nearest stellar neighbour, is about 4 light years away, and it may also have planets too small to be detected, even indirectly, with current techniques.

The Milky Way belongs to a class of galaxies which are described as *spiral*, with a shape roughly that of a disc with spiral arms. The diameter of the disc is about 100,000 light years and the sun is situated about 30,000 light years from the centre of the galaxy, in one of the spiral arms. The total mass of the galaxy has been estimated to be 150 billion times the mass of the sun. (Compare that to the

NGC7331. *Courtesy of NASA/JPL-Caltech.*

diameter of our solar system, which is about 10 light hours, or the diameter of the earth's orbit, which is 16 light minutes.) The image of NGC7331 (often referred to as the 'twin' of the Milky Way) shows how our galaxy might look to an extra-galactic viewer millions of light years away.

On a clear night, the Milky Way may be seen as a broad band of stars across the sky.

The Milky Way as seen from Death Valley, California.
Courtesy of Dan Driscoe, US National Park Service.

According to Albert Einstein, no information can travel faster than the speed of light. The news of the rise and fall of the Roman Empire has travelled barely 1/50 th of the way across our galaxy. If anyone is observing us from outside the Milky Way, they see the earth as it was before the arrival of Homo sapiens!

A historical interlude: Galileo Galilei (1564–1642)

Galileo was born in Pisa on February 15, 1564 (the same year as William Shakespeare). His family was of noble ancestry, but modest means. His father, Vincenzio, was a musician who made a number of discoveries about stringed instruments, which Galileo was later to incorporate into his own work. It was the wish of his father that he embark on a 'real' career as a doctor, rather than pursue his interest in mathematics, so the young Galileo enrolled in medical school at the University of Pisa. He never finished his medical studies, and left without obtaining a degree, reverting to teaching mathematics, privately in Florence.

Galileo was a determined and sometimes pugnacious character, red-haired and of stocky build. He was loyal to his friends, but was not known for suffering fools gladly. He was also known for writing verses about both his colleagues and his enemies. These verses were often funny, but sometimes also scurrilous.

At the age of 25 Galileo was appointed to a teaching post in mathematics in Pisa. It was badly paid and he did not enjoy the work. One of his duties was to teach Euclidean geometry to medical students, so that they could understand the then-standard *geocentric* model of the universe, in which the earth was stationary at the centre, with the sun and planets moving in complicated paths around it. Students of medicine were required to study astronomy so that they could apply this knowledge to astrology and use it in their medical practice.

One might safely assume that Galileo had no time for astrology, but what made the situation more ironic was that he did not believe in the geocentric model. Galileo followed the theories of Copernicus, at that time considered to be unsound and even heretical. In 1543, Copernicus had published *De Revolutionibus Orbium Celaestium*, a book in which he put forward a *heliocentric* theory of the universe, in which the earth and planets orbit a stationary sun.

In 1592, Galileo was appointed as Professor of Mathematics at Padua, where he spent the next 18 years of his life. Not only was he a mathematician, but he also had a great talent for practical

(Continued)

inventions. Some of these (such as a high precision balance to measure the density of precious stones and what would now be called an analogue computer for military use, invented in 1597) have been preserved at the Accademia Cimento in Florence. He lived with his common law wife, Marina Gamba, and their son and two daughters. Those were the best years of his life and it is thought that the foundations of much of his work were laid at that time.

Courtesy of An Post, Irish Post Office.

In 1610, Galileo moved to Florence, on foot, after receiving an invitation from Grand Duke Cosimo II, who had been one of his pupils, to become 'court mathematician and philosopher'. The duke had made him an offer he could not refuse. His salary was to be five times that at Padua and there would be very little teaching. Galileo abandoned Marina Gamba and placed his daughters in the care of a convent at Arcetri, some distance from Florence. Remarkably, his elder daughter, Virginia, was to become his greatest supporter during the troubled times in his old age.

Galileo pioneered the application of mathematical argument to observation and to experiment. Hitherto, the authority of Aristotle had been entrenched in philosophical discussions on the

Galileo facing the Roman Inquistion, painting by Cristiano Banti.

(Continued)

cosmos. It was believed that the *Ptolemaic system* of a fixed earth at the centre of the universe was endorsed by the Bible with certain passages. One example is Joshua 10: 12–14: 'God caused the sun to stand still in response to Joshua's prayer'. The church authorities in Rome went so far as to condemn the works of Copernicus, stating that 'the doctrine that the earth moves around the sun and that the sun is at the centre of the universe, and does not move from east to west, is contrary to the Holy Scriptures, and cannot be defended or held'.

In 1616 Pope Paul V, Cardinal Bellarmine, called Galileo to his residence, warning him not to defend the Copernican theory and not to discuss it orally or in writing. However, in 1624, having met the at first seemingly more liberal Pope Urban VIII, a compromise was reached in which Galileo was allowed to write about the theory, as long as it was treated purely as a mathematical hypothesis, and did not represent 'the actual construction of the universe'.

In 1632, Galileo published his *Dialogue Concerning the Two Chief World Systems*. Written in his typical satirical manner, it depicted a discussion between three people regarding the relative merits of the models of Ptolemy and Copernicus. The case for Copernicus was argued most convincingly by *Salviati,* an intellectual who spoke for Galileo. *Sagredo* was a wealthy nobleman, an interested onlooker who was seeking the truth. *Simplicio*, an Aristotelian philosopher who defended the theories of Ptolemy, was portrayed as not being very bright and using arguments which were not only weak but also included word-for-word statements made by Pope Urban VIII. The *Dialogue* was not well received in Rome, and Galileo was summoned to appear before a tribunal of the Inquisition in 1633. Although the Inquisition did not have the power to force him to comply, Galileo voluntarily went to Rome to appear before the tribunal.

The tribunal concluded that the idea that the earth moves was absurd and erroneous, if not actually blasphemous, and Galileo was ordered not to discuss the theory in writing or in speech. He was confined to his villa in Arcetri under what we now call house arrest,

(Continued)

ordered to recant his 'errors', and forbidden to publish any further writings. As he was by then 70 years of age, the inquisitors must have thought that he was unlikely to continue to give them further problems. They were underestimating the resolve of this man, who is said to have muttered as the verdict was announced *'Eppuor si muove'* — 'All the same it moves'.

Galileo's sharp style and cynical wit come through clearly, even in English translation. His scant regard for some of his adversaries is echoed in his words: '*I have taken the Copernican side of the discourse against the arguments of the Peripatetics[†]. These men indeed deserve not even the name, for they do not walk about, they are content to adore the shadows, philosophising not with due circumspection, but merely from having memorised a few ill-understood principles.*'

The three debaters

'Many years ago I was to be found in the marvellous city of Venice, in discussions with Signore Giovanni Francesco Sagredo and Signore Filippo Salviati, in the presence of a certain Peripatetic philosopher whose greatest obstacle in apprehending the truth seemed to be the reputation he had acquired by his interpretations of Aristotle. I have resolved to make their fame live on these pages by introducing them as interlocutors in the present argument.'

The discussion centres on the question whether the earth moves, or stands still at the centre of the universe. The argument of Simplicio is that if the earth were moving, this would be noticeable because falling objects would not fall perpendicularly, because thier motion would be compounded of 'both transverse and perpendioular'. Salviati proposes a very interesting experiment:

[†] (1) One who walks from place to place, an itinerant; (2) a follower of Aristotle, who gave his instructions while walking about in the Lyceum in Athens.

(Continued)

'Shut yourself up with some friend in the largest room below decks of some large ship, and there procure gnats, flies, and such other, small winged creatures. Also get a great tub full of water and within it put certain fishes; let also a certain bottle be hung up, which drop by drop lets forth its water into another narrow-necked bottle placed directly. Having observed all these particulars as long as the vessel stands still, how the small winged animals fly with like velocity towards all parts of the room, how the fishes swim indifferently towards all sides, and how the distilling drops all fall into the bottle placed underneath, now make the ship move with what velocity you please, so as the motion is uniform and not fluctuating this way and that. You will not be able to discern the least alteration in all the above-mentioned effects, or gather by any of them whether the ship moves or stands still.'

Many of the passages are entertaining, and often amusing, showing the fascination of the mysterious action of gravity. Salvatio: 'If from the top of a tower you let fall a dead bird and also a live one, the dead bird shall do the same as the stone does, it shall first follow the general diurnal motion, and then the motion of descent, just like a stone. But if the bird be let fall alive, the difference is that the stone is moved by an external projector, and the birds by an internal principle.'

The controversy did not affect Galileo's religious beliefs. He felt that the Bible was intended to be understood by common people, and that it was pointless to try to read advanced physical theories out of its pages. A few months before the final trial he wrote to his friend Ella Deliati: 'Nobody will maintain that Nature has ever changed in order to make its works palatable to men. If this be the case, then I ask why it is that, in order to arrive at an understanding of the different parts of the world, we must begin with an investigation of the Words of God, rather than his Works. Is then the Work less venerable than the Word?'

During the final period of his life Galileo was nominally under house arrest, but this did not debar him from settling at a place of his choice. He remained at Alcetri to be near his daughter Virginia, who

(Continued)

now proved to be a great support. Sadly she died in 1634, as he was becoming increasingly ill and feeble. Nevertheless, he still managed to write what turned out to be perhaps his greatest book, *Two New Sciences Dialogue.* Salvati, Sagredo and Simplicio remain as central characters, but they have changed. Simplicio is no longer the stubborn fool, but represents Galileo's own thinking when he was young, Sagredo his middle period and Salvati his mature reflections. They are discussing the motion of objects and the properties of matter; in fact many ideas form the starting points of what was to constitute Newtonian dynamics.

The two new sciences depicted by Galileo were the science of motion and the science of materials and construction. Galileo questioned the ideas of Aristotle, regarding objects falling to earth. Aristotle's 'explanation' was that this was the motion of things seeking their natural place in the centre of the earth. He also asserted that heavier objects fall faster as their attraction towards their natural habitat is more intense. In the course of the discussion, Salviati makes the following observation:

'I greatly doubt that Aristotle ever tested by experiment whether it be true that two stones, one weighing ten times as much as the other, if allowed to fall at the same instant, from a height of, say, 100 cubits, would so differ in speeds that when the heavier had reached the ground, the other would not have fallen more than 10 cubits.'

Since Galileo was not allowed to publish his writings in Italy, the manuscript was brought to Holland, and published in Leiden in 1638.

This was the beginning of a new era of science. The world is not constructed in the manner one might assert it should be constructed, even if that assertion is made by an authority, no matter how much that authority is revered. The laws of nature must be discovered and tested by experiment. Whether or not Galileo's experiments were carried out, as legend says, at the leaning Tower of Pisa, is not clear, but he did test the laws of gravity in many

(Continued)

carefully controlled experiments with smooth polished spheres of different weights rolling down inclined planes and found no discrepancy in the results.

Galileo died in 1642, the same year that marked the birth of Isaac Newton, who was to make the well-known statement 'If I have seen further than most men, it is by standing on the shoulders of giants'. Galileo Galilei was undoubtedly one of those giants.

Appendix 4.1 Mathematics of the ellipse

The ellipse can be defined mathematically in a number of ways, so we will choose the definition which is most relevant physically:

An ellipse is the locus of a point that moves so that the sum of its distances from two fixed points is constant. The two fixed points are called the *foci* of the ellipse.

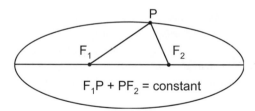

Figure 4.18 Properties of an ellipse.

This property can be used as a simple method to draw the ellipse. Fix two pins at the foci, and place a loop of thin string over them. With a pencil point pull the string taut, forming a triangle. As the pencil is moved over the paper around the pins, keeping the string taut, it will trace out an ellipse.

The eccentricity describes the elongation of the ellipse. The smaller the eccentricity, the nearer the foci are to each other, and the more it resembles a circle. For $e = 0$ the foci coincide at the centre of the circle.

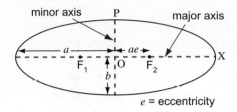

minor axis   P   major axis
e = eccentricity

Figure 4.19 The eccentricity of an ellipse.

From the definition of the ellipse (Figures 4.18 and 4.19),

$$F_1P + PF_2 = const = F_1X + XF_2 = (a + ae) + (a - ae) = 2a$$

From symmetry,

$$F_1P = PF_2 = a$$

In triangle OPF_2,

$$b^2 + (ae)^2 = a^2 \quad \Rightarrow \quad \frac{b^2}{a^2} = 1 - e^2$$

The ellipse is a closed conic section where the ratio

$$\frac{\text{semi-minor axis}}{\text{semi-major axis}} = \sqrt{1 - e^2}$$

Chapter 5

Light from the Past — Astrophysics

When Isaac Newton discovered the law of gravity, and that it applies equally to matter, whether on the earth, the moon or anywhere else in the universe, a completely new chapter was opened. It became possible to calculate the orbits of planets, even to predict the existence and motion of hitherto unseen planets. The science of *astrophysics* was born.

It now became possible, using the laws of physics, to trace the history of the universe back to the original *'big bang'*. It was later realised that the function of *light* as a messenger extended to *bringing news of the first moments of the universe*, when massive energy was released in the form of electromagnetic radiation which is still around us.

The laws of physics predict and explain other exciting things — among them the collapse and death of stars, seen as explosions of *supernovae*, and the existence of *pulsars* and even of *'windows'* out of our universe, or *'black holes'*.

5.1 The birth of astrophysics

So far we have been dealing with the science of *astronomy* — the study of the heavenly bodies and of their motion. Fascinating as has been the discovery of the order and organisation of the solar system, we have not asked the questions: *Why do the planets move as they do? Why do they move at all? Is there any explanation of Kepler's laws, in particular the extraordinary relation*

121

between the orbital radii and the orbital periods? The answers to these questions fall into the domain of *astrophysics.*

5.1.1 *Isaac Newton and gravitation*

Isaac Newton (1642–1727) is very well known, throughout the scientific world and outside it, for his formulation of the

mechanical laws of motion. His first law states that *an object continues in a state of rest, or uniform motion in a straight line,* provided that it is not subject to the action of a *force.*

We now know that the earth, the moon and the other planets are all in motion but *not* in a straight line. If we assume

Courtesy of La Poste Monaco.

that Newton's laws of motion hold not only for terrestrial but also for astronomical objects, then the earth, moon and planets must be under the influence of a *force*, which somehow seems to act at a distance. This force must be exerted on the earth and the other planets by the star they orbit, i.e. the sun. Similarly, the

An early experiment with gravity.

earth and other planets must exert forces on their respective moons. The existence of action at a distance cannot be denied.

The *force of gravity* may have lost some of its mystery, because it is so familiar from childhood. We start experiments on gravity at an early age by dropping things. Let it go, and it goes by itself. There are no strings, nothing visible is pulling it — truly a mysterious 'action at a distance'!

5.1.2 *Falling without getting nearer*

It is said that Newton got the brilliant unifying idea whilst watching an apple fall from a tree. Perhaps the force which causes the apple to fall is the same force as that which keeps the moon in orbit? Perhaps every particle of matter in the universe attracts every other particle according to some law which applies everywhere? Perhaps the moon is falling too, but *falling without getting nearer?*

The ski-jumper is falling from the moment he starts his jump, but at first his height above ground will not decrease as the slope makes the ground recede faster than he is falling, as illustrated in Figure 5.1. This will not last, of course, and after a short time he will begin to lose height more and more rapidly — until,

Figure 5.1 How can you fall without getting nearer?

hopefully, he makes a soft landing! How long he stays in the air, and how far he goes, will depend on his velocity at take-off.*

Now consider an imaginary 'super-athlete' who goes so fast that his jump takes him right over the horizon! If his speed is just right (equal to the *orbital velocity* at that particular orbital radius), his flight path will be such as to match exactly the curvature of the earth. The ground will keep 'falling away' from him at the same rate as that at which he falls towards it, as seen in Figure 5.2. His jump will never come to an end — he will literally 'go into orbit', and continue in orbit for ever! He will keep on falling, without getting nearer to the earth.

Newton realised that if the force of gravity extends to where the moon is located, it too would be falling. That it does not crash to the earth could be explained by the fact that it moves

Figure 5.2 The super-athlete keeps falling for ever.

---

* We are ignoring the buoyancy and viscosity of the air, which may be quite significant in ski-jumping but do not play a role in the main argument.

in orbit like our super-athlete. It has its own orbital velocity and its own orbital plane.

5.1.3 *The mystery of gravitation*

The earth seems to exert a mysterious *action at a distance*. Whether on the apple ripening on a tree or on the moon 384,000 km away, Newton made no hypothesis as to how this action might be transmitted, but the fact remained that a force was exerted by the earth on all objects.

Newton extended this notion beyond forces involving the earth, and made a general statement that *every mass in the universe attracts every other mass*. Just by existing, every piece of matter creates around it an area of influence or *gravitational field*, which extends indefinitely into space. This influence gives rise to a force of attraction on every other particle of matter which happens to be there. This influence never ceases, or becomes 'used up'. It is a mysterious property of matter in our universe.

5.1.4 *Newton's law of gravitation*

Newton then proceeded to construct a *universal* formula for the *gravitational force*, based on simple, logical argument. He postulated that the force of attraction between two point masses separated by a distance d is given by the expression

$$F = G \frac{m_1 m_2}{d^2} \quad \text{(Newton's universal law of gravitation)}$$

Things to note about Newton's law of gravitation:

1. The force is proportional to the masses, defined on the basis of Newton's second law of motion, i.e. on the basis of their *inertia*. For example, a mass which has twice the inertia of another mass will also exert twice the gravitational attraction.

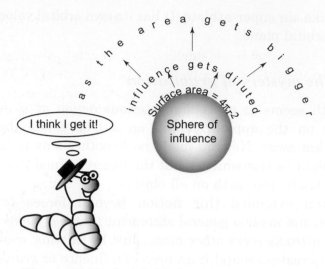

Figure 5.3 The logic behind an inverse square law.

2. The force obeys an *inverse square law*, i.e. decreases *inversely* as the square of the distance. This is a likely relation given a field of influence which spreads uniformly in three dimensions, as shown in Figure 5.3.
3. The force is independent of the nature of the mass. The attraction of a mass of aluminium is exactly the same as that of an equal mass of cheese.
4. The force cannot be shielded, and does not depend on whether there is a vacuum or an intervening medium between the masses, except in so far as the intervening medium itself will exert gravitational attraction.
5. Finally, and perhaps most important, the constant G is *universal* and applies equally to all masses in the universe.

5.1.5 *Testing the law*

Acceleration due to gravity at the earth's surface

Without doubt the best-known experimental result on gravitational attraction is that discovered by Galileo, namely that all objects fall with the same acceleration, $g = 9.8$ ms^{-2}.

The story (probably historically inaccurate) goes that Galileo dropped different weights from the leaning tower of Pisa, and found that all hit the ground at the same time, in contradiction with the Aristotelian belief — which had been widely held for more than a millennium — that heavier objects fall faster than light ones.

Galileo's result is a direct consequence of the universal law. At a given distance from the centre of the earth, for example at the surface, as shown in Figure 5.4, a mass m experiences a gravitational force of magnitude F (which we call its weight):

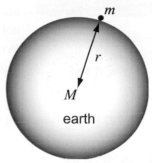

Figure 5.4 The earth's gravity.

$$F = G \frac{Mm}{r^2}$$

If g is the acceleration due to gravity, then

$$G \frac{Mm}{r^2} = mg$$

$$\Rightarrow g = G \frac{Mm}{r^2}$$

The law predicts that g is independent of m.

All objects at the same distance from the centre of the earth fall at the same rate.

David Scott takes samples
on the lunar surface.
Courtesy of NASA.

In a vacuum, a feather falls just as quickly as a stone. This was most memorably demonstrated in 1971 by David Scott, commander of the *Apollo 15* lunar mission, as he stood on the surface of the moon and let fall a falcon feather and a hammer. An almost equally convincing, if less spectacular, demonstration may be set up without going to the moon or even to Pisa, using an evacuated glass tube.

The acceleration of falling objects at the surface of any star or planet is equal to GM/r^2, where M is the mass and r is the radius of the star or planet. So, for example the mass of the moon, $M = 7.36 \times 10^{22}$ kg, and its radius $r = 1.74 \times 10^6$ m giving a value of $g = 1.62$ ms^{-2} on the moon's surface. Despite his massive spacesuit, Commander Scott walked quite freely; the hammer and feather fell together, but accelerated six times more slowly than on earth. On Jupiter, which is approximately

G is universal but g is a local condition.

318 times more massive than the earth and has a radius about 11 times larger, the value of g is about 25.9 ms^{-2}.

5.1.6 *Acceleration of the moon towards the earth*

As we have seen in the previous chapter, Newton knew that the radius of the moon's orbit about the earth, $R = 3.8 \times 10^8$ m, or approximately 60 times the radius of the earth.

Knowing that an apple near the surface of the earth falls with an acceleration $g = 9.8$ ms^{-2} he could now calculate the value of g', the acceleration due to earth's gravity at the distance of the moon:

$$\frac{g}{g'} = \frac{(60r)^2}{r^2} = 3600 \Rightarrow g' = \frac{9.8}{3600} = 2.7 \times 10^{-3}\,\text{ms}^{-2}$$

This is the rate at which the moon is 'falling' (accelerating) towards the earth.

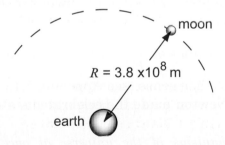

Figure 5.5 Value of the earth's gravity at the moon.

5.1.7 *The period of the moon's orbit*

Once he knew the rate at which the moon is falling, Newton could calculate its orbital period using his own mechanical laws of motion. An object in circular motion experiences an acceleration $= \dfrac{v^2}{R}$ towards the centre of the circle. Assuming that acceleration is the result of the gravitational force of the earth, $g' = \dfrac{v^2}{R}$,

$$v = 2.7 \times 10^{-3} \times 3.8 \times 10^8 = 1013 \text{ ms}^{-1}$$

The period of the moon's orbit

$$T = \frac{2\pi R}{v} = 2\pi \frac{3 \times 10^8}{1.013 \times 10^3} = 27.3 \text{ days},$$

in good agreement with observation.

5.1.8 *Explanation of Kepler's laws*

Newton realised immediately that all of Kepler's laws can be derived from his own universal law of gravitation.

All of the results of experimental observation were justified on the basis of Newton's mechanical laws. The relative periodic times of the planets, and of the moons of Jupiter, became understood.

The constant values of $\dfrac{T^2}{R^3}$ in Tables 4.3 and 4.4 were not a coincidence, but made sense (see Appendix 5.1).

No wonder Newton made the celebrated statement

> *All the mechanisms of the universe at once lay spread before me.*

5.2 The methods of astrophysics

5.2.1 *The moon and the falling apple*

By applying his laws of mechanics and gravitation to outside the earth, Newton established the principle that the same laws govern phenomena everywhere. The apple falling from the tree, the earth orbiting the sun, the moons of Jupiter, are all subject to exactly the same laws of nature. Gravitational forces keep us in orbit, just at the right distance from the sun to give conditions suitable for life. Gravitational forces govern the motion of all the stars in the galaxy, and even the motions of the galaxies themselves.

The same principle was soon extended to other forces and other physical laws. Electromagnetic and nuclear forces all have their parts to play in the running of the universe. The sun is powered by nuclear fusion, its heat energy transported away in all directions by electromagnetic waves, subject to the laws of electromagnetism.

5.2.2 *Predicting the existence of new planets*

In 1781 William Herschel (1738–1822) of Bath, England, using a homemade 10 ft telescope, discovered a new planet, *Uranus*, about twice as far from the sun as the then-outermost known planet, *Saturn*, and with an orbit of about 84 years. Perhaps the most interesting thing about Uranus was that subsequent careful measurements showed that it was 'misbehaving', by not quite following the path predicted by Newton's law of gravitation. Even when the perturbing forces of the other planets had been taken into account, Uranus did not follow exactly its schedule worked out by the most careful Newtonian calculations.

One possible suggested explanation was that Newton's law of gravitation did not hold exactly at distances as large as the distance from Uranus to the sun. Such an explanation could be accepted only as a last resort. Destroying, as it would, the belief in the universal nature of the law of gravitation, it would be a huge setback to *natural philosophy*.

Why is Uranus 'misbehaving'?

John Couch Adams (1819–1892), a student of mathematics at Cambridge, undertook a very difficult mathematical task based on the assumption that another, as-yet-undiscovered, planet was orbiting outside Uranus. One can imagine the magnitude of the task of explaining the observed perturbations of the orbit of Uranus on the basis of three-dimensional calculations involving an unknown mass in an unknown elliptical orbit.

The calculations were, nevertheless, completed by Adams in September 1845, at which time he presented his result to George Airy (1801–1892), his professor at Cambridge, and then to James Challis, director of the Greenwich Royal Observatory. With all the confidence of youth, he proposed that *"should they point their telescopes in a certain direction at a certain time they would observe a new planet hitherto unknown to man"*. As Adams was young and unknown, his suggestion was not taken seriously at Greenwich.

Neptune is there!

Shortly afterwards, Jean Joseph Le Verrier (1811–1877), working in France, quite independently published a result very similar to that of Adams. In his case, however, he wrote to Johann Gottfried Galle, the head of the Berlin Observatory, where he *was* taken seriously. Galle himself looked, apparently within an hour, and found the predicted planet. Thus another planet, *Neptune*, was added to the solar system in October 1846. This was a triumph for Newton's theory. Any doubt that the theory was universal evaporated.

This is a classic example of *predicting an as-yet-undiscovered phenomenon*. In this case, confirmation followed almost immediately. The existence of the planet was inferred independently by Adams and Le Verrier, with no instruments other than pen and paper. Le Verrier's work has been recognised with the naming of a lunar crater, Crater Le Verrier, and a street (Rue Le Verrier) in Paris.

Pluto

In more modern times, using the sophisticated telescopes of the 20th century, additional perturbations were found in the motion of both Uranus and Neptune, leading to the hypothesis that there was still another, as-yet-undiscovered, planet. Eventually, in 1930, the relatively small planet *Pluto*[†] was discovered at the Lowell Observatory in Arizona.

In Table 5.1, planetary masses and sizes, together with their respective orbital parameters, are expressed in terms of the corresponding quantities for the earth.

We can see that $T^2 = a^3 \left(\text{or } \dfrac{T^2}{a^3} = 1 \right)$ for all the planets, to a high degree of accuracy.

The final column gives the inclination of the plane of the orbit of the planet to the plane of the earth's orbit. With the exception of Pluto and, to a lesser extent, Mercury, all the planets orbit in approximately the same plane.

5.3 Other stars and their 'solar systems'

5.3.1 *Planets of other suns*

Do 'they' know about our planet earth?

If we imagine 'little green men' living on a planet similar to ours in some other solar system, it is interesting to speculate whether they see us with highly developed telescopes! The answer is that *their* observation technique would have to be more highly developed than ours, since the light from our sun would completely block out the relatively small intensity of light reflected from the earth. As mentioned before, unless they happen to be among our nearest neighbours in the Milky Way, their information about the earth would be out of date by tens if not hundreds of thousands of years.

Indirect evidence of planets which orbit stars other than the sun has recently become available, leading to considerable

---

[†] Following a decision of the International Astronomical Union (IAU), made in Prague on 24 August 2006, Pluto is no longer officially considered to be a planet. Its status has been changed to that of a dwarf planet (one of more than 40).

Table 5.1 Major members of the solar system (2006) and their properties.

Planet	Diameter (earth = 1)	Mass (earth = 1)	Sidereal period T (years)	Semi-major axis a (AU)	T^2	a^3	Inclin. of orbit to ecliptic
Mercury	0.38	0.055	0.2408	0.3871	0.0580	0.0580	7.00°
Venus	0.95	0.82	0.6150	0.7233	0.3782	0.3784	3.39°
Earth	1.00	1.00	1.0000	1.0000	1.00	1.00	0.00°
Mars	0.53	0.107	1.8809	1.5237	3.54	3.60	1.85°
Jupiter	11.2	317.8	11.86	5.2028	140.66	140.83	1.31°
Saturn	9.41	94.3	29.46	9.588	867.9	881.4	2.49°
Uranus	3.98	14.6	84.07	19.191	7.07×10^3	7.01×10^3	0.77°
Neptune	3.81	17.2	164.82	30.061	2.71×10^4	2.72×10^4	1.77°
Pluto	0.18	0.002	248.6	39.529	6.16×10^4	6.18×10^4	17.15°

excitement about what might be seen as a move from science fiction to scientific fact. Despite some over-enthusiastic early comments, the search for planets likely to support life (much less 'little green men' with whom we might communicate) is likely to be a lengthy one.

The most convincing evidence for extra-solar planets comes from the Hubble Space Telescope, which has measured a reduction in the intensity of light from a star as a result of a planet passing in front of it. In a press release on 4 October 2006, NASA reported evidence for 16 new extra-solar planets in a survey of part of the central region of the Milky Way. The candidates must be at least as massive as Jupiter to block out a measurable amount of light. An extrapolation of the survey data to the entire galaxy would seem to indicate the presence of about 6 billion Jupiter-sized planets in the Milky Way! Five of the 16 candidates belong to a new class, the 'ultra-short-period planets' (USPPs), which orbit their respective stars in about one earth day. They are quite different from planets in the solar system in that they are located only about one million miles from the parent star, a hundred times closer than the earth is from the sun.

Extra-solar planets may also be detected using the Doppler effect (a change in the frequency of light waves as a result of motion of the source, which is described in Chapter 6). This can be achieved by considering how planets and stars move relative to one another and therefore to us.

The planet and its parent star, bound together by gravitational force, orbit their common centre of mass in a manner similar to the atoms of a hydrogen molecule. (In this diatomic molecule, the two atoms rotate about their common centre of mass, midway along the line joining them.) However, the star, being much larger than the planet, is much closer to the centre of mass and the radius of its orbit is very small, as illustrated in Figure 5.6.

The star moves towards us and then away from us as it goes around in its tiny orbit. The star emits light characteristic of

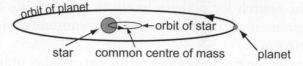

Diagram not to scale. Centre of mass of star-planet system much closer to centre of star and often inside star.

orbit of planet

orbit of star

star   common centre of mass   planet

Figure 5.6 Planets orbiting other stars — the Doppler shift.

its composition and the wavelengths of that light will wobble around their known values, oscillating between lower (blue-shifted) values as it approaches us and higher (red-shifted) values as it recedes from us. The planet itself is not visible but we can 'see' its effect on the light emitted by the star. The effect is so small that very sensitive means of detection are needed.

Probably the best-known of all USPPs orbits a star called *51 Pegasi*. It was first discovered in 1995, using the Doppler technique. The Jupiter-like planet races around the parent star in just four days, in an orbit about $\frac{1}{6}$th of the size of the orbit of Mercury.

5.3.2 *Other galaxies*

The Milky Way is by no means the only cluster of stars in the universe. In 1924 the astronomer Edwin Hubble (1889–1953) proved that what were then known as *extragalactic nebulae* or clouds, such as the Andromeda Nebula, are in fact vast assemblies of stars, similar to our own galaxy. It takes light more than 100,000 years to cross from one side of such galaxies to the other.

The *Large Magellanic Cloud*, at about 170,000 light years from the Milky Way, is the nearest neighbour. The largest member of our 'local group' of four galaxies is M31 — the Andromeda galaxy — at about 2.2 million light years away.

M31 — Andromeda galaxy. *Courtesy of NASA Marshall Space Flight Centre (NASA-MSFC).*

HCG 87 group of galaxies. *Courtesy of NASA/ESA.*

(The many large bright stars in the picture are from our own galaxy, and are much nearer.)

Hubble was the first astronomer to realise that, on the scale of the universe, the Milky Way is insignificant, one of tens of billions of galaxies. We now know that there are more galaxies in the universe than there are stars in our galaxy. There are galaxies at distances of more than 20 billion light years from us.

In 1929, Hubble made the remarkable discovery that there was a systematic change in the wavelength of characteristic spectral lines of basic elements in the light emitted by other galaxies. (He found that the measured wavelengths were longer than expected; the lines were 'red-shifted'.) Attributing this to the *Doppler effect*, he concluded that practically all the galaxies are moving away from us. The more distant the galaxy, the faster it is moving. Galaxies 10 or 20 billion light years away are receding with speeds approaching the speed of light.

Hubble derived an equation connecting the recession speed v of a distant galaxy and its distance r from us. He combined his own distance measurements with red shifts measured by another American, Vesto Slipher, and established *Hubble's law*:

$$v = Hr$$

H is Hubble's constant, which he estimated to be 150 km per second per kpc; more recent estimates average out at about 75 km per second per kpc. (1 kpc = 3.09×10^{22} m.) There is no commonly accepted value for H.

> *The more distant the galaxy, the faster it moves away from us and the larger the red shift.*

For stars which are closer to us, the red shift is smaller and is sometimes masked owing to the motion of the earth in its orbit about the sun.

A remarkable conclusion

The evidence that galaxies are moving away from us in such a systematic fashion points to an unavoidable conclusion:

> *The universe, i.e. the totality of space, mass and time observable by us, is expanding.*

5.4 Reconstructing the past

The laws of physics can be used to predict how a system will develop in the future. Equally, they can be used to reconstruct the past. By observing the universe as it is now, one can form models of what it was like in the distant past. This is a fascinating exercise, an epic detective story based on clues we observe around us.

In the middle of the 20th century, there were two conflicting theories on the origin of the universe.

5.4.1 *The steady state cosmological model*

Originally put forward by James Jeans (1877–1946) in about 1920, and revised in 1948 by Thomas Gold (1920–) and Hermann Bondi (1919–) and later by Fred Hoyle (1915–2001), this model assumed that the universe is and has always been effectively homogeneous in space and time. There was no beginning, there will be no end, the universe remains in equilibrium, always essentially the same. One of the predictions of this model is that matter is being continuously created out of the vacuum. The creation is extremely slow, of the order of one hydrogen atom per cubic metre every 10^{10} years; nevertheless we are continuously getting something out of nothing, and such a model must violate the principle of conservation of energy.

5.4.2 *The 'big bang' theory*

This cosmological model, somewhat controversial at the time, incorporated the expansion of the universe into its framework. If the universe is expanding now, there is no reason why it should not have been expanding in the past. Its expansion might have been slower before, because of the gravitational attraction of matter in the then smaller and denser universe, but the expansion was always there.

George Gamow (1904–1968) and his colleagues reversed the expansion mathematically and found that, projecting back about 15 billion years, the universe started as a point infinitely dense, and infinitely hot.

Presumably, at that first instant the universe was born, and exploded as a 'big bang'. This implies an instant of creation, beyond the laws of physics, and is one of the reasons why the theory was controversial. At the same time, it is difficult to see why continuous creation, as depicted in the steady state model, should be any easier to accept.

Assuming that the laws of physics immediately after the big bang were the same as they are now, the first moments can be reconstructed, right back to a fraction of a second. Thus, in the first one hundredth of a second, the temperature was about 10^{11} °C (one hundred thousand million degrees Celsius). Over the first 3 minutes the universe cooled down rapidly

to one thousand million degrees (10^9 °C). Fundamental particles such as electrons, positrons, neutrinos, and later protons and neutrons, were in violent motion. Pairs of each were continually being created out of energy, and quickly annihilated. Nuclear forces produced pairs of particles of opposite charge, and electric charges in motion gave rise to electromagnetic waves. Eventually the universe was filled with light.

The study of what happened during the very early history of the universe is a most exciting intellectual adventure. Mathematical models based on the laws of physics can be developed, and to some extent tested at high energy particle accelerators. Another approach is to search for evidence of 'cosmological relics' of these early moments.

5.4.3 *A blast from the past*

In 1960, the Bell Telephone laboratory built a 'horn' antenna on a hill in Holmdel New Jersey, to be used for communication with the *Echo* satellite. As it turned out, within a couple of years the *Telstar* satellite was launched and the Echo system became obsolete. Arno Penzias (1933–) and Robert Wilson (1936–), who had joined Bell Labs as radio astronomers, had had their eye on the horn and jumped at the opportunity to use it as a radio telescope when it became free to use for pure research.

As soon as they began to use the antenna, they noticed a troublesome background of '*microwave noise*' which, no matter how they tried, they were unable to eliminate. The background had a wavelength of 7.35 cm, far shorter than the communication band, and had therefore not interfered with the Echo system. They pointed the horn in the direction of New York City — it was not urban interference. They even removed the pigeons which were nesting in the horn, but the noise persisted. There was no seasonal variation. The only conclusion they could come to was that it was a

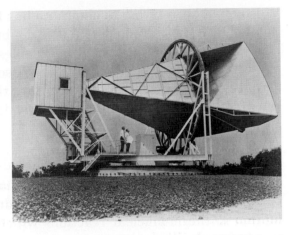

Echo horn antenna. *Courtesy of NASA.*

genuine electromagnetic signal, which appeared to be coming from every direction.

Unaware of these 'technical' problems in New Jersey, P.J. Peebles (1935–), a Princeton theorist, gave a talk at Johns Hopkins University in the spring of 1965. In his lecture, he presented work by Robert Dicke (1916–1997) and himself, which predicted the existence of electromagnetic radiation left over from the early universe, still present as a 'cosmological relic' of the big bang. Such radiation was necessary to carry off energy from nuclear interactions during the formation of matter in the first few minutes. The theory even predicted that as the universe expanded, the wavelength of the radiation would become longer and would by now be in the microwave region.

Since 1965, the 'troublesome' radiation which Penzias and

Wilson had been trying to eliminate has been studied and documented by radio astronomers. Little doubt remains that it is precisely the radiation predicted by Peebles, a genuine relic from the very distant past. Penzias and Wilson had found gold — a signal coming from the first moments of the universe. They were rewarded with the Nobel Prize for Physics in 1978.

5.5 The life and death of a star

The central core of a star is its 'boiler-house'. In a typical star, such as our sun, hydrogen nuclei are driven together to form helium in thermonuclear reactions which liberate energy. As a result the core exerts pressure outwards, and there is a delicate balance between that pressure and gravitational forces which tend to collapse the star into itself.

5.5.1 *White dwarfs*

Subrahmanyan Chandrasekhar (1910–1995), an Indian astrophysicist born in Lahore, was one of the first to combine the laws of quantum mechanics with the classical laws of gravitation and thermodynamics, in a physical model of the evolution of a star. During a voyage from India to England he developed the basics of a theory of the 'death' of a star, and in particular the dependence of the process on the original size of the star.

As the hydrogen 'fuel' becomes used up, the pressure within the 'boiler-house' drops, and the balance is upset. The centre, now mostly helium, is compressed further by gravitational forces until its density becomes far greater than the density of the heaviest material found on the earth.

Further nuclear fusion of helium into carbon and heavier elements takes place, producing more energy and more outward pressure, but eventually the forces of gravitation win and the star collapses. During the collapse, gravitational potential energy is converted into radiation and heat. Material outside the core is blasted off. If the mass of the star is below 1.44 times the mass of the sun (the *Chandrasekhar limit*), the star dies relatively gently. It becomes a *white dwarf* and fades away over billions of years. Such a fate eventually awaits our sun, which will finally end up as a cold dark sphere.

> *A sugar cube of white dwarf material, if brought to the earth, would weigh about 5 tons.*

5.5.2 *Supernovae*

If the mass of the star is greater than the Chandrasekhar limit, a much more violent death awaits it. The end, when it comes, comes dramatically. The collapse becomes sudden and the core shrinks until the electrons are forced to combine with the protons present in the nuclei of the atoms. In the core of the star gravitation scores a victory over both *electric* and *nuclear* forces. The rest of the star is ejected at speed into space. The star becomes a *supernova*, expelling as much energy in a matter of weeks as it had previously radiated in its lifetime.

Only the core of the star remains, made up of matter entirely composed of neutrons: a *neutron star*. Its density is equal to the density of an atomic nucleus, about 10^{17} kg/m^3. A sugar cube of neutron star material would weigh about 100 million tons. To put this into perspective, the heaviest tank to see action in World War II was the German King Tiger 2, weighing 70 tons. The weight of the sugar cube of concentrated neutrons would equal the weight of over a million

If you thought White Dwarf matter is heavy listen to this!..

Tiger tanks — in fact, more than the combined armoured vehicles of all armies participating in wars in the 20th century!

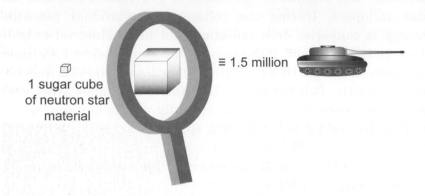

1 sugar cube
of neutron star
material

≡ 1.5 million

Five such supernova explosions have been recorded in our galaxy in the last 1000 years. The first dates back to 4 July 1054:

> *'In the first year of the period of Chih-ho, the fifth moon, the day Chi-ch'ou, a guest star, appeared south-east of T'ien-kuan. After more than a year it gradually became invisible.'* (Chinese manuscript of the Sung Dynasty.)

We now know that what the Chinese saw was the death and not the birth of a star.

Supernovae in other galaxies

With the aid of modern telescopes, several hundred supernova explosions are now being observed in distant galaxies every

Table 5.2 Supernovae in our galaxy.

Year	Supernova
1054	Chinese
1151	
1172	Tycho
1604	Kepler
1667	Cassiopei (deduced from its remnant)

year. Most of these are so faint that they can be studied only with difficulty.

In 1987, a supernova was observed in the Large Magellanic Cloud. Named SN1987a, it marked the death of the star Sanduleak–69 202, with a mass approximately 20 times that of the sun. The first observation was made on the night of 23 February 1987 in Chile and, as the earth rotated, there were further observations in New Zealand, Australia and southern Africa.

The picture on the right was taken by the Hubble Space Telescope in 1991. It shows a gaseous ring around SN1987, which became the progenitor star several thousand years before the supernova explosion. The light from the far edge of the ring arrived at the earth nearly one year after the arrival of light from the forward edge. This gives a very accurate

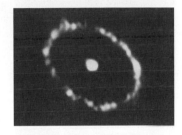

Ring of gas around SN187a. *Courtesy of NASA/ESA.*

value of the physical diameter of the ring, which, when compared to its angular diameter of 1.66 arc seconds, enabled astronomers to calculate the distance from the earth to the Large Magellanic Cloud as 169,000 light years (to within 5%). As the explosive expansion continues over the next few years, the envelope will become more transparent, enabling astronomers to carry out the most detailed study to date of such a remnant.

5.5.3 *Pulsars*

Neutron stars do not 'shine' like other stars, emitting a steady stream of light. They emit electromagnetic waves, which come in characteristic pulses, and hence neutron stars are often called 'pulsating radio stars', or *pulsars*.

All spinning objects have angular momentum and stars are no exception.

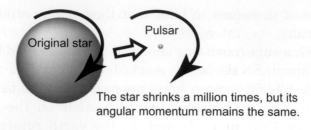

The star shrinks a million times, but its angular momentum remains the same.

Jocelyn Bell Burnell

A pulsar acts like a rapidly spinning magnet emitting a rotating beam of electromagnetic radiation like a 'lighthouse in the sky'. Its mass is made from the core of the original star and is greater than 1.44 times the mass of the Sun, but its radius has shrunk to about 10 km. If the earth happens to be in the line of the light beam, we will see a source pulsating every time the beam sweeps past.

The first evidence of pulsars came in 1967 at the 4.5-acre radio telescope array in Cambridge, England. Jocelyn Bell Burnell (1943–), a postgraduate

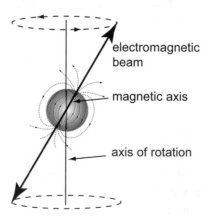

Figure 5.7 Lighthouse in the sky.

student born in Belfast, Northern Ireland, and working under the supervision of Antony Hewish (1924–), discovered a source of radio waves which came in bursts 1.3 seconds apart. The pulses were regular, like a signal — perhaps from another planet?

The source was given a provisional name, LGM1 ('Little Green Man 1'), and kept under quiet surveillance. Within 1 month, a second source was discovered, by which time it had become 100% clear that it was not extraterrestrial intelligence but a neutron star behaving like a rapidly spinning magnet.

An extract from Jocelyn Bell's original chart records is reproduced as shown on the right. The bottom trace marks out one-second time pips and so LGM1 is pulsing every one and a third seconds.

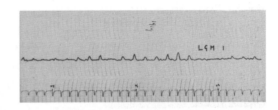

Courtesy of Jocelyn Bell Burnell.

The signal is weak and often drops below the detection threshold but it is still in phase (on-beat) when it reappears.

The 1974, Nobel Prize for Physics was awarded to Anthony Hewish *'for the discovery of pulsars'*, and Martin Ryle (1918–1984) *'for his pioneering work in radio-astronomy'*. Due credit must also be given to Jocelyn Bell, who built the radio telescope and was the first to notice the signals.

The *remnant* of the Chinese *supernova* was discovered in 1963. It is situated in the Crab Nebula 5500 light years away. Calculations tracing back its expansion showed that it should have been observable from the earth in the year 1054. The actual explosion occurred 5500 years earlier, about 4500 BC.

5.5.4 *Black holes*

Game, set and match to gravity!

A neutron star is not necessarily 'the end of the line' in the life and death of a star. Gravitational forces are still at work, compressing matter more and more tightly. The more concentrated the mass, the greater the inward pull. If the original star was big enough, the resulting neutron star is merely an intermediate stage, itself unstable under gravity. Chandrasekhar calculated that if the original star had a mass in excess of about 3 solar masses, the resulting 'electron degenerate matter' (neutron star) can no longer exist. It will collapse even further into itself — the ultimate sacrifice to gravity. With so much mass concentrated in such a small volume, the force of gravity dominates, and nothing can stop it. The further the collapse proceeds, the greater the gravitational force and the density becomes infinite. The known laws of physics no longer apply. We reach a window in our universe to another world beyond our own, with which we cannot communicate. Matter can be sucked in but nothing can get out, not even light. We have a *black hole*.

Robert Oppenheimer (1904–1967), perhaps better known for his later work on the Manhattan Project, was the first to apply the theory of general relativity to what happens when a massive star collapses. The end result is much the

same as predicted by classical Newtonian mechanics, but the mechanism is more difficult to visualise. The curvature of space–time (which has four dimensions and is described in Chapter 15) increases dramatically. This is not easy to visualise, but a useful two-dimensional analogy might be to push a knitting needle against a flat membrane. The membrane becomes more and more distorted and, in the end, it is punctured. The two-dimensional space ceases to exist.

Robert Oppenheimer

A black hole is a *singularity in space–time.* Space and time, as we know them, cease to exist. The rules of physics no longer apply.

Oppenheimer's work has become a rich source for theoretical research by many physicists and mathematicians. Stephen Hawking, Roger Penrose, John A. Wheeler and, of course, Chandrasekhar are some of the more prominent names associated with the subject.

More than 200 years ago, Pierre Simon Laplace (1749–1827) postulated:

'A *luminous star of the same density as the Earth, whose diameter be 250 times larger than that of the Sun, would not, in consequence of its gravitational attraction, allow any of its rays to arrive at us. It is therefore possible that the largest luminous bodies in the Universe may, through this cause, be invisible.'*

5.5.5 *Escape velocities*

How to get away from a large mass

The first obstacle faced by prospective space travellers is the gravitational attraction of the earth. The energy to overcome the gravitational force can be provided in the form of combustible rocket fuel. Alternatively, one *could* imagine building up speed

Table 5.3 Some escape velocities.

To escape from:	Speed (km/s)	Speed (mph)
Earth	11.2	25,000
Moon	2.4	5,400
Jupiter	58	130,000
Sun	620	1.4 million
Neutron star	$\approx 150,000 \approx 0.5c$	3.4×10^8
Black hole	$> c$	

c is the speed of light; 1 km/s = 2236.9 mph.

horizontally and then flipping the rocket upwards and away from the earth, turning off the motors and using kinetic energy to overcome gravitation. The speed required to do that is called the *escape velocity*. Table 5.3 gives some relevant values of escape velocities.

(Appendix 5.2 shows how to calculate the escape velocity for planets such as the earth.)

5.5.6 *How to 'see' the invisible*

Black holes are no longer in the realm of science fiction. Even though light cannot escape from inside a black hole, we can detect its existence from what is happening around it. Matter close to a black hole gets sucked in and spirals around it like water in a

plug hole. As electric charges spiral, they emit electromagnetic waves — a 'last gasp' signal back to the universe. X-rays and other radiation are emitted, not from the black hole but from the *accretion disc* of matter spiralling into it.

Data reported in February 2000 from the Chandra X-ray observatory show strong evidence of the existence of a supermassive massive black hole, Sagittarius A*, near the centre of the

Artist's impression of a massive black hole accreting matter. *Courtesy of NASA/JPL-Caltech.*

Milky Way. According to recent estimates, this black hole is about 3.7 million times more massive than the sun and is very compact, with a radius of about 45 AU, at most. It is generally believed that most galaxies harbour supermassive black holes at their centres.

5.5.7 *A strange event in the Milky Way*

In 2002, V838, a star in the Monoceros constellation, 20,000 light years from the sun, suddenly brightened. Its maximum luminosity was about 1 million times the luminosity of the sun and it was one of the brightest stars in the Milky Way at that time. Figure 5.8 shows two images of Monoceros, the first taken

Figure 5.8 V838 explodes. *Courtesy of NASA, USNO, AAO and Z. Levy (STScI).*

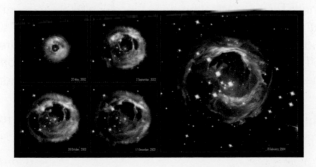

Figure 5.9 Evolution of an explosion. *Courtesy of NASA, Hubble Heritage Team (AURA / STScI) and ESA.*

in 1989, when the star appeared quite 'normal', and the second in 2002, during the outburst.

The sequence of events was recorded by the Hubble Space Telescope and is illustrated in Figure 5.9. An interesting result of the examination of the spectrum of the emitted radiation is that it shows a strong enrichment of Li, Al, Mg and other elements. This leads to the hypothesis that these elements came from planets which had been orbiting the star and were vaporised by the explosion. In about 5 billion years our own solar system may meet a similar fate!

5.5.8 *Time stands still*

There are other extraordinary properties predicted for black holes. According to the general theory of relativity, time slows down in a strong gravitational field. In the extreme environment of a black hole, we must conclude that *time will stand still relative to us*. In an 'experiment of the mind', let us imagine a space traveller approaching a black hole. To us, his last movements will appear to be very slow. His pulse rate slows down, and his heart beats once every hundred years. Eventually he 'freezes in time'. His final image appears for ever, as he was, just before being sucked in!

A historical interlude: Isaac Newton (1642–1727)

Isaac Newton was born on Christmas Day, 25 December 1642*, at his mother's house at Woolthorpe in Lincolnshire, three months after the death of his father, who was a farmer. When Newton was two years old, his mother remarried and left her son in the care of her mother. That Newton resented being abandoned by his mother and stepfather (Barnabas Smith) is apparent in his later admission

---

* According to the Gregorian calendar adopted in England in 1752, this date becomes 4 January 1643.

(Continued)

'Threatened my father and mother Smith to burn them and the house over them.'

On returning to his father's farm in 1656, Newton spent most of his time tinkering with mechanical gadgets, solving mathematical problems and devising his own experiments. His maternal uncle, William Ayscough, noticed that young Isaac showed little interest in farming. Realising that Newton's talents lay elsewhere, Ayscough arranged to send him to his own *alma mater*, Trinity College, Cambridge, in 1661.

Isaac Newton

Despite the fact that his mother appeared to be financially well off, Newton had to work for his keep and college expenses as a *sizar* acting as a servant for wealthier students. He kept a diary, from which we know that he had little interest in student life, and pursued a solitary existence, reading whatever books he could find on logic, philosophy and mathematics. He was particularly fascinated by the brilliant logic of Euclid's geometry, which led to conclusions which he felt were undeniably true. This may have prompted him to write a thesis entitled *Questiones Quaedam Philosophicae*, which he prefaced with the enlightening statement in Latin: *'Plato is my friend, Aristotle is my friend, but my best friend is truth.'*

Newton obtained his BA degree in Mathematics in 1664. The university was closed for the following two years because of the Great Plague. In London alone, more than 30,000 died from the Black Death over a two-year period. Newton returned to Lincolnshire and spent the years 1665 and 1666 studying at home in isolation. These years turned out to be the most productive of his life. Not only did he read widely in mathematics, but he also made a series of discoveries. It was then, while still under the age of 25, that he developed his ideas on optics and, perhaps more important, the laws of motion, and universal gravitation. He also

(Continued)

laid the foundations of the mathematical methods of differential and integral calculus. He called this the 'method of fluxions' and used it as a general technique to calculate seemingly unrelated things such as areas, tangents to curves, and maxima and minima of functions.

Newton did not publish any of his results until many years later. Apart from some letters to friends and reference to some of his work on optics, he kept his ideas to himself. This was to create a controversy on priority with the German philosopher Gottfried Leibnitz (1646–1716).

When Cambridge reopened in 1667 after the plague, Newton applied for a fellowship and was soon awarded the title of Senior Fellow. This allowed him to dine at the fellows' table — a considerable rise in status compared with that of a *sizar*. More rapid promotion was to follow almost at once, when Barrow resigned the Lucasian Chair of Mathematics in 1669 and Newton was appointed in his place, at the age of only 27 years.

In January 1670, Newton began his first lecture course as Lucasian Professor. The course described experiments which he had carried out on his own during the years of the plague. Alone in his room, using only the simplest equipment, he had discovered things about light which were new and in conflict with conventional opinion. He published his first scientific paper on light and colour in 1672, describing his experiments in graphic style:

> '*I procured me a Triangular glass-Prisme, having darkened my chamber made a small hole in my window-shuts, to let in a convenient quantity of Suns light.... It was at first a very pleasing divertisement to view the vivid and intense colours thereby, but after a while I applied myself to consider them more circumspectly*'.

The old 'explanation' of the coloured spectrum produced when light passes through a prism was based on an assumption that a light ray is modified gradually as it travels through a medium like glass. The change manifests itself as a change of colour. The light is modified slightly at the thin end of the prism and becomes red; it is darkened

(Continued)

a little more when it passes through the middle, and becomes green; it is darkened even more when it passes through the widest part, and becomes violet. This could hardly be called an 'explanation' since it said nothing about the nature of light; moreover, it was clearly wrong, as could easily be shown by sending light through blocks of different thickness. However, there had been no climate for experimentation to test the laws of nature. Hypotheses were not tested, but simply accepted on the authority of philosophers such as Aristotle.

Newton showed that light was not *modified* by passing through a substance like glass, but was *physically separated*. According to him, white light was composed of different types of ray, each one producing a different colour. These rays were diffracted at slightly different angles and, as they entered the glass, the components separated, then continued onwards undisturbed. He also discovered that colour had nothing to do with light and dark.

Newton's painstaking methods can be described in his own words: *'I keep the subject of my inquiry constantly before me, and wait until the first dawning opens gradually, little by little, into a full and clear light.'*

Newton did not rely on hypotheses, but persisted in confirming his findings by further experiments. Using a second inverted prism, he was able to recombine those colours and reproduce the original white light. *'Hence therefore it comes to pass, that Whiteness is the usual colour of Light; for, Light is a confused aggregate of Rays indued with all sorts of Colors, as they are promiscuously darted from the various parts of luminous bodies.'*

Newton then tried yet another experiment and showed that once a colour had been separated it could not be changed any further: *'Then I placed another Prisme, so that the light might pass through that also, and be again refracted before it arrived at the wall... when any sort of Ray hath been parted from those of other kinds, it hath afterwards obstinately retained its Color, notwithstanding my utmost endeavours to change it....'*

New ideas which were contrary to established opinions were not acceptable to many of Newton's contemporaries. One can well imagine a certain reluctance to admit that established doctrine did not just

(*Continued*)

have to be modified, but was completely wrong. Another Cambridge physicist, Robert Hooke (1635–1703), involved Newton in a bitter dispute. The controversy became bitter and personal, with no punches pulled by either side. So much so that Newton's often-quoted words *'If I have seen further than other men it is because I have stood upon the shoulders of giants'* are suspected to contain a hidden barb. Hooke was a small, somewhat deformed man with a hunched back.

The criticism and fighting caused Newton to withdraw within himself, and in particular not to publish any more of his ideas. His book *Opticks* did not appear until some 30 years later, in 1704. In a letter to his then-friend Gottfried Leibnitz he confided: *'I was so persecuted with discussions arising from the publication of my theory of light that I blamed my own imprudence for parting with so substantial a blessing as my quiet to run after a shadow.'*

Newton's greatest achievement, his formulation of the laws governing force and motion, dates back to his solitary work during the plague and remained unpublished for many years afterwards. The story of the falling apple originated in a biography by his friend William Stukeley, who describes having tea with Newton under some apple trees. The mystery of action at a distance, by which objects somehow exert a force on each other without touching, was the subject of discussion. Could this same kind of force be exerted on the moon and on a falling apple?

Newton continued where Galileo had left off. Galileo had discovered that all objects at the earth's surface experience the same acceleration under gravity; Newton formulated a general law which explained why. Not only that, but he calculated why the moon should take about 27 days to orbit the earth. The famous *Philosophiae Naturalis Principia Mathematica* finally appeared in 1687. It seems very likely that Newton's frustration caused by arguments about the light spectrum was the prime cause of the delay. Newton became chronically ill about 1692, suffering from insomnia and nervous depression and irritability. He decided to abandon academic life and to look for some other kind of job. However, the offer of the position of headmaster of Charterhouse, an aristocratic London school, was not exactly what he wanted. He wrote somewhat undiplomatically: *'...besides a coach, which I consider*

(*Continued*)

not, it is but £200 per annum, with a confinement to the London air, and to such a way of living as I am not in love with....'

He was appointed Master of the Royal Mint in 1699, at the then-enormous salary of £1500 per year. He took this position quite seriously, and used his talents and imagination to invent methods of preventing counterfeiting of coins. In 1703 he was elected president of the Royal Society, a position of great power in the realm of science in England, and re-elected every year until his death. Newton was knighted in 1704 by Queen Anne — the first scientist to be so honoured.

Newton's controversy with Gottfried Leibnitz (1646–1716), regarding who was the inventor of calculus, dominated much of the later part of his life. Leibnitz published his paper '*Nova Methodus pro Maximis et Minimis itemque Tangentibus*' in 1684. It contained the basic ideas on what we now call differential calculus. These ideas, particularly his methods of calculating tangents, were based on the same principles as the methods of Newton. Two years later, in 1686, Leibnitz published a paper on integral calculus, which saw the first appearance of the now-familiar integral notation.

To put this into historical context: Newton's unpublished work on fluxions was written in 1671 but did not appear in print until 1704, and then as two mathematical appendices in his book *Opticks*. John Collins produced it in English translation in 1736, nine years after Newton's death.

In the meantime John Keill wrote an article in *The Philosophical Transactions of the Royal Society* in 1710 which accused Leibnitz of having plagiarised Newtons's work, essentially changing only the method of notation. When Leibnitz read this, he took grave exception, and immediately wrote to the secretary of the Society, saying that he had never heard of fluxions and demanding an apology. Keill's reply quite possibly made with the collaboration of Newton himself, was to restate the charge in even more aggressive language. Leibnitz immediately demanded a retraction and the stage was set for a major controversy.

The Royal Society set up a commission to investigate the claims. With Newton as its president it could hardly have been impartial, and, not unexpectedly, its report concluded in favour of Newton. To copper-fasten the matter Newton anonymously wrote a

(Continued)

favourable review of the report, which appeared in *The Transactions of the Royal Society* in 1712.

Newton was straightforward and honest but, while scrupulously just, he was unwilling to compromise on his convictions or to be generous to his adversaries. He is said to have been the originator of the statement *'Tact is the knack of making a point without making an enemy'*, and yet he was easily irritated, and tended to take offence. Many of his friends eventually became enemies.

As one might expect, he was absent-minded and frequently became lost in his thoughts. It is said about him that on occasion he left his dinner guests to their own devices by disappearing from the table to work on some problem and forgetting to come back.

It is clear that Newton was fascinated by his discoveries and by deep questions as to how the laws of Nature came into being: *'Did blind chance know that there was light and what was its refraction, and fit the eyes of all creatures after the most curious manner to make use of it? These and other such like considerations, always have, and always will prevail with mankind, to believe that there is a Being who made all things, who has all things in his power, and who is therefore to be feared.'*

Appendix 5.1 Kepler's third law, derived from Newton's law of universal gravitation

To derive Kepler's third law, that $\dfrac{T^2}{r^3}$ is constant for a circular orbit,

$$F = \frac{mv^2}{R} = G\frac{Mm}{R^2}$$

$$v^2 = \frac{GM}{R}$$

$$T = \frac{2\pi R}{v} \Rightarrow T^2 = \frac{4\pi^2 R^2}{v^2} = \frac{4\pi^2 R^3}{GM}$$

$$\frac{T^2}{R^3} = \frac{4\pi^2}{GM}, \text{ which is constant}$$

Appendix 5.2 Escape velocity

As one gets further away from the earth, its gravitational force diminishes and the work to cover each extra metre gets smaller. One must therefore integrate the work done, as follows:

Total work done:

$$W = \int_R^\infty \frac{GMm}{r^2} = \left[-\frac{GMm}{r} \right]_R^\infty = \frac{GMm}{R}$$

W = kinetic energy = $\dfrac{1}{2}mv^2$

$$\frac{1}{2}mv^2 = \frac{GMm}{R}$$

$$v = \sqrt{\frac{2GM}{R}} = \sqrt{\frac{2 \times 6.668 \times 10^{-11} \times 5.976 \times 10^{24}}{6.378 \times 10^6}}$$

Escape velocity $v = 1.12 \times 10^{-4}\,\mathrm{ms}^{-1}$ (about 400,000 km hr^{-1})

Constants used in the calculation:

$G = 6.668 \times 10^{-11}\ \mathrm{Nm^2\ kg^{-2}}$ (universal constant)
$M = 5.976 \times 10^{24}\ \mathrm{kg}$ (mass of the earth)
$R = 6378\ \mathrm{km} = 6.378 \times 10^6\ \mathrm{m}$ (radius of the earth)

Chapter 6

Introducing Waves

What is a wave? Waves are so diverse that we might tend to regard different types of waves as separate entities, unrelated to one another. Do events such as the devastation of a region by an earthquake and the melodic tones of a musical instrument have anything in common? In this chapter we explore the common ground between these and many other phenomena.

Waves carry energy from one place to another. They also act as messengers, transmitting information. We look at different types of waves, how they are created, what they are 'made of' and how they carry out their functions of carrier and messenger.

To deal with some aspects of wave behaviour, we need to express the properties of the waves mathematically. We can then make quantitative predictions regarding many wave-related phenomena seen in nature. In a mathematical analysis of waves, it is most convenient to use the simplest type of wave form — a continuous sine wave. It is gratifying, if rather startling, to find that even the most complicated periodic waves can be constructed simply by adding a number of these sine waves.

6.1 Waves — the basic means of communication

When talking about *waves*, we most likely picture a seaside scene with ocean waves approaching the shore or perhaps a ship at sea tossed in a storm. A survey asking the question 'What serves as the most common means of communication?' would be unlikely to favour the answer '*Waves*'. Yet, heat and light from

Waves coming in on a beach. *Courtesy of Johntex.*

the sun are carried by waves. We could neither see nor speak to one another were it not for light waves and sound waves. Even the sensation of touch relies on the transmission of nerve impulses, which are composed of wave packets.

Electromagnetic waves, which pervade all space, and without which the universe could not exist, are the basic theme of this book. *Light* is just one member of that family. These waves propagate in a mysterious way, as we shall see in later chapters.

We have learned to produce electromagnetic waves and to put them to a variety of uses, many of which are now taken almost for granted. Radio waves facilitate the remote communication that enables us to hear and to see the latest news and to watch our favourite television programmes. We can receive information about the lunar surface from space probes. In medicine, lasers produce waves for keyhole surgery, X-rays are used in diagnostic imaging and radiation therapy; infrared waves heat muscles and relieve pain. Microwaves cook food; radar waves guide planes and ships. Last but not least, a tiny infrared beam from our remote control allows us to change channels without leaving our armchair in front of the TV set!

Once we become aware of the seemingly endless variety of waves, it begs questions such as: Do all waves have something in common? How can we deal with waves mathematically?

What is the justification to say that light behaves as a wave?

6.1.1 *Mechanical waves in a medium*

When a medium is disturbed by a wave, individual particles oscillate about their equilibrium positions and transmit energy by their mutual interactions.

A stone dropped into the middle of a pond gives rise to waves spreading out on the surface of the water in ever-increasing circles. The expanding waves show that something definitely propagates, but it is not the water itself; ducks sitting in the path of the wave bob up and down but do not necessarily move in the apparent direction of propagation. Energy con-tained in the local oscillations of indi-

Surface waves created by dipping a stick into water.
Courtesy of Roger McLassus.

vidual water particles is transmitted with no net motion of these particles from the middle of the pond to its edge.

> *Waves transmit energy, momentum and information from one place to another by means of coordinated local oscillations about an equilibrium position.*

Waves, whether naturally occurring or artificially produced, can be classified according to the type of disturbance which is transmitted and also by the relationship between the direction in which things change locally and the direction of energy transfer.

When molecules are stimulated by an external force which varies periodically (repeats itself at regular intervals), they vibrate and a mechanical wave is set in motion. Individual mol-ecules remain localised, but energy is transferred from molecule to molecule and hence from one place to another. The energy is transmitted through the material by molecular interactions but the precise nature of the interaction depends on the material. On the macroscopic scale, we refer to the level of the molecular response as the elasticity of the material.

Mechanical waves are propagated by the interaction of molecules with their neighbours and do not exist in a vacuum.

6.1.2 *Transverse waves*

Energy may be transmitted by particles oscillating in a direction either perpendicular or parallel to the direction of energy transfer. In the disturbance caused by the stone, the water surface moved up and down, but the wave moved horizontally. This sort of wave is called a transverse wave.

A transverse wave is a wave for which the local disturbance is perpendicular to the direction of propagation of energy.

We need some form of *bonding* to support a transverse wave. When molecules are bonded together, transverse waves can propagate. The bonding must be such that, when a molecule moves up and down, it tends to drag neighbouring molecules with it. Force is required to disturb the molecule and the phenomenon is technically known as *resistance to shearing stress*. Such bonding exists in solids and liquids.

In a row of children holding hands, if one child jumps up, the children on each side will be pulled up. The bonding is in the holding of hands.

There is no bonding if the children do not hold hands and the jumping motion of one child does not affect the others.

The 'Mexican wave', often seen at sporting events, is another example of a transverse wave.

Where is the bonding in a Mexican wave?

How about bonding of hearts and minds?

6.1.3 *Longitudinal waves*

Transverse waves cannot propagate in gases, because gases have practically no resistance to shearing stress. Air molecules are much further apart than the molecules of a solid or liquid, and as a result the viscous forces tending to drag adjacent layers of air are too small to propagate a transverse disturbance.

Courtesy of The Hovercraft Museum Trust.

There are, however, forces between the molecules which resist compression. A fully laden bus is supported by the compressed air in its tyres; a hovercraft rides on a cushion of compressed air.

In 1887, John Dunlop developed the pneumatic tyre for his son's tricycle, and patented it in 1889. The patent description read: 'A device covering the circumference of a wheel. It cushions the rider from a bumpy road, reduces wear and tear on the wheels and provides a friction bond between the vehicle and the ground.'

A pulse of compressed gas or a series of such pulses is transmitted through a gas at a speed which depends on the elastic properties of the gas and on its temperature. The human ear is a delicate instrument which can detect such compressions. We call this sensation *sound*.

The mechanism of transmission of a longitudinal pulse can be shown by the example of an orderly queue of people waiting for a bus.

As the bus approaches, the people at the back of the queue begin to push the people in front of them. By the time the bus stops, the compression has been transmitted to the front of the queue. Individual people move in the same direction as the compression.

Sound waves propagate as a series of compressions. A vibrating membrane such as a guitar string exerts a varying pressure on the adjacent air and the pressure variations are transmitted as a longitudinal wave.

A longitudinal wave is a wave where the local disturbance is parallel to the direction of propagation of energy.

In a solid each molecule is bonded to *all* of its neighbours, so the application of a force in *any direction* will distort the equilibrium arrangement. This means that solids can transmit both transverse and longitudinal waves. An underground explosion, for example, generates both transverse and longitudinal seismic waves which are transmitted through the earth. The production of simultaneous longitudinal and transverse waves can be demonstrated using a slinky, as illustrated in Figure 6.1.

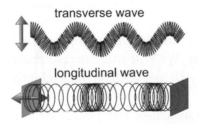

Figure 6.1 Transverse and longitudinal waves on a slinky.

Water waves can be a mixture of transverse and longitudinal vibrations. Individual elements of water move in circles or ellipses, oscillating both perpendicular and parallel to the surface of the water.

As sea waves approach a beach the motion of the particles changes as the depth of the water decreases. At a certain point the friction of the sand causes the wave to 'break'.

Sea storm in Pacifica.
Courtesy of Mila Zinkova.

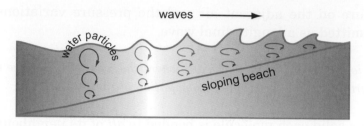

The waves appear to be coming in but they do not bring in the sea!

6.2 The mathematics of a travelling wave

6.2.1 *The making of a wave*

We base the mathematical representation of a wave on the assumption that travelling waves are made up of basic components oscillating continuously about given positions in space.

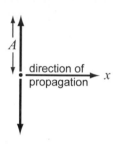

Let us choose a simple transverse wave in a material medium. We will call the maximum displacement of each oscillating particle the amplitude A.

6.2.2 *From the sine of an angle to the picture of a wave*

The *sine* of an angle is defined in terms of the ratio of two sides of a right angled triangle. Many physical phenomena, such as refraction, are described in terms of the sines of angles (e.g. Snell's law in Chapter 3). The function which expresses the value of $\sin(\theta)$ in terms of θ is called the *sine function*. It is the basic function which represents the physical properties of periodic waves.

Generating the sine function

Let us increase the angle θ in Figure 6.2a by rotating the hypoteneuse of the triangle ABC anti-clockwise. As θ increases

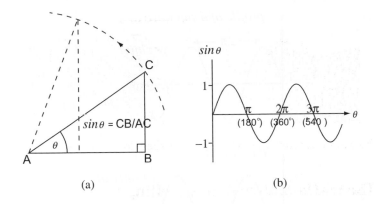

Figure 6.2 (a) $\sin(\theta)$ = opposite/hypoteneuse. (b) The sine function.

from zero to 90° the value of $\sin(\theta)$ rises from 0 to 1, falls back to zero at $\theta = 180°$ and then becomes negative as the side BC points in the negative y-direction. Finally, when θ reaches 360° the function $\sin(\theta)$ returns to its initial value at $\theta = 0$. As we continue to rotate the arm AC in an imaginary circle, the function $\sin(\theta)$ continues to repeat in a periodic cycle as illustrated in Figure 6.2b.

If we rotate the radius at a constant angular speed ω (radians per second), θ changes at constant rate, and we can write $\theta = \omega t$. The graph then shows how the lateral displacement of a particle in a given place in the medium varies with time. Such a particle oscillates with *simple harmonic motion* represented by the function $y = \sin(\omega t)$.

In a continuous medium which can transmit transverse waves, any one particle will not oscillate in isolation but will transmit its motion to neighbouring particles resulting in the creation of a *simple harmonic wave*. The upper part of Figure 6.3 depicts a 'photograph' of such a wave giving the transverse displacements from equilibrium (y) of successive particles at a given moment in time. The profile of the wave is the function $y = \sin(x)$, where x is the distance in the direction of propagation (x). Note

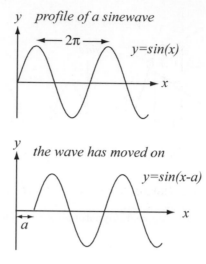

Figure 6.3 Advancing sine wave.

that the pattern repeats after a distance $\lambda = 2\pi$, the *wavelength* of the wave.

The lower curve is another snapshot, taken when the wave has advanced a distance a in the x direction. We represent this wave by introducing a *phase difference* $= -a$ into the original expression. The first null point, which had previously been at the origin is now at the position $x = +a$. All other points on the wave have advanced by the same distance.

6.2.3 An expression for a sine wave in motion

So far our mathematical expression represents a wave frozen in time. It is a 'snapshot' which shows the wave at a given instant. We can represent a moving wave by letting the phase shift change with time at the rate $a = vt$ where v is the speed of the wave.

Making adjustments to the scale
We adjust the equation to give the correct wave amplitude by multiplying the sine function by the factor A. We can also adjust

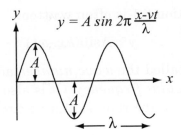

$$y = A \sin 2\pi \frac{x - vt}{\lambda}$$

Figure 6.4 The final product.

the scale so that the distance between wave crests is λ instead of 2π by writing it in the form:

$$y = A \sin\left(\frac{2\pi x}{\lambda}\right)$$

The value of this function is $+A$ for $x = 0, \lambda, 2\lambda, 3\lambda, \ldots$, etc. Combining the above with a phase shift which changes with time we obtain:

$$y = A \sin\left\{2\pi\left(\frac{x - vt}{\lambda}\right)\right\} \tag{6.1}$$

This represents a simple harmonic wave of amplitude A and wavelength λ travelling with speed v from left to right along the x-axis.

6.2.4 Wave parameters

The time for any particle to complete one oscillation is called the period T while the number of oscillations per second is called the frequency $f = \frac{1}{T}$.

The wave travels a distance λ in a time T so the speed of the wave $v = \dfrac{\lambda}{T} = \lambda f$.

The wave equation 6.1 is often written:

$$y = A\sin(kx - \omega t)$$

where $k = 2\pi/\lambda$ is called the *wave number* and $\omega = 2\pi v/\lambda = 2\pi f$ is known as the *angular frequency*. (It is also the angular speed of the radius vector in the representative circle in Figure 6.2a).

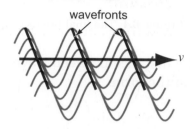

It is often convenient to picture a wave in terms of wave fronts, which are the surfaces joining points equally displaced by the wave at any one time.

6.3 The superposition of waves

6.3.1 *The superposition principle*

Raindrops. *Courtesy of Piotr Pieranski.*

The superposition principle states that the total displacement of any particle, simultaneously disturbed by more than one wave, is simply the linear sum of the displacements due to the individual waves.

When droplets of rain fall on the surface of a pool, they create circular surface waves which expand and overlap one another. Each wave is unaffected by the presence of the others, and each independently displaces particles of water. To get the total displacement of any particle, we simply add the displacements due to individual waves.

6.4 Applying the superposition principle

6.4.1 *The superposition of two waves travelling in the same direction*

Two identical sinc waves travel in the same direction:

Waves in phase: the individual waves combine to give a wave with a total amplitude twice that of either wave.

Waves out of phase: the individual waves combine to give a total amplitude of zero.

6.4.2 *Path difference and phase difference*

If two sources emit periodic waves in phase, the total amplitude of the disturbance at any point where the waves overlap depends on the phase difference between them.

This phase difference depends on how far each wave has travelled from its source. Waves from the two sources will be in phase provided that the difference in the length of the path is zero or some whole number of wavelengths (λ, 2λ, 3λ, etc.). The

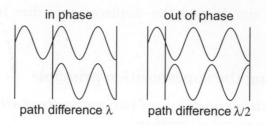

in phase   out of phase

path difference λ   path difference λ/2

Figure 6.5 Path difference and phase difference.

waves will be completely out of phase if the path difference is one half wavelength or any odd number of half wavelengths $\left(\dfrac{\lambda}{2}, \dfrac{3\lambda}{2}, \dfrac{5\lambda}{2}, \text{ etc.}\right)$, as illustrated in Figure 6.5.

6.4.3 *When two waves travelling in opposite directions meet*

If a transverse pulse is sent down a string tied at one end, it will be reflected and come back upside down. This is according to Newton's third law of motion, which states that action and reaction at the point of reflection are opposite. (The fact that it is reflected upside down is not particularly important as far as the argument that follows is concerned; what is more relevant is that the pulse comes back with the same speed with which it was sent.) If the support is rigid, very little energy is absorbed and the amplitude of the pulse will not be significantly diminished.

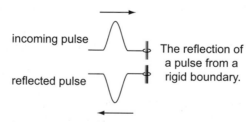

incoming pulse

reflected pulse

The reflection of a pulse from a rigid boundary.

If, instead of a single pulse, a continuous wave is reflected from a rigid boundary such as a wall, a reflected wave is generated. We have something which seems hard to imagine — two equal waves travelling in opposite directions in the same string.

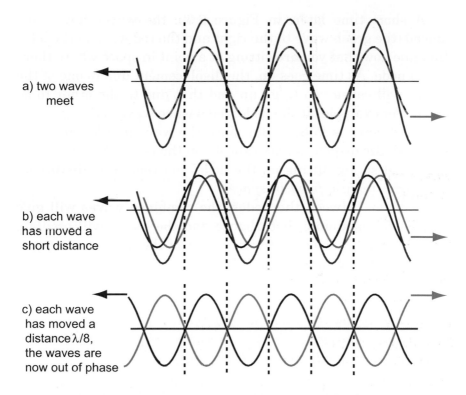

a) two waves meet

b) each wave has moved a short distance

c) each wave has moved a distance λ/8, the waves are now out of phase

Figure 6.6 Two waves meet and move out of phase.

The incident and reflected waves will be superimposed on one another, as seen in Figure 6.6. The resultant pattern is shown at three instants in time and tells the story of two waves meeting. The story of two waves meeting. The right-travelling wave is represented by a green line and the left-travelling wave by a red line. The waves are identical and travel with the same speed in opposite directions.

Figure 6.6a is a 'snapshot' taken at the instant when they happen to coincide. The resultant wave *at that instant* (dotted line) has the same wavelength and twice the amplitude. At the points where the disturbance due to both waves is zero, the resultant is of course also zero. These points are indicated by the dashed lines and are called *nodes*.

A short time later, in Figure 6.6b, the waves have each moved (the green wave to the right and the red wave to the left). Imagine now that you are sitting at a point in space where there was a node. As time goes on, the displacement due to one of the waves will cause you to go up and that due to the other to go down. For example, at the node, the dashed wave is rising, while the solid wave is falling. But since the wave profiles are symmetrical, the displacements cancel and the resultant remains at zero. And so, by 'sitting on the node' you remain undisturbed! *The nodes remain fixed in space.*

Halfway between the nodes, the combined waves will give rise to the maximum disturbance and the string will oscillate with an amplitude equal to twice that of the individual waves. These points are called *antinodes*.

The argument could be stated even more simply on general principles: since the waves moving left and right are identical, there is no preferred direction in which nodes can move, and therefore they will remain where they are.

Figure 6.6c completes the story. The waves have moved on, each by a distance of $\lambda/8$. At this instant the green and red waves are completely out of phase, and the resultant is zero everywhere. The distance between successive nodes or successive antinodes is $\lambda/2$.

(The analysis effectively applies only to a string of infinite length. In practice it will only last until the returning wave is again reflected, at the other end, and yet another wave enters the scene.)

6.4.4 *A string fixed at both ends*

The situation described above is rather unreal, in that we are ignoring what will happen when the reflected wave reaches the other end of the string which must be held by somebody, or attached to something. When the reflected wave comes back to that point it will be reflected again and the wave will continue bouncing back and forth — in principle *ad infinitum*! Nodes and antinodes will be set up, this time in different places, and the vibrations will soon die out.

If the length of string is equal to some multiple of the inter-node distance, nodes and antinodes set up by the wave reflected from the far end of the string will be in exactly the same positions as the nodes and antinodes produced by the original wave and reflected wave. In these circumstances the waves combine to give a standing wave.

The particles of the string continue to vibrate in the same sort of way, but with the one important difference that *all the particles oscillate in phase at the frequency of the wave*, as illustrated in Figure 6.7. The dots show successive positions of string particles at equal intervals of time, indicated by the different shades. Particles situated at antinodes such as A vibrate with maximal amplitude and those at nodes such as B do not vibrate at all. The contemporaneous positions of adjacent particles lie on curves of the same shade. The unique feature of standing waves is that energy is not transmitted but stored as vibrational energy of the particles.

A standing wave stores energy in the oscillations of the particles disturbed by the two waves.

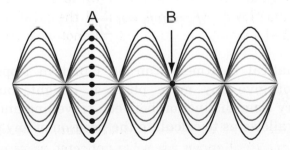

Figure 6.7 Time sequence of the positions of the particles of a vibrating string.

(In practice, there is no such a thing as a completely rigid support and a small amount of energy will 'escape' from the string at each reflection.)

6.4.5 *Standing waves*

A string of fixed length has a number of normal modes of vibration corresponding to exact numbers of half wavelengths which 'fit exactly' into that length of string. The amplitude of vibration must be zero at both ends of the string.

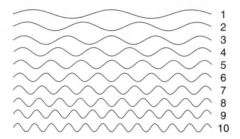

The normal modes of a vibrating string.

The frequencies of the normal modes are called the *natural* frequencies (f_n) of the string, and

$$f_n = nvf_1 = n\,\frac{v}{2L}$$

A mathematical treatment can be found in Appendix 6.3. It does not give any more physical insight, but it does make a very clear distinction between travelling and standing waves and allows us to calculate the position of any particle at any time.

6.5 Forced oscillations and resonance

We can use an oscillator to produce waves on a string or in other materials, such as air. If the string or other system has natural frequencies of vibration, its subsequent behaviour depends on whether the frequency of the oscillator matches one of these frequencies.

6.5.1 *Forced oscillations*

Suppose that an oscillator drives a string to vibrate. The wave will be directed down the string and reflected at the far end, which is a node because it is fixed. As the incoming and reflected waves meet, nodes separated by distances of $\lambda/2$ will be formed. The reflected wave returns to the oscillator end, which is also a node (assuming the amplitude of the waves generated by the oscillator to be small). In Figure 6.8 the positions of nodes created when reflections take place at each of the ends illustrate the condition for the formation of standing waves on a string.

If the oscillator frequency is not one of the natural frequencies of the string, the fixed end at B is not exactly one or more half wavelengths away from A and the condition for standing waves is not met. The reflected wave, which travels back up the string, is out of phase with the waves pumped in by the oscillator. So, although the oscillator keeps

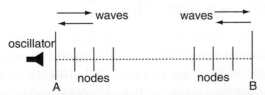

Nodes must match for standing waves to form.

Figure 6.8 Nodes and the condition for standing waves.

pumping energy into the string, this energy is absorbed mainly at the ends of the string. The waves themselves are travelling waves of small amplitude. This type of behaviour can also be seen in electrical circuits which have characteristic modes of vibration. The energy is absorbed by the resistance of the circuit.

6.5.2 *Natural frequencies of vibration and resonance*

When someone plucks a stretched string, the disturbance propagates in both directions and is reflected from the ends. The incident and reflected waves may combine to give a standing wave at one of the natural frequencies of the string. Now suppose that an oscillator is coupled to one end of the string. If the oscillator frequency matches one of the natural frequencies, each new wave from the oscillator reinforces the wave reflected from the far end of the string. The total amplitude of the wave is higher than at other frequencies; we refer to this as *resonance*. The amplitude of vibration is limited by elasticity and by other constraints on the system.

All flexible mechanical systems which are constrained in some way, like suspension bridges, guitar strings or the air in an organ pipe, have natural frequencies at which they will vibrate if prompted by an impulse.

6.6 Resonance — a part of life

We make use of resonant behaviour in many ways. For example, resonance chambers are used to amplify sound in acoustic instruments; resonant behaviour of electrical circuits is used in the transmission and reception of the wireless communications of radio, television and cellphones; lasers operate on the principle of resonance.

Resonant systems can sometimes create situations over which we have no control. There have been some devastating consequences of uncontrolled resonant oscillations.

6.6.1 *The Tacoma Narrows bridge disaster*

One of the most spectacular mistakes in the history of civil engineering was the design of the Tacoma Narrows bridge, built at a cost of seven million dollars. It was the third-longest suspension bridge in the world when it opened and the supporting towers were separated by more than half a mile.

Official opening on 1 July 1940. *Courtesy of University of Washington Libraries.*

Well in advance of the opening, spectacular motions of the bridge had earned it the nickname of Galloping Gertie. Oscillations with an amplitude of about 1 metre were a regular feature and attracted sightseers.

On 7 November 1940, in steady winds of about 40 mph, the bridge began to twist violently and eventually collapsed, shedding a 183-m-long section into the valley below. The only casualty was a dog.

The 1941 report of the investigating committee attributed the disaster to some form of resonance. It was not a simple case of forced resonance (such as when the periodic motion of soldiers marching in step over a

A violet twisting motion. *Courtesy of University of Washington Libraries.*

The aftermath.
*Courtesy of University of
Washington Libraries.*

suspension bridge sets it into resonant oscillation). The wind could not have maintained a sufficiently large regular stimulus for simple resonant oscillation.

There has been a lengthy debate as to the exact cause of the collapse, but what *is* certain is that a full understanding of what happened can only be gained through a rigorous mathematical analysis of the sequence of events.

6.6.2 *The Mexico City earthquake*

'Considerable liquefaction and damage to new buildings occurred in Mexico City during the Great Earthquake of 19 September 1985. Although the epicentre was more than 300 km away, the

Mexico City earthquake.
*Courtesy of George Pararas-
Carayannis.*

valley of Mexico experienced acceleration of up to 17% *g*, with peaks concentrated at 2-sec periods. The extreme damage in Mexico City was attributed to the monochromatic type of seismic wave, with this predominant period causing 11 harmonic resonant oscillations of buildings in downtown Mexico City. The ground accelerations were enhanced within a layer of 30 ft of unconsolidated sediments underneath downtown Mexico City, which had been the site of a lake in the 15th century, causing many buildings to collapse.' (*Pararas-Carayannis, G. 1985**)

---

* drgeorgepc.com.

6.7 Diffraction — waves can bend around corners

All waves can spread around corners — a phenomenon called diffraction. A classic example of diffraction is where incoming waves pass between rocks or other obstacles close to the shore. The photograph shows a series of stone breakwaters along the shore at Kingsmill in Chesapeake Bay, Virginia. The breakwaters were constructed as part of a coastal erosion control programme in the Bay. Incoming wave energy is sapped by frictional effects, refraction and diffraction. The semicircular way in which the sea has eroded the shoreline behind the breakwaters is remarkably clear evidence of the effect of diffraction.

Courtesy of Chesapeake Bay Breakwater Database, Shoreline Studies Program, Virginia Institute of Marine Science — www.vims.edu.

6.8 The magic of sine and the simplicity of nature

A question which might spring to mind at this stage is why we should have chosen the simple sine wave as the means of describing natural waves which are frequently far from sinusoidal in their behaviour. The mathematics of a simple sine wave is relatively easy to handle, but there is a much more compelling reason.

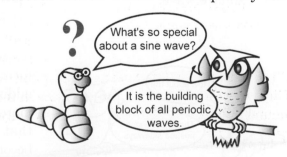

Music from an unexpected duet.

A young cellist, apprehensive about giving her very first performance, decides to practise in a secluded corner of a local park, early one morning.

The wave form is far from sinusoidal but surprisingly enough, provided that it is periodic, we can describe it as the sum of a number of sine waves of different amplitudes and frequencies.

6.8.1 *The sum of a number of sine waves*

To illustrate how complex wave forms may be produced we can take the example of a square wave form. We can graphically illustrate how adding the amplitudes of sine waves of certain frequencies and amplitudes results in a wave which becomes increasingly like a square wave.

Sets of 2, 3 and 7 sine waves are shown in Figure 6.9 together with the wave form which results from adding them. As we add more sine waves, we can see that the resultant wave becomes closer and closer to a square wave.

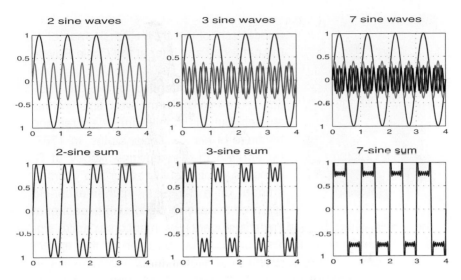

Figure 6.9 The evolution of a square wave.

The frequencies of the component sine waves are all harmonics of the frequency of the square wave (see Appendix 6.3). The mathematical method for the decomposition of any harmonic wave into its component sine waves is called *Fourier analysis*.

So the simple sine wave proves to be the basis for the mathematical description of any periodic wave, even complex ones such as the sound waves from musical instruments. The key to the process of describing such waves is in knowing the relative amplitudes and frequencies of the component sine waves. This is the essence of the true harmony and simplicity of nature, which becomes more apparent as we learn what to look for.

A historical interlude: Jean Baptiste Joseph Fourier (1768–1830)

Jean Baptiste Joseph Fourier was born in Auxerre, a small cathedral town in the Burgundy region of central France. He was educated at

(Continued)

a local school run by the music master of the cathedral and later at the Ecole Royale Militaire.

Fourier had intended to take the tradition route into the army after finishing his studies but his application was refused because of his low social status. In 1787, he entered the Benedictine Abbey at St Benoit sur Loire, but left after two years to embark on a career in mathematics. He returned to Auxerre, becoming a teacher at the school where he had been a student. His intention to make significant contributions to mathematics may be seen in this extract from a letter written to Bonard, a professor of mathematics in Auxerre: *'Yesterday was my 21ˢᵗ birthday; at that age Newton and Pascal had already acquired many claims to immortality.'*

In the turbulent years following the French revolution in 1789, Fourier developed an active interest in politics and joined the local revolutionary committee in 1793. A political idealist, he wrote: *'As the natural ideas of equality developed it was possible to conceive the sublime hope of establishing among us a free government exempt from kings and priests and to free from this double yoke the long-usurped soil of Europe. I readily became enamoured of this cause, in my opinion the greatest and most beautiful which any nation has ever undertaken.'*

French politics at that time were extremely complex. There were numerous small groups of 'liberators' who fought bitterly amongst themselves, despite having the same general aims. 'Citizen' Fourier was arrested and imprisoned in 1794 because of his support for victims of terror. At this time, during the 'Reign of Terror' imposed by Robespierre, the interval between imprisonment and being guillotined was often very short. Fortunately (for Fourier), Robespierre himself was guillotined shortly after Fourier was imprisoned and the political climate changed. Fourier was freed and went to the Ecole Normale in Paris, where he studied

(Continued)

under the eminent mathematicians Lagrange, Laplace and Monge. Fourier was appointed to a teaching post at the newly established Ecole Polytechnique. He continued his research until he was arrested and imprisoned once more (for the same offences), but he was soon released and was back at work before the end of 1795. Within two years, Fourier succeeded Lagrange as professor of analysis and mechanics at the Ecole Polytechnique.

In post-revolutionary France, Fourier was no longer debarred from a military career as he had been in his youth. In 1798, he was drafted and joined Napoleon's expeditionary force to Egypt as a *savant* (scientific advisor). The campaign began well. The occupation of Malta in June 1798 was rapidly followed by the fall of Alexandria and the occupation of the Nile delta some three weeks later. Napoleon's fleet was decimated by the English fleet, under the command of Lord Nelson, at the battle of the Nile and he was cut off from France by the English fleet. Nevertheless Napoleon proceeded to set up an administrative structure modelled on the French system. Fourier was appointed governor of Lower Egypt and was one of the founders of the Institute d'Egypte in Cairo. He began to develop ideas on heat, not only in the mathematical sense but also as a source of healing.

Napoleon left about 30,000 troops stranded in Egypt and returned secretly to France. Fourier himself returned in 1801 hoping to continue his work as a professor at the Ecole Polytechnique. He was not allowed to remain there for long. Napoleon, having noticed his remarkable administrative abilities in Egypt, appointed him as prefect of the Departement of Isere, where he was responsible for the drainage of a huge area of marshland and the partial construction of a road from Grenoble to Turin.

Whilst in Grenoble, Fourier found the time not only to further his mathematical studies but also to write a book on Egypt. He was made a count and appointed prefect of the Rhone shortly before Napoleon's final defeat in the Battle of Waterloo in 1815. At that point, Fourier resigned the title and the position and returned to Paris, where he had a small pension but no job and a rather

(Continued)

undesirable political pedigree. One of his former students secured him an undemanding administrative job. Fourier finalised his most important mathematical work during this time. His major achievement was to describe the conduction of heat in terms of a differential equation (similar to the wave equation) without trying to figure out what heat actually was. The conclusion of his work was that these equations could be solved in terms of periodic functions which themselves could be constructed from simple sinusoidal functions. Fourier died in 1830 in Paris.

Appendix 6.1 The speed of transverse waves on a string

We can show that the speed of a transverse wave on a stretched string depends on the tension and the mass and length of the string. Take a single pulse which travels along a string at a constant speed v:

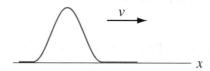

When the pulse arrives at a point on the string, a component of the tension suddenly acts at right angles to the string and displaces the string at that point. Owing to 'bonding', successive particles are displaced from their equilibrium positions and the pulse moves along the string. *The faster the response of each particle*, the larger its acceleration, the more rapidly it rises and falls and *the faster the wave progresses. Increasing the tension increases the wave speed. Increasing the mass (a measure of the resistance to motion) reduces the wave speed.*

Calculating the speed

We look at a very small segment of length $2L$ around the peak of the pulse and make a number of approximations.

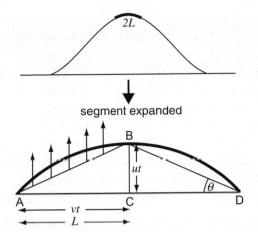

Figure 6.10 The peak of the pulse passes along the string.

Let t be the time for the left hand half of the pulse to move from A to C in Figure 6.10 and let the average transverse speed of the particles be u.

For small L, arc AB ≈ chord AB.

$$AC = vt \qquad BC = ut \text{ and } \frac{BC}{BD} = \frac{u}{v} = \tan\theta$$

For small values of θ

$$\sin\theta \approx \tan\theta = \frac{u}{v}$$

The force towards the centre of the circle $F = T\sin\theta = \dfrac{Tu}{v}$, where T is the tension of the string.

Impulse $Ft = \dfrac{Tut}{v} = mu$, the momentum transferred

$m = \dfrac{Tt}{v} = \mu L$ (μ is the mass per unit length, a property of the string)

since $\qquad \dfrac{L}{t} = v \; \Rightarrow \; T = \mu v^2 \quad$ and $\quad v^2 = \dfrac{T}{\mu}$

The speed of a transverse pulse on a string

$$v = \sqrt{\dfrac{T}{\mu}} \tag{6.2}$$

The speed of the wave depends only on the characteristics of the string, and not on the frequency of the wave.

There is an equivalent expression for all mechanical waves.

The dimensionless analysis in Appendix 6.2 shows that the speed of any transverse wave in any medium is related to the ratio of the appropriate elastic constant divided by a mass-related constant.

Appendix 6.2 Dimensional analysis

A dimensional analysis will confirm that the form of Equation (6.2) for the speed of a transverse pulse on a string has the correct form.

We found that

$$v^2 = \dfrac{T}{\mu}$$

Dimensions of the left hand side:

$$\left(\dfrac{L}{T}\right)^2 = \dfrac{L^2}{T^2}$$

Dimensions of the right hand side:

$$\dfrac{ML}{T^2} \bigg/ \dfrac{M}{L} = \dfrac{L^2}{T^2}$$

confirming that Equation (6.2) has the correct form.

A dimensional analysis is independent of the type of wave and the nature of the material, so we can extrapolate and say that the general expression for the speed of a wave in a material will be

$$v \propto \frac{\text{elastic constant}}{\text{mass-related constant}}$$

Appendix 6.3 Calculation of the natural frequencies of a string fixed at both ends

We can write the displacement due to a right-travelling wave as $y_1 = A \sin(kx - \omega t)$ and due to a left-travelling wave as $y_2 = A \sin(kx + \omega t)$. (At all places on the string, there are times when $y_1 + y_2 = 0$, which justifies using these expressions for y_1 and y_2.)

Using the superposition principle, $y = y_1 + y_2$,

$$y = A[\sin(kx - \omega t) + \sin(kx + \omega t)]$$

$$\Rightarrow \qquad y = 2A \sin(kx)\cos(\omega t) \qquad\qquad (6.3)$$

$$\left[\text{using the trignometric identity} \right.$$

$$\sin(A + B) = 2\sin\left(\frac{A + B}{2}\right)\cos\left(\frac{A - B}{2}\right)\left.\vphantom{\frac{A}{2}}\right].$$

There is no $(kx - \omega t)$ term in Equation (6.3), which means that the wave does not propagate. If we look at the particle at $x = a$, for example, the amplitude of its vibration is $2A \sin(ka)$, which is constant. Each particle performs simple harmonic motion at the frequency of the waves and all the particles vibrate in phase.

In simple harmonic motion, the total energy of any particular oscillating particle is constant — a standing wave stores energy.

An oscillator attached to one end of a string generates waves of very small amplitude, so the 'oscillator' end of the string can be taken to be more or less fixed. If the other end of the string is also fixed, then standing waves can be set up. Suppose that the oscillator is situated at $x = 0$ and the other end of the string at $x = L$. The displacement will always be zero at $x = 0$ and $x = L$,

$$\text{i.e.} \quad y = 2A[\sin(kL)\cos(\omega t)] = 0$$

This means that $\sin(kL) = 0$ and $kL = 0$, π, $2\pi = n\pi$, where n is an integer.

By comparing the expression $y = A\sin(kx - \omega t)$ with Equation (6.2), we see that $k = \dfrac{2\pi}{\lambda}$, giving $\lambda = \dfrac{2L}{n}$. The value of the wavelength is fixed solely by the length of the string; it is a *natural* wavelength.

The natural wavelengths have a series of values, $\lambda_n = \dfrac{2L}{n}$.

The corresponding natural frequencies are $f_n = n\dfrac{v}{2L}$.

The frequency of the first harmonic $f_1 = \dfrac{v}{2L}$ (also called the *fundamental* frequency). The frequency of the second harmonic $f_2 = \dfrac{v}{L}$. The frequency of the third harmonic $f_3 = \dfrac{3v}{2L}$ and so on.

Since v depends on the material itself $\left(v = \sqrt{\dfrac{T}{\mu}} \text{ for a wave on} \right.$

a string $\left.\vphantom{\sqrt{\dfrac{T}{\mu}}}\right)$, a vibrating string will not have the same natural frequencies as a vibrating rod, or indeed a vibrating air column of the same length.

Chapter 7

Sound Waves

No account of waves and their properties can be complete without the inclusion of a discussion of sound waves. Sounds in the frequency range of 20–20,000 Hz are associated with human hearing. Many animals and birds can produce and detect sounds above and below this frequency range, providing them with a means of communication and an aid to survival.

The perception of sound follows general biological laws which relate stimulus to perception. We will look at the relationship between real and perceived changes and how loudness, pitch and tone quality relate to measurable properties of sound waves.

Music at an open-air concert sounds very different from the same music played indoors, where reflections can modify the sound by distorting the wave form. The design of good auditoria is both a science and an art.

The apparent change in the frequency of sound due to motion of the source or the listener — the Doppler effect — provides useful and sometimes vital information. It is used by some animals and birds to navigate and locate food. The familiar 'sonic boom' heard when an aircraft moves faster than the local speed of sound in air is also associated with the Doppler effect.

7.1 Sound and hearing

7.1.1 *Sound as a pressure wave*

Sound waves in air are longitudinal. They can be produced, for example, by vibrations of the membrane of a loudspeaker. When

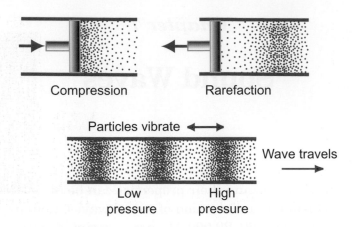

Figure 7.1 Pressure waves travel from source to listener.

a membrane starts to vibrate, it produces to-and-fro motions of neighbouring air molecules, creating alternate regions of high pressure (compressions) and low pressure (rarefactions), as illustrated in Figure 7.1. The compressions and rarefactions are pushed along according to the principles of propagation of any mechanical disturbance. High pressure regions are shown as being more densely populated with dots than low pressure regions. (The mean pressure is usually atmospheric pressure.)

The amplitudes of vibration of the diaphragm, the air and the eardrum are all different. In low pressure regions, particles are more easily displaced than in high pressure regions, so low pressure means high displacement, and the converse.

Sound waves exert a varying pressure on the eardrum, causing it to vibrate. The vibrations are communicated to a chamber called the auditory canal, then transmitted along a series of small, interlinked bones, and eventually received as nerve impulses by the brain.

7.1.2 *The speed of sound*

The velocity of sound varies from one material to another. Contrary to what one might expect, sound can travel faster in

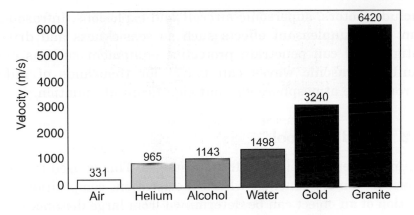

Figure 7.2 Measured values of the speed of sound.

liquids and in solids than in gases, because the particles in⤴ solids and liquids are closer together and can respond more rapidly. Some measured values may be seen in Figure 7.2. Sound propagates as a longitudinal wave in gases but can also propagate as a transverse wave in solids because solids resist *any* form of shearing stress.

The speed of sound in a given medium is found to be independent of its frequency.

7.1.3 *Ultrasound and infrasound*

Sound waves with frequencies in the range of *20–20,000 Hz* are audible to humans. Frequencies *above 20,000 Hz* are called *ultrasound*. Ultrasonic waves can be produced when a quartz crystal is subjected to an oscillating electric field and changes shape, rather like a diaphragm (piezoelectric effect). Many animals use ultrasonic waves as a means of communication. Ultrasound is also used extensively in medicine and in industry. *Infrasound* is the name given to sound waves with frequencies *below about 20 Hz*. They are usually accompanied by low frequency sound audible to humans. Natural sources include ocean waves, lightning and large mammals; man-made sources include motors, supersonic aircraft and explosions. Infrasound can cause unpleasant effects such as seasickness and driver fatigue and can penetrate protective equipment such as ear-muffs. Infrasonic waves can travel for thousands of miles through the atmosphere without significant attenuation.

7.2 Sound as a tool

Sound waves are reflected from obstacles. This is used to great effect in remote sensing, where the location, composition and motion of an object can be determined from large distances.

Detailed analysis of the intensities and frequencies of reflected sounds forms the basis for a whole range of imaging techniques. In most applications the sound is emitted as a series of pulses. The distance between the source and the reflecting surface is calculated from the time between the emission of a pulse of sound and the detection of the echo. The term *echolation* has been coined to describe the use of echoes to determine distances. Echolation is routinely used to measure the depth of water beneath a ship.

We can map the contours of reflecting surfaces by measuring the times for echoes to return from different parts of the surface. Echoes vary in intensity, loud for a highly reflecting surface and soft for a highly absorbing surface, enabling us to determine the nature of the surface. We can also measure the velocity of a moving surface from the difference in frequency between emitted and reflected sounds (using the Doppler effect).

Sound navigation and ranging (SONAR)

Sound travels large distances in water. Sound-based underwater communication systems were used as early as the 19th century when lightships equipped with underwater bells acted as navigation aids. The ringing sound was detected in passing ships using stethoscopes. These were the forerunners of today's passive sonar systems.

⨉*Passive sonar.* A listening device which does not radiate any sound, it may used in studies of marine life and by submarines to minimise the risk of detection. Wartime use of passive sonar revealed many unexpected sources of underwater noise, such as the so-called 'snapping' of shrimps. There are numerous sources of noise: man-made (such as ships' propellers, engines and seismic drilling), biological (whales and other marine life) and natural (such as tides and seaquakes). Echoes from the seabed increase the underwater noise level. Many modern passive sonar systems have very advanced signal discrimination and extended receivers help to improve the direction and range finding.

Active sonar. Most sonar systems are active and gather information by echolation. The sound is emitted as a narrow pulsed beam of high intensity, often at ultrasonic frequencies.

The first active sonar system was an ultrasonic submarine detector developed in 1917 by Paul Langevin (1872–1946). The ultrasonic waves were emitted and detected by the same device, called a transducer. The reflected wave was detected using the inverse piezoelectric effect, whereby the incoming ultrasonic wave 'squeezes' a quartz crystal in the detector, producing an oscillating electrical signal. It was the first practical application of the inverse piezoelectric effect.

Current uses of active sonar are numerous. There is extensive military use, such as the detection of submarines and sea mines, navigation and acoustic missiles. Civilian applications include accurate underwater surveying, marine communication and the location of shoals of fish, sunken aircraft and shipwrecks. The

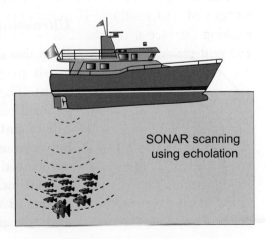

SONAR scanning using echolation

transducer may be fixed to a rotating platform mounted on the keel of a boat or towed along in a small 'towfish', just above the sea bed (side-scan sonar).

This 'ghostly' image of the sunken cruise liner *Mikhail Lermonotov* was recorded using side-scan sonar operating at a frequency of 675 kHz, purchased as part of a research project to map marine habitats.

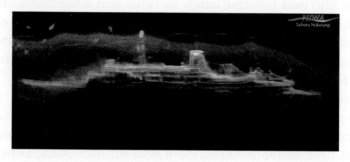

Sunken cruise liner. *Courtesy of Ken Grange National Institute of Water and Atmospheric Research, Nelson, New Zealand.*

Ultrasound is absorbed more rapidly than audible sound but has higher resolution and is ideal for searching for small objects such as mines. Submarine-hunting equipment operates at lower frequencies (mostly high frequency audible sound and very low frequency ultrasound).

Ultrasound in nature

Bats use ultrasonic waves to distinguish prey from obstacles when in flight. They can isolate echoes from general background sound. In a cave full of bats, a bat can differentiate between its own signals and those associated with other bats. The time resolution of bat sonar apparatus (about 2 millionths of a second) and the precision with which bats locate and identify prey is the envy of manufacturers of sonar equipment.

Ultrasound in medicine

Ultrasonic energy is absorbed more rapidly in tissues such as bone, tendons and tissue boundaries which have a high concentration of a protein called collagen. This forms the basis for diagnostic and therapeutic procedures in medicine. Ultrasound does not produce the toxic side effects associated with ionising radiations such as X-rays.

Low intensity ultrasound as a diagnostic tool

The imaging of tissue and the velocity measurement of fluids within the body have become invaluable tools in medicine. The intensity of diagnostic ultrasound is kept as low as possible to minimise heat absorption. The times of arrival and relative intensities of the echoes are used to construct a detailed image of the tissue structure. The level of detail in ultrasonic images increases with frequency but the level of penetration decreases. As an example, body tissue absorbs about 50% of the energy of 1 MHz ultrasonic waves in about 4 cm but the same amount of energy at a frequency of 3 MHz will be absorbed in 2–2.5 cm. This limits the efficiency of the technique.

Sonography is a pulse-echo imaging technique which is widely used in medicine. It is almost 50 years since ultrasound was first used to produce images of a developing foetus and it is now standard medical practice. A transducer is placed on the patient's skin, close to the area being examined. A coupling gel applied to the skin ensures that about 99.9% of the ultrasonic energy enters the body. The

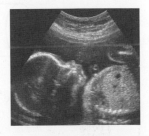

Foetal image. *Courtesy of Wessex Fetal Medicine Unit, St Anne's Hospital, Southhampton.*

ultrasound probe is moved over the abdomen of the patient to obtain different perspectives of the foetal image.

Doppler ultrasound

Pulsed ultrasonic waves are beamed into the body and are partially reflected by body tissue and fluids, as illustrated in Figure 7.3. The frequency of waves reflected from moving blood particles is different from the frequency of the waves emitted by the source. The difference between the two frequencies is directly proportional to the velocity of the blood (see Appendix 7.1 — Doppler effect). The velocities of blood particles in the aorta, the carotid artery, the umbilical chord and numerous other blood vessels can be calculated from measurements of this frequency shift.

The velocity of the blood particles depends on the diameter of the blood vessel and the uniformity of the walls at the point of measurement. It is possible to measure velocity profiles and detect abnormal blood flow patterns caused, for example, by the build-up of fatty deposits on the walls of arteries using this technique.

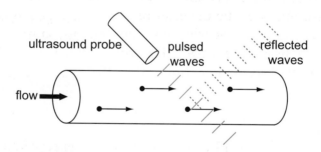

Figure 7.3 Doppler blood flow monitoring.

Heat treatment

As ultrasonic waves pass through the body, tissue expands and contracts thousands of times each second, producing heat. High-power, well-focused ultrasound can create temperatures in excess of 100°C in targeted tissue without damaging surrounding tissue, in the same way that a magnifying glass can be used to focus sunlight and burn holes in paper. Focused ultrasound can kill cells and is used to treat some forms of cancer.

Shock wave lithotripsy

Shock wave lithotripsy is the non-invasive destruction of kidney stones using ultrasound. Shock waves generated outside the body reduce the stones to a harmless powder. To avoid damaging surrounding body tissue, the shock waves are focused on the stone itself rather than being simply directed at the kidney. The ultrasound source S is situated at one focus of a silvered ellipsoidal reflector and the patient is positioned so that the kidney stone is at the other focus of the ellipsoid. Ultrasonic waves reflected from the silvered surface are focused on the stone and pulverise it. Waves not reflected from the silvered surface simply spread out and become weaker. The shaded areas in the diagram represent wavefronts converging on the kidney stone (only waves travelling to the stone are shown).

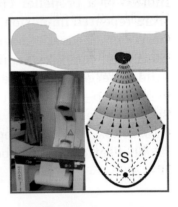

Lithotripter. *Courtesy of Mark Quinlan, UCD Medical School, Dublin.*

Acoustic cavitation

A liquid may be boiled by heating it to the appropriate temperature or by lowering the ambient pressure. In either case, the result is cavitation, the formation of bubbles of vapour within the liquid. Localised pressure changes, such as those caused by waves, can cause bubbles to form. Acoustic cavitation occurs when a liquid is exposed to intense ultrasound, causing bubbles to form, grow and collapse repeatedly. Energy is absorbed from the waves as the bubbles grow, and released when they collapse. The collapse or implosion of the bubbles causes localised heating. Research indicates that the internal temperature of almost completely collapsed bubbles may reach 10,000 K. When low frequency ultrasound bubbles expand and contract, flashes of blue light may be seen. This phenomenon is called *sonoluminescense* (from the Greek words for sound and light).

Surface erosion and cleaning

The erosion of ships' propellers is one of the earliest-known examples of surface erosion due to acoustic cavitation. The motion of a propeller creates sound waves and bubbles. Heat from repeated implosion of the bubbles erodes the surface of the propeller. Acoustic cavitation is an efficient and non-toxic method of cleaning. Equipment, immersed in a bath containing mild detergent, may be irradiated with high intensity ultrasound for 5–10 seconds to remove surface grime. It is often possible to see dirt being lifted from the immersed objects. In *liposuction*, another application of cavitation, sound from a probe inserted into the body attacks and erodes fat cells.

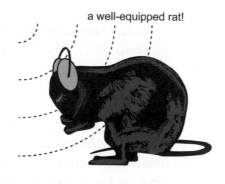
a well-equipped rat!

Pest control

There are pest control devices which emit intense blasts of ultrasound at frequencies which totally disrupt the nervous systems of rodents, such as rats, but are harmless and often inaudible to humans and other mammals, such as cats and dogs.

Infrasound in nature

Birds may travel thousands of miles in summer or winter migrations — a remarkable feat of navigation. One of the best avian navigators is the pigeon, once used to carry messages.

The pigeon's extraordinary homing instinct is the basis for the popular sport of pigeon racing. Pigeons can fly home in cloudy and windy conditions and can be trained to navigate at night. It is well known that pigeons use the earth's magnetic field as a compass, but a sense of direction is not sufficient to return them to their home loft. They need to be able to relate their position to their destination — they need a map. It is thought that infrasound plays a part in this very precise homing process. Infrasonic waves,

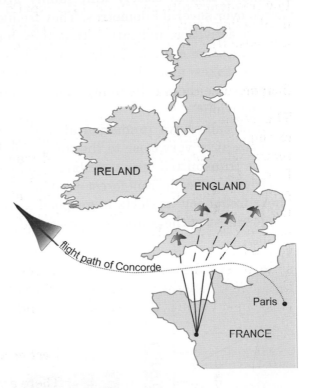

generated by natural sources such as seismic activity and ocean waves, propagate in the air. They are reflected off geographical features such as steep hillsides and may serve as a navigation aid for the pigeons. There have been instances where a large number of birds were inexplicably lost or delayed. In June 1997, 60,000 English pigeons were released in France and about one third of them failed to return to England. It has been suggested* that the birds were disoriented after crossing the path of low frequency shock waves generated by the Concorde supersonic aircraft.

---

* J.T. Hagstrum, The Journal of Experimental Biology, 203: 1103–1111 (2007); Proc. 63rd Annual Meeting of The Institute of Navigation, 2007, Cambridge, Massachusetts.

Elephants and infrasound

Elephants live in closely knit family groups sometimes dispersed over several kilometres. They frequently engage in coordinated activities, indicating that they have a well-developed communication system. Their sight is poor and remote communication takes place by means of sound. Conservation programmes use these calls to monitor elephant numbers, location and behaviour.

Most calls have fundamental frequencies in the range of 5–30 Hz, at or very close to the infrasonic frequency band, and may be transmitted over very large distances. A characteristic spectrogram of calls from forest elephants is shown in Figure 7.4, where frequency in Hertz is graphed against time in seconds. The loudness of the sound is represented by the darkness of the display.

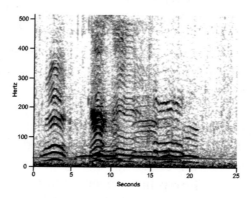

Figure 7.4 Elephant calls. *Courtesy of The Elephant Listening Project — www.birds.cornell.edu.*

Listening for danger

Networks of listening stations, originally set up to monitor explosions from underground testing of nuclear weapons, are now important sources of information on potentially disastrous events such as hurricanes, earthquakes and meteors.

In addition to furthering scientific understanding, such networks have the potential to become early warning systems in the event of impending natural catastrophies.

In 1999, infrared noise from the explosion of a meteor in the atmosphere was detected by scientists at the Royal Netherlands Meteorological Service.

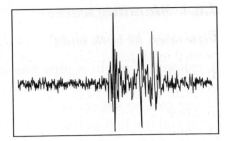

Courtesy of Hans Haak, Royal Netherlands Meteorological Service.

7.3 Superposition of sound waves

The sounds we hear are usually the result of the superposition of a number of sound waves. Superposition of waves can result in amplification or in reduction of the intensity of the sound. Passive aids, such as earmuffs, which reduce background sound are inefficient for low frequency noise. *Noise cancellation*, i.e. active suppression of noise, is widely used by airline pilots to

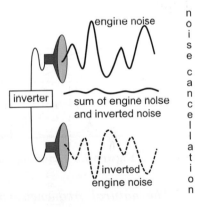

suppress high levels of low frequency engine noise. Small microphones mounted on a pilot's headset convert the ambient noise in the cockpit to an electrical signal, which is inverted and re-emitted as sound by small speakers. The noise experienced by the pilot is the sum of the ambient and inverted noise wave forms and is substantially lower than the ambient noise in the cockpit. The technique is most effective against the steady low frequency sounds created by engines.

7.3.1 *Standing waves*

Pipe open at both ends

A column of air in a pipe has natural resonant frequencies, in much the same way as does a string fixed at both ends. Standing waves can be set up in a pipe open at one or both ends. At an open end (or just outside), the air molecules are free to vibrate, constrained only by the elasticity of the air, so the open ends are anti-nodes. A closed end is an almost perfectly rigid boundary, so it must be a node. Figure 7.5 shows the first two harmonics for a pipe of length L, open at both ends.

The displacement is maximal at both ends of the pipe, so the *basic symmetry is the same as a string fixed at both ends* (the positions of the nodes on a string are the same as the positions of the anti-nodes in a pipe of the same length). Any whole number of half wavelengths can be fitted into the pipe.

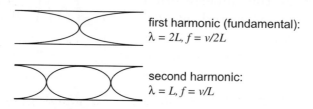

first harmonic (fundamental):
$\lambda = 2L, f = v/2L$

second harmonic:
$\lambda = L, f = v/L$

Figure 7.5 Standing waves in a pipe open at both ends.

The natural frequencies of a pipe open at both ends are $f_n = \dfrac{nv}{2L}$ *where* $n = 1, 2, 3, \ldots$.

Pipe closed at one end

In Figure 7.6 we see the first and second allowed modes of vibration in a pipe closed at one end. The situation is *not* the same as for the open-ended pipe — *the symmetry has been broken* because there is now a node at one end and an anti-node at the other end.

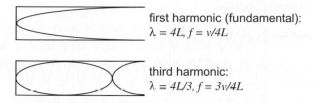

first harmonic (fundamental):
$\lambda = 4L, f = v/4L$

third harmonic:
$\lambda = 4L/3, f = 3v/4L$

Figure 7.6 Standing waves in a pipe closed at one end.

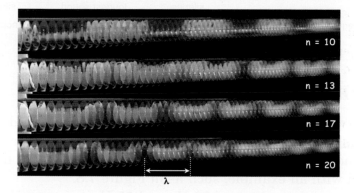

Figure 7.7 Standing waves in a tube. *Courtesy of Stefan Huzler, Trinity College, Dublin.*

Only odd harmonics are found in the sound from a pipe closed at one end.

Breaking the symmetry removes half the natural frequencies.

The anti-nodes of sound waves in a perspex tube may be made visible by the presence of equally spaced soap films, as seen in Figure 7.7. The coloured soap films reveal the positions of the anti-nodes. The photograph is part of a novel experiment for the visualisation and measurement of standing waves in a tube.*

---

*F. Elias, S. Hutzler and M.S. Ferreria, Visualisation of sound waves using regularly spaced soap films. *Eur. J. Phys.* 28: 755–765 (2007).

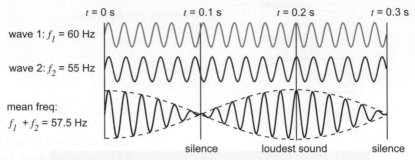

beat frequency = 60 Hz - 55 Hz = 5 Hz, the difference between the two frequencies

Figure 7.8 Beats — an example of amplitude modulation.

7.3.2 *Beats*

When two notes of almost the same frequency are played simultaneously, the result is a throbbing note with periodic intervals of sound and silence called *beats*. The effect is best demonstrated with an example, such as in Figure 7.8.

Suppose that waves of 60 Hz and 55 Hz are superimposed. The 60 Hz wave produces 6 vibrations in 0.1 seconds but the 55 Hz wave produces only 5.5 vibrations in the same time interval. If the two waves are in phase at some particular time they will be completely out of phase 0.1 seconds later. After a further 0.1 seconds they are in phase again, and so on, giving intermittent periods of sound and silence.

If several notes with almost the same frequency are played simultaneously, the bursts of loud sound are shorter than with just two notes. In between, there are fainter sounds when only some of the waves are in phase. When the number of waves is extremely large, there is only one noticeable burst of sound at a time when all the waves are in phase. The number of waves in phase at any other time becomes very small.

7.4 Sound intensity

The intensity of sound is the power (energy per unit time) transmitted through an area of 1 m^2 perpendicular to the direction in which the wave is travelling.

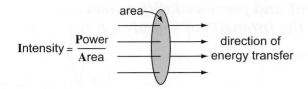

If sound from a point source moves at the same speed in all directions, the wave fronts form a series of expanding spheres. As the distance from the source increases, the area over which the energy is spread gets larger and the energy gets 'diluted'.

Sound waves from a point source

The wave fronts form concentric spheres with intensity

$$I = \frac{P}{A} = \frac{P}{4\pi r^2}$$

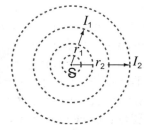

The total power of the expanding wave front remains the same, so the intensity at a distance r_1 from the source

$$I_1 = \frac{P}{4\pi r_1^{\,2}}$$

and the intensity at a distance r_2 from the source

$$I_2 = \frac{P}{4\pi r_2^{\,2}}$$

The intensity falls off as $\dfrac{1}{r^2}$, where r is the distance from the source.

When we double our distance from a source of sound, the intensity falls by a factor of 4.

7.4.1 *Real and perceived differences in the intensity of sound*

Things that do not make sense

Any kind of perception is subjective and cannot be quantified directly. For example, nurses in hospital often ask patients to quantify the pain they feel on a scale of one to ten. Answers vary widely from patient to patient. Even for the same patient, the numbers must be taken purely as an indication. Suppose that one injection gives rise to a level 2 pain. Will two identical injections given together produce pain at level 4?

this radio is twice as loud

this soup tastes twice as good

I love you ten times more

my toothache is twice as bad

These four statements attempt to quantify a perception without defining units. The numbers are essentially random, 'I love you ten times more' could be replaced by 'I love you one hundred times more' without changing the meaning. The tests are not repeatable.

Things that do make sense

- You have turned up the radio so that I can just hear it.
- Salt has made this soup just about edible.
- I just about prefer Lisa to Jane.
- I think the aspirin is beginning to work.

These statements also quantify changes in perception, but this time a standard minimal step has been introduced as a unit.

For example, in the second statement we are comparing the taste of the soup with and without salt. The test could be repeated any number of times and the same result would be obtained. The change in perception is just noticeable. The size of the minimal step will depend very much on circumstances. A single candle will light up a dark room but will not be perceived in the bright midday sun. A cough which would disturb a solo violin recital would seldom be heard at a rock concert.

7.4.2 *Quantifying perception*

The concept of the just-noticeable difference was introduced by Ernst Weber (1795–1878). He set out to describe the relationship, for various human sense organs, between the physical magnitude of the stimulus and the perceived intensity of sensation. In one of his first experiments he increased, in very small steps, the weight held by a blindfolded person, in one hand and not the other, and recorded the steps of 'just-noticeable' difference.

Figure 7.9 illustrates such an experiment. The weightlifter is training with dumbbells of equal weight on each side. When the weights are light (top diagram) he notices that a small extra mass has been added to one side, but when he is lifting heavier weights (lower diagram) he does not notice the addition of the same small mass. The Weber–Fechner law deals with sensation and does not apply when he reaches the physical limit.

It is the perception of any sensation that is important to us because in the end we do not respond to differences in the stimuli themselves but to our perception of those differences.

What does that mean?

It means the change you perceive depends on the original stimulus.

both weights same　　left weight heavier

both weights same　　both weights same

Figure 7.9 Equal difference in weight but equal difference in perceived weight.

Weber observed that, for all senses, the *smallest detectable* change in sensation ΔS in a given stimulus is directly proportional to the existing intensity of the stimulus I:

$$\Delta S = k\,\frac{\Delta I}{I}\ \ (k \text{ is a constant})$$

Gustav Theodore Fechner (1801–1887) studied medicine under Weber. He suggested that *the magnitude of a sensation S could be measured in terms of the number of just-perceptible steps above threshold required to attain it.*

Integrating each side of Weber's equation,

$$S = \int \Delta S = k \int \frac{\Delta I}{I} = k \log_e I \qquad \text{(The Weber–Fechner law)}$$

Sensation is proportional to the logarithm of the stimulus.

7.4.3 *Intensity level (loudness)*

The term *intensity level* or *loudness* is sometimes used as a measure of the perception of sound. The ear is a remarkable organ, capable of detecting an enormous range of sound intensity. The Weber–Fechner law tells us how we *perceive* sounds, how *loud* they are.

Intensity level can be expressed in terms of a dimensionless unit called the decibel (dB).

The decibel is named after the Scot Alexander Graham Bell (1847–1922). Bell had a life-long interest in the education of deaf people and is generally credited with the invention of the telephone, which he patented in 1876. A recent US Congress resolution has however drawn attention to the contribution

Alexander Graham Bell.
*Courtesy of An Post,
Irish Post Office.*

made by an Italian-American inventor, Antonio Meucci, who had demonstrated a rudimentary communications link in 1860, which had been described in New York's Italian language press.

Unable to afford a formal patent application for his *'teletrofono'*, Meucci filed a renewable one-year notice of an impending patent on 28th December 1871, but was not in a position to pay the $10 needed to maintain the caveat after 1874, and the patent was granted to Bell. Meucci took legal action and in January 1887, the US government initiated proceedings to annul the patent issued to Bell. The case was remanded for trial by the Supreme Court but the legal action terminated with the death of Meucci in 1889.

The role played by Meucci was formally recognized by the US Congress on 11th June 2002.

Resolved: 'That it is the sense of the House of Representatives that the life and achievements of Antonio Meucci should be recognized and that his work in the invention of the telephone should be acknowledged.' *Thomas Jefferson Papers, US library of Congress.*

Calculating intensity levels

Applying the Weber–Fechner law we can write:

$$\text{Intensity level} = 10 \, \log_{10} \frac{I}{I_0} \, \text{dB}$$

The minimum intensity audible to the human ear, I_0, is about $10^{-12} \, \text{Wm}^{-2}$, and corresponds to 0 dB.

An intensity 10 times as large corresponds to 10 dB, 100 times as large to 20 dB, and so on.

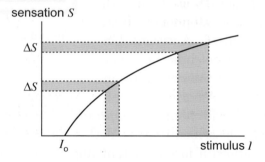

Figure 7.10 Sensation and stimulus.

In Figure 7.10, a graphical representation of the Weber–Fechner law, sensation S is plotted against stimulus I and ΔS represents a just-noticeable increase in sensation. The figure shows that as the stimulus increases it takes a much larger change in stimulus to produce a just-noticeable increase in sensation.

The 'threshold of hearing' corresponds to an intensity level of 0 dB and the 'threshold of pain' corresponds to an intensity level of 120 dB. Prolonged exposure to intensity levels in excess of 100 dB can permanently damage hearing, and at around 150 dB the eardrum may rupture.

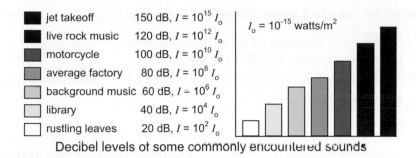

■ jet takeoff	150 dB, $I = 10^{15}\, I_o$	$I_o = 10^{-15}$ watts/m^2
■ live rock music	120 dB, $I = 10^{12}\, I_o$	
■ motorcycle	100 dB, $I = 10^{10}\, I_o$	
■ average factory	80 dB, $I = 10^8\, I_o$	
■ background music	60 dB, $I = 10^6\, I_o$	
□ library	40 dB, $I = 10^4\, I_o$	
□ rustling leaves	20 dB, $I = 10^2\, I_o$	

Decibel levels of some commonly encountered sounds

Example — crying babies

Two crying babies will produce double the intensity of sound of one baby, but will they be 'twice as loud'? How about three babies?

1 baby $\qquad L_1 = 10 \log I_1 \text{ dB}$

2 babies $\qquad L_2 = 10 \log I_2 = 10 \log 2I_1 = 10 \log I_1 + 10 \log 2$
$\qquad\qquad\quad = L_1 + 3.01 \text{ dB}$

3 babies $\qquad L_3 = L_1 + 10 \log 3 = L_1 + 4.77 \text{ dB}$

Making the reasonable estimate that the loudness of one baby is equivalent to the average background noise of about 80 dB in a factory, twins will only increase this to 83 dB and triplets to approximately 84.8 dB.

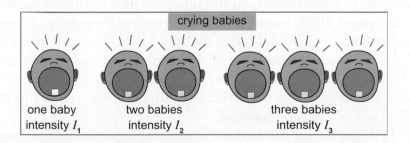

crying babies

one baby	two babies	three babies
intensity I_1	intensity I_2	intensity I_3

7.4.4 The 'annoyance factor'

Loudness is not the only factor that affects our tolerance of sound waves.

Which do you find more annoying?

(a) 58 dB for 5 seconds or 55 dB for 5 hours?
(b) 58 dB at 3 pm or 55 dB at 3 am?
(c) Someone scraping their fingernails over a blackboard at 55 dB?
(d) Handel's *Water Music* at 58 dB?

The amount of annoyance created by a noise should not be measured simply in terms of decibels. Nobody can tell us that we are not annoyed by some sound simply because the decibel reading indicates that we should not be annoyed. This is very relevant, because laws governing the level of noise may well be based only on the decibel level of the sound and not our perception.

7.5 Other sensations

7.5.1 *Pitch*

The pitch of a sound is our perception of its frequency. It depends, to some extent, on the intensity and harmonic content of the sound. Pitch increases with increasing frequency. A 'pure' tone is a sine wave and has a definite pitch. The vast majority of sounds are combinations of different tones and do not give a single value of pitch. Complex tones produced by musical instruments give the sensation of a single pitch, related to the frequency of the fundamental.

Pythagoras did the earliest acoustical experiments. A woodcut, dated 1492, in the Theorica Musice in Naples, shows him establishing the relationship between mathematical frequency ratios and harmonious musical sounds.

A *musical interval* is the difference in pitch between pairs of tones. It is related to the ratio of the frequencies of the tones

↲ and not to the frequencies themselves (for example, the interval between tones of 200 Hz and 400 Hz is the same as between tones of 2000 Hz and 4000 Hz). Harmonious tones have intervals with small integer ratios such as 2:1 or 3:2 and form the basis of all music. Pairs of tones with equal intervals sound similar when played together. For example, a pair of tones at 440 Hz and 660 Hz sounds similar to a pair at 330 Hz and 495 Hz (both intervals are 3:2).

2:1 (octave)   3:2 (fifth)

Frequency ratios 2:1 and 3:2.

The most important interval is the *octave*. If notes an octave apart are played simultaneously, the difference in their tones is barely noticeable. A musical scale is a set of tones which are used to make music. The octave is normally the interval of repetition. (In the diatonic major scale *do-re-mi-fa-sol-la-ti-do´*, the interval between *do* and *do´* is *one octave*). The ear recognizes notes in a scale as being related to one another

Ancient Irish pipes

A set of six prehistoric pipes was discovered in 2004, during archaeological work on a Bronze Age site in County Wicklow, on the east coast of Ireland. The site was being surveyed to comply with planning regulations, prior to the construction of a residential development. Six pipes made from yew wood were

Ancient Pipes.
Courtesy of Margaret Gowan & Co.
www.mglarc.com.

found at the bottom of a wood-lined trough, and radiocarbon dating has established that a wooden peg used in the construction of the lining dates back to somewhere between 2120 BC and 2085 BC. The pipes are not perforated but there is evidence to suggest that they originally formed a set. The photograph shows the pipes, together with a marker graduated in centimetres.

7.5.2 *Tone quality*

Each musical instrument has its own characteristic tone. Individual notes of exactly the same pitch sound different if played on different instruments, and we say the notes have a different tone quality. Terms like 'throaty', 'mellow' and 'full-bodied' are used to describe tone quality. We can often identify someone we know from the tone quality of their voice.

The tone quality of a sound depends on the shape of the wave form. The simplest wave form, a *sine wave*, is called a *pure tone*. All other periodic wave forms (*complex tones*) can be expressed as the sum of a sine wave and some of its harmonics, as we saw in Chapter 6. The number and the relative intensities of the harmonics determine the shape of the wave form. In terms of tone quality, we say that *complex tones are combinations of pure tones*. The ability to make distinctions between different tones is more enhanced in some people than in others.

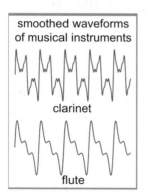

smoothed waveforms of musical instruments

clarinet

flute

7.5.3 *Propagation of sound in open and confined spaces*

Perfect reproduction of sound in an open space

In open spaces, the sound waves travel directly from the source to the listener, who hears an almost perfect reproduction of the sound, as illustrated in Figure 7.11. The loudness of the sound

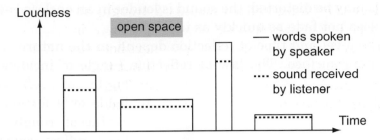

Figure 7.11 Comparison of words 'spoken' and 'heard' in an open space.

depends on the separation of the musician and the listener (except when the two are close together). As the listener moves away from the source, the loudness decreases.

In confined spaces, reflection from the surroundings can cause considerable distortion of the sound. A well-known example is the echoing of sound in a large cave. The echo from any one individual word reverberates in the cave and a whole string of words may result in a completely garbled sound.

When we listen to instruments and voices in rooms, halls and other auditoria, we are bombarded by 'surround' sound from all directions due to multiple reflections from walls, ceilings, floors and other surfaces. Figure 7.12 illustrates how individual

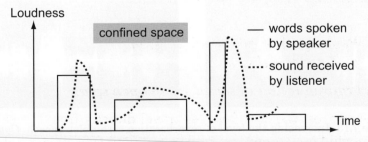

Figure 7.12 Comparison of words 'spoken' and 'heard' in a confined space.

words may be distorted; the sound is louder in an enclosed space and does not fade as quickly as in the open.

The level and type of reflection depend on the nature of the reflecting surface. The law of reflection (angle of incidence = angle of reflection)

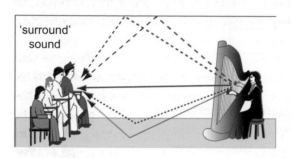

holds only for smooth surfaces; rough surfaces reflect sound at all angles. The level of sound which reaches a listener in an auditorium depends on the nature and distribution of sound-absorbing surfaces. We can identify three different parameters — the time to reach a more or less steady level of loudness, the value of this loudness, and the time for the sound to become inaudible.

✘ The decay of sound is called *reverberation* and is a very important parameter. The reverberant sound from one note (or word) can often be heard when the next one is produced. We quantify reverberation by defining reverberation time as the time taken for the loudness to decrease by 60 dB. Figure 7.13 shows a time sequence of the amplitude of a string of words for two different values of the reverberation time (the left hand image was recorded at the lower value).

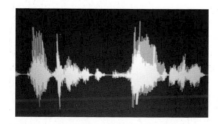 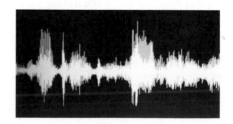

Figure 7.13 Increasing the level of reverberation scrambles speech. *Courtesy of James Ellis, School of Physics, UCD.*

7.6 Strings and pipes in music

String instruments

Stretched strings may be made to vibrate at characteristic frequencies. The surface area of a vibrating string is too small to compress the surrounding air effectively and string instruments use sound boards to amplify the vibrations.

Instruments which have strings mounted paralled to the soundboard and a body which resonates are generally known as lutes. The earliest lutes were long-necked and made from materials such as hide and turtle. There is evidence for the existence of lutes in Egypt and Mesopotamia (modern Iran) as early as about 2,500 BC. The European lute was modelled on the classical Arab instrument, *Aloud* (meaning a wooden stick), a short-necked instrument with a body of lightweight wood. The Europeans modified the Arabic design by introducing frets (metal strips embedded in the neck of the instrument).

The earliest surviving classical (six-string) guitars were made in Naples towards the end of the eighteenth century. The body shape of the modern guitar was largely standardized by the Spanish guitarist and guitar maker Antonino de Torres (1817–1892) in Seville in the mid to late nineteenth century.

The *acoustic guitar* does not use any external means of amplification. Vibrations produced by plucking the strings are transmitted to the body of the guitar, which resonates at the frequency of the vibrating string. Sound generated in the chamber is projected into the air through the sound hole. Strings are stretched from the bridge to the top of the neck. The vibrating lengths of a string

Acoustic guitar. *Courtesy of Massimo Giuliani.*

may be shortened by pressing it against the frets to produce sounds of different pitch.

The modern form of the *violin* dates from the mid sixteenth century. These early violins were made in northern Italy, By 1600, the centre of violin making was Cremona, where Nicolo Amamti (1596–1684) and Antonio Stradivari (1644–1737) worked. It is remarkable that violins made centuries ago are now cherished for their beautiful sound. Violinmakers such as Stradivari, knowing only practical physics and acoutics, were able to produce instruments so desirable that they are highly prized. The Stradivarius violin known as 'The Hammer' sold at public auction in 2006 for more the $3.5M.

Wind instruments

A column of air, like a stretched string, can be made to vibrate at a series of frequencies whose values depend on the length of the column and on whether it is open at one or both ends (see 'standing waves').

Woodwind instruments such as the clarinet and the flute have a long, thin column of air with a series of tone holes. The pitch is changed by opening or closing the holes. The lowest (fundamental) note is produced when all the holes are closed.

The *clarinet* is a reed instrument. Standing waves are excited by vibrations of the reed. The mouthpiece is closed during play and even harmonics are strongly suppressed, as would be expected for a column of air closed at one end. The *flute*, on the other hand, is open at both ends. Standing waves are excited when air is blown across the open mouthpiece. Adjacent odd and even harmonics are fairly equally represented.

7.7 The Doppler effect

Sound from the siren of a vehicle rushing to the scene of an emergency may appear to have a higher frequency as the vehicle approaches the scene and a lower frequency as it leaves the scene. This apparent change in the frequency of sound, which can be due to motion of the source or the observer, is called the *Doppler effect.*

The Doppler effect is named after the Austrian physicist Christian Johann Doppler (1803–1853), who developed formulae to express the change in frequency as a function of the change in velocity. These were verified, only three years later, in 1845 by a Dutch meteorologist, Christoph Heinrich Dietrich, who used trumpets as the source of sound and trains to transport the trumpeters and observers, in turn.

Christian Johann Doppler. *Courtesy of Austrian Post.*

Fans of Indy car or Formula One racing may be familiar with apparent changes in frequency as the cars speed past them. If the cars are approaching the spectators, they hear an increase in frequency, and as the cars speed away the frequency appears to decrease.

7.7.1 *A moving observer*

If an observer runs towards a source of sound, the number of vibrations per second increases. It is like running into the sea. The number of breakers you cross in one second (the *frequency*) is higher. Running away from the breakers has the opposite effect. The distance between breakers (the *wavelength*) is the same in both cases, which means that there is an apparent change in frequency. The frequency is higher when the observer approaches the source and lower when he recedes from it.

7.7.2 *A moving source*

A point source emits spherical waves. When the source is stationary, the wave fronts are concentric spheres, centred on the source. As the source starts to move, it essentially chases its own wave fronts, compressing the sound waves it has produced.

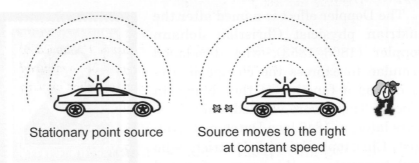

Stationary point source   Source moves to the right
                              at constant speed

The drawing shows how successive wave fronts are pushed along to the right. The wave velocity is unchanged but the wavelength is smaller in front of the source. An observer in front of the source will receive more waves per second and measures a higher frequency than an observer behind the source.

> *In general, the frequency is higher when the source and the observer approach each other, and lower when they move away from each other.*

7.7.3 *Two Doppler effects?*

An observer approaches a stationary source at one half the speed of sound in air and is subsequently approached, when stationary, by a source moving at the same speed. Is the apparent frequency the same in both cases?

Does it matter who is moving?

Let's do our sums and see.

When the observer moves: $f' = f(1 + 0.5) = 1.5f$

When the source moves: $f' = \dfrac{f}{0.5} = 2f$ [†]

> *The answer is most definitely no! This is because the speed of a sound wave and the speed of an observer can be measured independently, relative to a medium. This means that we can tell which is moving and which is at rest in the medium.*

7.7.4 *Moving away from a source at almost the speed of sound*

The nose of an aircraft creates pressure waves in the same way as a source of sound. If an aircraft travels at almost the speed of sound, the nose almost 'catches' its own waves. The apparent wavelength becomes extremely small and the frequency very high.

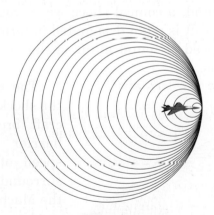

Wave pattern emitted by a source travelling at about 0.9 times the speed of sound in air.

[†] These equations are derived in Appendix 7.1.

7.7.5 *Shock waves*

If an aircraft moves faster than the speed of sound, it 'overtakes' its own wave fronts and shock waves are created. The

overlapping spherical wave fronts reinforce one another to form a conical surface of very high pressure, propagating outwards. The pressure is much greater than in individual compressions and the intensity can reach 100 million W/m^2 (corresponding to about 200 dB). The sonic boom is the audible sound created by this pressure wave, which is confined to the surface of a cone with the aircraft at its apex. This is called the *Mach cone*, after Ernst Mach.

Ernst Mach.
*Courtesy of
Austrian Post.*

Ernst Mach (1838–1916), an Austrian physicist, was the first to realise that if an object were to travel through air at a speed greater than the local speed of sound, a conical shock wave would be produced. The speed of aircraft is often designated in terms of *Mach number*. Mach 1 = speed of sound in air in the prevailing environmental conditions.

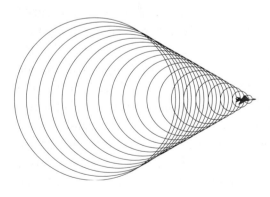

Shock wave at Mach 2.

The cone created by the shock wave as an aircraft breaks the sound barrier (exceeds the speed of sound in the surrounding air) is called the Mach cone.

As the pressure wave created by a supersonic aircraft sweeps along the ground in the wake of the aircraft, the pressure variations can be large enough to cause significant damage. The Concorde passenger jet was only allowed to travel at supersonic speed over uninhabited areas and oceans.

The half angle of the Mach cone becomes smaller as the speed increases.[‡]

Shock waves.

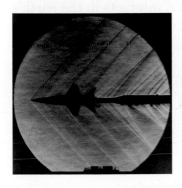

X15 Jet. *Courtesy of NASA.*   Vigorously paddling duck. *Courtesy of Piotr Pieranski.*

7.7.6 *Shock waves and light*

Optical shock waves may be formed when high energy electrons and protons travel through materials at speeds in excess of the local speed of light. These waves are referred to as *Cherenkov light.*

Pavel Alekseyvich Cherenkov (1904–1990) was one of three Russian physicists who shared the 1958 Nobel Prize for Physics. In 1934, whilst a PhD student, Cherenkov discovered that there was a faint glow of blue light when high speed electrons passed through a transparent liquid. Igor Tamm and Ilya Frank, co-recipients of the Nobel Prize, interpreted his discovery.

Cherenkov light is emitted from the cores of nuclear reactors. A reactor may produce more than 10^{15} fissions per second whilst operating at full power. Each fission generates a huge amount of energy, mostly carried away by fission products.

[‡] See Appendix 7.1.

These decays often produce high energy electrons travelling at speeds in excess of the speed of light in the water. This results in emission of characteristic intense blue light.

VERITAS imaging array.
Courtesy of Veritas Collaboration.

Astrophysical objects such as neutron stars and black holes emit high energy gamma radiation which interacts in the earth's upper atmosphere and generates electrons travelling faster than the local speed of light. Cherenkov light radiated by these electrons is detected by arrays of ground-based imaging telescopes, such as VERITAS in southern Arizona.

A historical interlude: The sound barrier

The speed of sound in air at sea level is about 350 ms⁻¹, or 760 mph. At 40,000 ft, where the air is colder and more rarified, it is about 660 mph.

As planes became faster and faster, mysterious things seemed to happen when they reached speeds approaching Mach 1. Some pilots in World War I reported that controls froze in fighter planes when they were executing steep dives. There were even rumours, subsequently found to be false, that controls reversed at such speeds.

In the late 1940s, propeller aircraft began to be superseded by jets capable of flying speeds of over 600 mph, and it became crucial to find out if it was safe to fly as fast as, or faster than, the speed of sound. What happens if the plane overtakes its own sound waves? The interference pattern of the sound waves could of course be predicted, as for any waves created by a source travelling faster than the speed of the waves. They would form a 'Mach cone', like the bow wave of a ship travelling at a speed greater than the speed of waves

(Continued)

in water. Nobody could tell, however, if the plane would encounter new and extreme stress loads. Such stresses are normally measured in wind tunnels, but these could not go above 0.85 Mach.

There were speculations that a plane would encounter a 'sonic wall' which it could not penetrate and which indeed could destroy it. In 1946 Geoffrey de Havilland had tried to take one of his father's DH 108 jets through Mach 1. Tragically the plane disintegrated and he was killed. The mystery of the *sound barrier* had become another challenge for exploration.

In 1947, a new rocket plane, called the *X-1*, was being constructed in the US which it was believed might exceed Mach 1. The centre for testing this plane was established at Muroc Air Base in the desert in California. Test pilots had to have the courage to fly into the unknown, nerves of steel, and the ability to react instantly to any unforeseen problems as the speed of the plane came closer and closer to Mach 1.0. Chuck Yeager was the chief test pilot, whose skills in flying were legendary. As

Chuck Yeager. *Courtesy of NASA.*

soon as he entered the cockpit, he was said to become part of the plane, irrespective of whether it was new to him or well known. Chuck had an outstanding record as a fighter pilot. He had been shot down over France, but was helped by the French underground to escape into Spain and return to England. Military rules prohibited his return to active combat, but exceptionally, by special permission from General Eisenhower, he returned to his squadron to continue with fighter missions. Chuck was a West Virginian, with a laid-back southern drawl, relaxed and confident, commanding the highest respect among his

(Continued)

flying colleagues. If anyone could guide the *X-1* through the sound barrier, it was he.

The plan for the test flights and for the final record attempt was rather primitive compared to modern computerised techniques. In order to give it a flying start, the *X-1* would be hooked under the bomb bay of a B29 bomber and lifted to an altitude of about 25,000 ft. At that point Yeager would climb down the ladder in the bomb bay and slide into the *X-1* cockpit, lock the hatch, and wait to be dropped like a bomb. It literally was a bomb, filled to capacity with highly volatile fuel and liquid oxygen. As soon as he was dropped, and clear of the mother plane, Yeager would fire the rocket engines and shoot forwards from the mother ship. If the engines failed to ignite, the *X-1*, which was loaded with fuel, would not glide and would probably end up in a fatal spin.

On 14 October 1947, Yeager was scheduled to bring the plane up to Mach 0.97. Typically, in his nonchalant way, he had gone horse-riding with his wife, Glennis, two days before the test flight.

The horse stumbled and threw him, resulting in two broken ribs. Yeager could hardly raise his right hand, but kept this fact to himself, knowing full well that if it became known that he was thus handicapped, he would be grounded. He was determined to go ahead with the test, particularly

The *X-1* beside its mother plane.
Courtesy of NASA.

since it might be his last flight for a while. He confided in his close friend Jack Ridley, the flight engineer who promised to keep mum and

(Continued)

to help him as far as possible. A minor but critical problem was that Yeager would have to pull hard with his right hand in order to lock the canopy above his head after he had lowered himself into the cockpit of the *X-1.* Jack's ingenuity provided a broken broom handle which Yeager could use as a lever with his left hand. With the problem of closing the cockpit solved, firing the rocket engine when in free fall did not seem to worry him as much.

As the critical moment arrived on the day of the test, and the plane began to fall freely from under the bay of the bomber, Yeager turned on the rocket engines in sequence. The machometer needle rose: 0.88, 0.92,…, 0.96 Mach. There was some buffeting, but the faster he flew, the smoother the ride. Since he still had some fuel left, he accelerated further. The needle went off the Mach scale, which was calibrated only up to Mach 1.0, and stayed off for about 20 seconds. The tracking station reported a distant rumble, the first sonic boom from an airplane. The sound barrier had been broken.

X-1 begins to drop. *Courtesy of NASA.*

Yeager wrote later: 'The real barrier was not in the sky, but in our knowledge and experience of supersonic flight.' Modern jets can take off on their own power and reach speeds high above Mach 1.0.

Once the sound barrier had been broken, progress was fast in making faster and faster planes. Planes could take off under their own power and accelerate up to Mach 1.0 and beyond.

The *F-104 Lockheed Starfighter* was the first to reach Mach 1 in a climb. It could reach an altitude of over 100,000 ft and travel at

(*Continued*)

X-15 wreckage. *Courtesy of NASA.*

more than twice the speed of sound. On 10 December 1963, Chuck Yeager took the *F-104* on a routine test flight. Travelling at a speed better than Mach 2 he fired the thrust rocket in its tail to bring it up to an altitude of 104,000 ft, when it went out of control and rapidly began to fall. The data recorder later showed that the plane had made 14 flat spins before it hit the desert floor. Yeager stayed with it for 13 of those spins before he ejected. As he wrote in his autobiography: 'I hated losing an expensive airplane, but I couldn't think of anything else to do.' He managed to eject successfully, although he was badly burned on his face by the rocket motor of his ejector seat. Yeager was not the only pilot to survive a crash of an X-series test plane. The picture shows the wreckage of an *X-15* after Jack McKay had been forced to make an emergency landing at Mud Lake, Nevada.

Almost exactly 50 years after Yeager's pioneering flight, on 15 October 1997, the sound barrier was broken on land. A jet-powered car, driven by British Air Force pilot Andy Green, made two runs, in opposite directions across Black Rock Desert in Nevada. The car accelerated over a distance of 6 miles, then was timed over a measured mile, and then followed a further 6 miles to stop. The runs were timed at 759 and 766 mph, both faster than the speed of sound, which under the prevailing conditions was calculated at 748 mph. When the hard desert ground was later inspected, it was found to be pulverised by the Mach wave.

(Continued)

Andy Green's Thrust Super Sonic Car was 54 ft long and 12 ft wide. It was powered by a pair of Rolls-Royce jet engines from a Phantom fighter producing 50,000 pounds of thrust and 110,000 horsepower.

The aerodynamic design had to be faultless. If the front of the car were to lift by as little as half a degree, all the weight would come off the front wheels. The car would then nose up and flip over backwards. Alternatively, if the nose were to dip, it would somersault forwards. As one of the engineers remarked at the time: 'If the nose comes up, you're going flying. But equally, if it goes down, then you're going mining.'

Appendix 7.1 Derivation of Doppler frequency changes

Apparent frequency measured by a moving observer

In Figure 7.14, S is a point source of sound and O is an observer.

The observer runs a distance v_0 in one second, so he meets $\dfrac{v_0}{\lambda}$ extra waves per second.

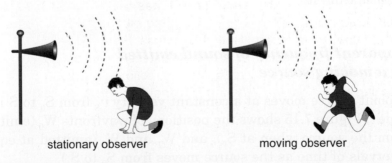

stationary observer   moving observer

Figure 7.14 Doppler effect — moving observer and stationary source.

The source emits waves at a frequency $f = \dfrac{v}{\lambda}$.

The observer measures a frequency $f' = \dfrac{v + v_0}{\lambda}$.

$$f' = f + \frac{v_0}{\lambda} = f\left(1 + \frac{v_0}{v}\right)$$

We can write this as

$$\frac{f'}{f} = \frac{v + v_0}{v}$$

If the observer 'runs away' from the source at the same speed, he measures a frequency:

$$f' = f\left(1 - \frac{v_0}{v}\right)$$

As we inferred from common sense, the frequency is higher when the observer approaches the source and lower when he recedes from it.

Apparent frequency of sound emitted by a moving source

A point source moves at a constant velocity v_S from S_0 to S in a time t. Figure 7.15 shows the positions of wavfronts W_0 (emitted from the source when at S_0), and W_1 and W_2 (emitted at equal intervals of time as the source moves from S_0 to S.)

$$S_0 X = vt \qquad S_0 S = v_S t$$

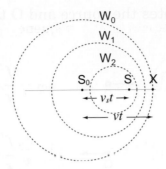

Figure 7.15 Doppler effect for a moving source.

The waves reaching an observer in front of the source are compressed into the distance $S_2X = (v - v_S)t$ and we can calculate their wavelength as folows:

$$\lambda' = \frac{vt}{(v - v_S)t}$$

The speed of sound v is constant, so $f\lambda = f'\lambda'$, giving $\lambda' = \dfrac{f\lambda}{f'}$

Substituting for λ' from Equation (7.3) $\dfrac{f'}{f} = \dfrac{v}{v - v_S}$,

$$f' = \frac{fv}{v - v_S}$$

If the source is moving away from the listener at the same speed, $f' = \dfrac{fv}{v + v_S}$.

Non-mathematical reasoning gives the same results, that the frequency is higher when the source and the observer approach each other and lower when they move away from each other.

Summary (S denotes the source and O the observer)

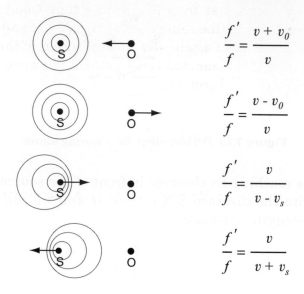

$$\frac{f'}{f} = \frac{v + v_0}{v}$$

$$\frac{f'}{f} = \frac{v - v_0}{v}$$

$$\frac{f'}{f} = \frac{v}{v - v_s}$$

$$\frac{f'}{f} = \frac{v}{v + v_s}$$

Calculating the Mach angle

The cone created by the shock wave as an aircraft exceeds the speed of sound is called the Mach cone. We can calculate φ, the

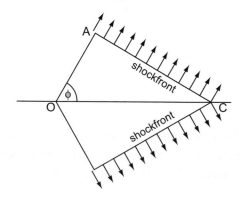

Figure 7.16 The Mach cone.

Pressure waves created by the jet expand as the aircraft moves along.

If the aircraft moves from O to C in a time t and the wave travels from A to C in the same time, then OA $= vt$ and OC $= v_A t$, where v is the speed of sound and v_A is the speed of the aircraft. The Mach cone is the common tangent to all the wave fronts, so OAC is a right-angled triangle.

$$\sin \theta = \frac{vt}{v_A t} = \frac{v}{v_A}$$

The half angle of the Mach cone becomes smaller as the speed increases.

Chapter 8

Light as a Wave

We have spent some time looking at waves — and by now should be able to recognise a wave when we see one! In this chapter we look at evidence that a light beam behaves like a train of waves.

To help us predict how the wave properties of light should manifest themselves, we use a construction originally proposed by the Dutch mathematician Christiaan Huygens and based on the hypothesis that a train of waves continuously regenerates itself, each point on the wave acting as if it were a new source of *secondary wavelets*. Snell's law of refraction can be derived using Huygens' construction — circumstantial if not over-whelming evidence that light behaves a wave.

A more convincing argument is available. If we really believe that light is a wave, it should be possible to arrange an experiment in which light waves interfere destructively — in other words, light plus more light sometimes makes darkness.

When we send a beam of light through a very small aperture such as a thin slit, Huygens' construction gives a curious prediction. The walls of the slit will cut off some of the secondary wavelets and create an imbalance in the emerging beam. Not only will there be a sideways spread (*diffraction*), but the unmatched wavelets will interfere constructively and destructively, producing a characteristic pattern.

In a practical way, the wave nature of light sets a limit to the resolving power of optical instruments. It is not possible to make microscopes and telescopes of unlimited magnifying power. No matter how sophisticated the optical system is, we cannot use light waves to examine structures finer than the structure of the waves themselves.

In Young's experiment two thin slits act as coherent light sources. When their light comes together we get constructive and destructive interference, a pattern of dark lines within the diffraction pattern — an example of light + light = dark !

Interference between light from diffraction gratings, which contain very large numbers of slits, gives us the ability to investigate atomic structure. We can probe even more deeply using X-rays and we were fortunate that nature provided us with perfect diffraction gratings, in the form of crystalline solids.

8.1 Light as a wave

Over a number of centuries, 'natural philosophers' were faced with two contradictory theories, one of which proposed that light is a wave and the other that light is made up of a stream of particles. This chapter presents the evidence for the *wave theory of light*.

8.1.1 *The mystery of waves in nothing*

Waves transmit energy and information from place to place; light certainly does just that. Energy is transferred from the sun to the earth in the form of light, and information about everything we see is transmitted to our eyes by light. Of course, carrying energy and information is not a feature of waves alone. Particles also carry energy and could be used to transmit information!

There is a strong argument that light could not possibly be a wave. Light can travel through empty space. Sunlight travels across 100 million miles in which there are virtually no particles of matter, no atoms and no molecules. Sound waves and other waves in gases, liquids and solids require a medium, as vibrations are transmitted from one particle to another. If there is no medium, how can vibrations be transmitted when there is nothing to vibrate?

The protagonists of the wave theory attempted to resolve this paradox in a simple, but not very satisfactory, way by proposing the existence of an invisible medium which permeates the universe. This hypothetical medium, with no known properties other than the ability to carry light waves, was given the name 'ether'. No theory is satisfactory if it applies to one phenomenon and that phenomenon only. To give the ether hypothesis credibility, one must have independent evidence, and many clever experiments were devised to try to observe some sign of the ether.

A series of such experiments was carried out by Albert Michelson and Edward Morley in Los Angeles in the 1880s. Their idea was to attempt to detect an *'ether wind'* blowing past the surface of the earth as it rotates and travels in its orbit around the sun. (We will look at their experiment in greater detail in Chapter 15.) Despite the best efforts of Michelson and Morley, no evidence was found for an ether wind and they were forced to conclude that either the ether does not exist or, if it does exist, it is invisible and undetectable! We must leave it to the philosophers to decide whether something invisible and undetectable can be said to exist.*

8.2 Wave properties which do not make reference to a medium

8.2.1 *Superposition*

We can bypass the whole argument about ether by considering those wave properties which are independent of the medium. Waves in general obey the *principle of superposition*, which states that the net effect of two or more overlapping waves is simply the sum of the effects caused by each wave individually. This applies to waves in a string, waves in

---

* George Berkeley (1685–1753) wrote a treatise entitled 'The Principles of Human Knowledge', which formed the basis of a philosophy of *immaterialism*, concerning the reality of physical objects and the existence of things which cannot be observed.

Figure 8.1 Nothing interferes with love!

water, and all kinds of waves both transverse and longitudinal. It means that waves pass through each other and then continue on independently, unaffected by their mutual interaction. An example of this was shown in the picture of overlapping waves created by drops of rain falling on the surface of water, in Chapter 6.

We do not need to set up a special experiment to illustrate that light behaves in similar fashion. A common domestic situation is shown in Figure 8.1. Father's watching of the television is not interrupted by the visual messages of affection that travel across his line of sight.

The world is full of light. Light beams of all wavelengths criss-cross every bit of space in the universe. Now, mankind has added radio and television waves, radar, infrared and other radiation, all independently superimposing on top of each other and on top of the natural background. While the phenomenon of superposition may not be conclusive proof that light is a wave, it is consistent with the wave hypothesis.

8.2.2 *Huygens' principle*

Christiaan Huygens (1629–1695), a Dutch astronomer and mathematician, believed that light is a mechanical wave produced

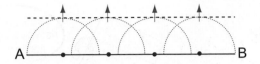

Figure 8.2 Huygens' construction at work.

when luminous objects vibrate and set the surrounding ether in motion.

In his *Traité de la Lumière*, published in 1690, Huygens showed how he could derive the laws of geometrical optics using a geometrical method of wave front reconstruction, now known as *Huygens' principle*.

According to Huygens' principle, every point on an advancing wave front can be considered as a source of spherical secondary *wavelets*. The forward-going surface of a wave front at any instant in time corresponds to the envelope of wavelets which had emanated at some previous time and are expanding at the speed of the wave. The wavelets themselves are simply a mechanism for wave front reconstruction and it does not matter whether or not they have a physical reality.

We can apply Huygens' construction to a plane wave travelling in a uniform medium. In Figure 8.2 the line AB represents a cross-section of the wave front of an infinite plane wave. The envelope of secondary wavelets is in this case their common tangent, representing a plane wave moving forwards at the speed of the original wave.

In the case of an 'infinite plane wave', we can tell from symmetry that the tangent to the wavelets will be a straight line moving directly forwards.

8.2.3 *Huygen's principle and refraction*

Light changes direction when it enters from a less dense into a denser medium. We have seen in Chapter 3 that this is governed

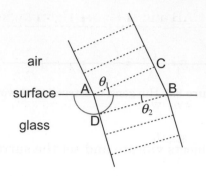

Figure 8.3 Huygens' principle and refraction.

by *Snell's law of refraction*, which can be derived on the basis of the assumption that light always chooses the path which takes the shortest time. We will now show using Figure 8.3 that the same law can be derived on the basis of the wave theory, using Huygens' construction.

A and C are points on an incoming plane wave front approaching the surface of a denser, yet transparent, medium such as glass or water. The spherical secondary wavelet travelling into the glass from point A has travelled a distance AD by the time point C on the wave front reaches the surface. The common tangent to all the secondary wavelets between A and B at this instant passes through points D and B and this, according to Huygens' principle, forms the new wave front. *Since the wave travels more slowly in glass than in air*, AD < CB, the angle $\theta_2 < \theta_1$ and the wave is bent towards the normal.

Snell's law follows directly from the construction in Figure 8.3:

If light travels at a speed v_1 in air and at a speed v_2 in glass,

$$CB = v_1 t \quad \text{and} \quad AD = v_2 t$$

where t is the time for the light to travel from C to B (or from A to D).

The triangles CAB and DAB are right-angled, so

$$\sin\theta_1 = \frac{CB}{AB} \quad \text{and} \quad \sin\theta_2 = \frac{AD}{AB}$$

$$\frac{\sin\theta_1}{\sin\theta_2} = \frac{v_1}{v_2} = n, \text{ where } n \text{ is the refractive index.}$$

$$\frac{\sin\theta_1}{\sin\theta_2} = n \quad \text{(Snell's law)}$$

The fact that light bends inwards towards the normal on entering a denser medium supports the argument that it behaves as a wave and not as a particle. A bullet, for example, when it enters water at an angle, slows down but keeps its original direction. Perhaps we could say that 'on the balance of probability' refraction can be taken as evidence for the wave theory. However, it is hardly evidence beyond reasonable doubt for the wave theory, to the exclusion of the particle theory!

8.2.4 *Diffraction*

We have seen in Chapter 6 that sea waves spread, bending around obstacles and penetrating into inlets. Such bending of waves around corners is a common feature of waves in water. Characteristically, when a wave front comes to a narrow opening or to the entrance of a harbour, it spreads out from this new point of origin in a semicircular manner. Somehow the wave seems to 'forget' that it had a plane wave front before it came to the opening and proceeds to spread sideways, at least in part!

The ripple tank experiment

In a controlled experiment in a water tank we can get a clearer view. Plane waves are sent through a narrow opening (by 'narrow' we mean 'not much larger than the wavelength of

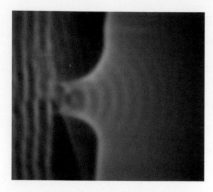

Water waves from narrow
slit ripple tank.
*Courtesy of James Ellis UCD
School of Physics, Dublin.*

the waves). The emerging waves spread out as did the waves on the beach. If we look closely, we can see the apex of a V made up of two diverging lines along which the waves appear to have been 'rubbed out'. Along these lines the waves *interfere destructively*, giving regions of calm water. If we had a moving picture we would see that these regions are stationary in space, like the nodes in the one-dimensional case of waves in a string.

8.2.5 *Huygens' principle and diffraction*

Let us apply Huygens' construction to the waves emerging from a slit.

As we see in Figure 8.4, the spherical wavelets emerging near the edge of the slit have no neighbouring wavelets to maintain a parallel wave front. As a result the envelope of wavelets bends slightly around the edge. The waves spread into the 'geometrical shadow' behind the slit.

We can use Huygens' method to reconstruct the pattern made by water waves emerging from the slit. In the

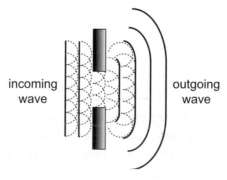

incoming
wave

outgoing
wave

Figure 8.4 A plane wave passes through a narrow slit.

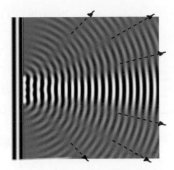

Figure 8.5 Computer reconstruction of water waves.

computer reconstruction of Figure 8.5, we see a striking resemblance to what is seen in the ripple tank experiment.

The V-shaped regions of calm water are clearly visible bordering the expanding wave crests. Most interestingly, we see a secondary and a third region of waves on each side of the central region.

8.3 Specifically light

8.3.1 *Diffraction of light*

To try the experiment with light we change the set-up slightly by shining a beam of light through a thin vertical slit. Some distance behind the slit we place a screen at right angles to the light beam. If the light is diffracted as it passes through the slit and produces an interference pattern in the same way as the water waves, we should see the evidence in the final image on the screen.

We can predict the form of the image using Huygens' method in the same way as we did for water waves. Looking at Figure 8.5, let us imagine a screen on the right side of the picture. Bright bands occur when the overlapping wavelets

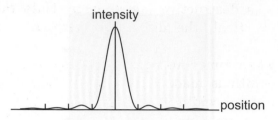

Figure 8.6 Distribution of intensity in the image of light diffracted from a slit.

interfere constructively as they arrive at the screen. Dark areas on the screen correspond to regions of calm water. The central bright band is brighter than the others because more light goes forwards than in any other direction. According to the rules of diffraction we have a bright band in the middle. The secondary maxima to either side are much less bright and about half the width of the central maximum.

The width of the diffraction pattern depends on the ratio of the wavelength to the width of the slit. The pattern of intensity variation represented in Figure 8.6 is taken from the mathematical treatment in Appendix 8.1 where Huygens' principle is used to predict the form of the image.

8.3.2 *The experiment with light*

The experimental set-up is quite simple. The slit and laser beam are schematic drawings, but the image is an actual photograph of what appears on the screen. There is a central bright spot of maximum intensity with smaller peaks to the left and right. These peaks occur where the secondary wavelets interfere constructively and are separated by nodes

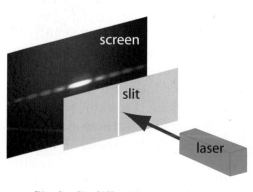

Single slit diffraction experiment.

where there is destructive interference. Note that whereas the slit is vertical, the diffraction pattern is spread out horizontally.

Contrary to common instinct, as the slit width is made *narrower* the angular width of the central maximum *increases*. Eventually we reach a limit where we effectively have a point source and the central maximum expands to the whole width of the screen. The three images show how the image expands laterally as the slit width is decreased (top to bottom in the picture).

Diffraction patterns for different slit widths.

8.3.3 *Other apertures*

The diffraction pattern of a square aperture is a superposition of vertical and horizontal single slit patterns. The symmetrical pattern reflects the proportions of the aperture.

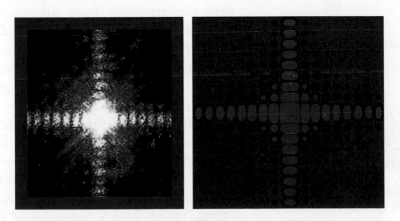

Diffraction pattern of a square aperture: real (left)
and computer reconstruction (right).

Diffraction pattern of
a circular aperture.

The diffraction pattern of a circular aperture (hole) is a series of concentric bright and dark rings, as we might expect from symmetry.

Diffraction images on pp. 248–250, *Courtesy of James Ellis, USD School of Physics, UCD.*

8.3.4 *The curious case of the opaque disc*

It would be perfectly reasonable to expect the diffraction pattern of a small opaque disc of about the same size as the hole to have the same symmetry as the diffraction pattern from the hole with a dark spot at the centre, but that is not what wave theory predicts.

In 1818, Augustin Fresnel (1788–1827) submitted a paper on the wave theory of diffraction for a competition sponsored by the French Academy of Sciences. Simeon Poisson (1781–1840), a member of the judging committee, applied Fresnel's theory to

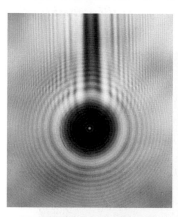

Poisson's Spot: Shadow of a ball bearing suspended on a needle. *Courtesy of Chris Jones, Union College, Schenectady, NY, USA.*

the diffraction of light by a small circular disc. It follows from symmetry that the wavelets from all sources around the circumference of the disc will all be in phase when they meet at some central point downstream from the disc. Thus the theory predicts that at the very centre of the diffraction pattern there should be a bright spot! Poisson, who did not believe in the wave theory of light, deemed this to be absurd. Fortunately Dominique Arago (1786–1853), also a member of the judging committee, soon verified the existence of the 'absurd' spot with a simple experiment. Fresnel won the competition and the wave theory triumphed again. It is ironic that the spot is generally referred to as Poisson's spot.

8.4 Is there a limit to what we can distinguish?

The extent to which optical systems can resolve closely spaced objects is limited by the wave nature of light.

8.4.1 *Images may overlap*

The resolving power of an optical instrument is its ability to produce separate images of adjacent points. Light entering an optical instrument through some sort of aperture (the pupil of an eye, the lens of a camera and the objective lens of a telescope are some examples) is diffracted. The image of a point source of light is never itself a point but is a pool of light accompanied by concentric bright rings. This is clearly illustrated in the image taken with the Nicmos camera on the Hubble Space Telescope.

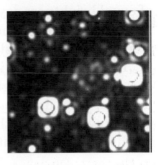

Images of stars at the galactic centre.
Courtesy of NASA JPL-Caltech.

8.4.2 *The Rayleigh criterion*

John William Strutt, Lord Rayleigh (1842–1919), received the Nobel Prize for Physics in 1904 for his discovery of the rare gas argon. His work, both experimental and theoretical, spanned almost the entire field of physics. Rayleigh developed the theory which explains the blue colour of the sky as the result of scattering of sunlight by small particles in the atmosphere. Craters on Mars and the moon are named after him.

Lord Rayleigh

When a telescope is pointed at a group of stars, light from each star passes through the aperture and is independently diffracted. The extent to which the images of individual stars can be resolved depends on how much their diffraction patterns overlap.

The amount of overlap depends on the size of the angle between the beams of light from the two stars. As this angle becomes smaller the individual diffraction patterns merge into one.

Rayleigh's empirical criterion for resolving two point sources of equal brightness says that two objects are 'just resolved' when the central peak of the diffraction pattern from the first object coincides with the first minimum of the pattern from the second.

In Figure 8.7 the images of the two stars are just resolved according to Rayleigh's criterion. At that point, the angle between the light beams from the two stars has a value equal to one half of the angular width of the main maximum of the diffraction pattern (Appendix 8.1). This then is the

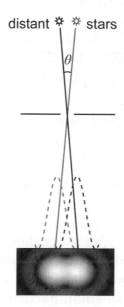

Figure 8.7 Image of a pair of distant stars.

minimum angular separation of objects which are just resolved by the telescope. It is called the angular resolution or diffraction limit.

$$\text{Angular resolution } \theta = \frac{\lambda}{a}$$

$$\left(\text{For circular apertures } \theta = \frac{1.22\lambda}{a}. \right)$$

To improve the angular resolution we can either increase the size of the aperture or decrease the value of the wavelength.

8.5 Other electromagnetic waves

There are other electromagnetic waves which travel at the speed of light and turn out to have the same wave structure as light but which are produced in different ways and can have enormously different wavelengths, as indicated. Radio waves have the lowest energies and gamma rays the highest.

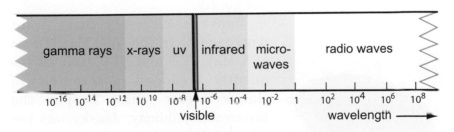

Spectum of electromagnetic waves.

We are all familiar with non-visible electromagnetic waves in certain contexts. Radio waves might be associated with radio and television, infrared waves with night vision binoculars, ultraviolet light with sunburn, microwaves with cooking, and X-rays and gamma rays with medical procedures.

8.5.1 *Message from the stars*

As we shall see in Chapter 10, electromagnetic waves are produced whenever an electric charge is accelerated. The universe is full of such radiation, ranging in wavelength right across the electromagnetic spectrum. It comes from natural sources and crosses billions of light years of interstellar space, carrying information about a variety of astrophysical phenomena.

Radio waves were the first cosmic radiations to be detected other than visible light.

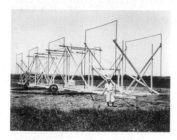

Jansky's antenna.
Courtesy of National Radio Astronomy Laboratory / AUI / NSF.

Karl Guthe Jansky (1905–1950) was employed by Bell Telephone Laboratories as a radio engineer. He was asked to investigate a mysterious sort of static in the form of a steady hiss of unknown origin that was interfering with the short wave radio signals used by Bell Laboratories for their transatlantic radio telephone system. Jansky found that the hiss was coming from space — from the centre of our galaxy.

Jansky's discovery made front page reading in the *New York Times* of 5 May 1933 and NBC radio subsequently broadcast a recording of 'Jansky's star noise'. Despite the favourable publicity, Jansky was not permitted by his employers to continue his investigations into the extra-terrestrial radio waves. Much later, in 1998, Bell Laboratories recognised his work by erecting the Karl Jansky monument on the site where the cosmic radio waves were first detected.

Green Bank Telescope.
Courtesy of National Radio Astronomy Laboratory / AUI / NSF.

The accidental discovery of cosmic radio emissions was not really followed up until after the end of World War II, when a number of strong sources of cosmic radio waves were identified. The largest individual radio telescope operating today is a fully steerable 100-m-diameter dish operated by the National Radio Astronomy Observatory at Green Bank, West Virginia. More than 30 years later, and less than a mile from the site of Jansky's aerial, Penzias and Wilson first detected the cosmic microwaves which are the tell-tale remnants of the big bang (see Chapter 5).

8.5.2 *Other windows on the universe*

The earth's atmosphere absorbs high energy electromagnetic radiation from space. Many important first results in astronomy were established with equipment sited on mountain tops or borne by balloons and rockets. The launch of the satellite *Explorer XI* in 1961 with equipment for detecting cosmic gamma rays marked a new era. The era of long term observations of electromagnetic radiation in all regions of the spectrum had begun.

Radio waves are emitted by the relatively low energy electrons in an antenna. An antenna does not emit light or X-rays, or indeed any other form of higher energy electromagnetic radiation. The particles that emit radio waves cannot 'pool' their energy to produce light or any other electromagnetic radiation. The detection of electromagnetic waves of a particular energy implies that the source of the waves contains individual particles with at least that energy.

Modern technology affords us the opportunity to monitor the emission of electromagnetic radiation of different energies. Such 'multi-wavelength' observations are crucial in the study of complex systems, whether they are millions of light years away or right beside us in the laboratory.

It is remarkable to think that gamma rays with energies more than 10^{13} times the energy of visible light have been detected in the radiation from the remnant of a supernova called the Crab Nebula.

8.6 Light from two sources

The English doctor and physician Thomas Young (1773–1889) demonstrated interference between light waves at an historic

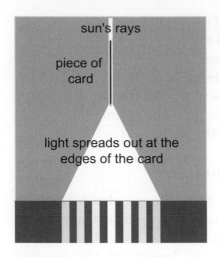

sun's rays

piece of card

light spreads out at the edges of the card

Two light waves add to give bright and dark bands.

meeting of the Royal Society of London in 1803. Young inserted a thin piece of card endways into a beam of sunlight entering the room through a hole in a window blind. This effectively created two sources of light. The coloured bands seen on a screen placed behind the card were evidence that interference had taken place.

8.6.1 *Young's experiment*

Waves which leave two different points, S_1 and S_2, in phase will generally not be in phase as they arrive at a point P. The light intensity is greatest at points where the overlapping waves are in phase and interfere constructively. Light waves from the slits

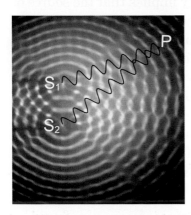

Young's fringes.
*Courtesy of Chris Phillips,
Physics, Department Imperial
College, London.*

travel along different paths. If the path difference between waves travelling from slits S_1 and S_2 is zero or any whole number m of wavelengths to a point P on the screen, the waves will be in phase at that point.

In the more familiar version of Young's apparatus, the piece of card is replaced by a pair of narrow slits. In the typical set-up illustrated in Figure 8.8, light from a lamp is diffracted at a single slit and illuminates two secondary slits. Light from these slits overlaps and interference bands appear

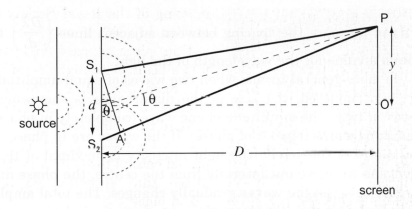

Figure 8.8 Double slit interference.

on a screen behind the slits. We can obtain a value for the wavelength of light by measuring the spacing between the bands.

$$\text{Path difference} = S_2A$$

If the separation of the slits is very much smaller than the distance between the slits and the screen (d ≪ D),

$$\angle S_1AS_2 = 90° : \qquad \sin\theta = \frac{S_2A}{S_1S_2}$$

$$S_2A = d\sin\theta \approx d\theta$$

For constructive interference

$$d\theta = m\lambda$$

m is the 'order' of interference and is an integer ($m = 1,2,3 \dots$).

The bright points on the screen form of a series of narrow, equally spaced bands.

In Figure 8.8, $\theta \approx \tan\theta = \dfrac{y}{D}$. Bright lines are found at

distances $y = \dfrac{D\lambda}{d}, \dfrac{2D\lambda}{d}, \dfrac{3D\lambda}{d}$ from the centre of the screen.

We can measure the spacing between adjacent lines $\left(\dfrac{D\lambda}{d}\right)$ to obtain a value for the wavelength of the light.

In more general terms, when two waves of equal amplitude are superimposed, the size of the disturbance is somewhere between twice the amplitude of one wave (completely in phase) and zero (completely out of phase). If the waves are in phase at the centre of the screen the light intensity is maximal at that point. As we move out laterally from the centre, the phase difference between the waves gradually changes. The total amplitude decreases smoothly towards zero, where the light intensity is minimal, and subsequently increases smoothly to its maximum value. The cyclic variation of amplitude is repeated across the screen, giving rise to the broad, regularly spaced, alternately bright and dark bands. Figure 8.9 shows the bands as they are seen on the screen and also (underneath) how the light intensity varies laterally across the bands.

Sunlight (as used in Newton's experiment) creates multicoloured bands because it contains a more or less continuous spectrum of wavelengths and it is very difficult to obtain anything other than a

central bright fringe

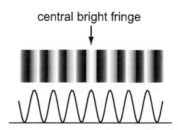

Figure 8.9 Double slit interference pattern.

very poor estimate of wavelength. If the light is monochromatic (has one wavelength) there is a single set of well-defined bands. The same is true of a conventional lamp if most of the light is emitted at one wavelength. Laser light is as monochromatic as you can get and produces very clear interference patterns.

What happens if the separation between the slits is changed?

The condition for constructive interference is $d\theta = m\lambda$. When d is increased, the angle θ between adjacent bright bands must decrease if $m\lambda$ is to remain constant. This means that as slit separation increases, the bands become closer together. The light intensity diminishes uniformly with increasing lateral distance from the central bright band. This is addressed in the next section.

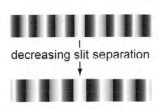

decreasing slit separation

From the middle of the 19th century, the wave theory of light was generally accepted.

The stature of Isaac Newton was one, if not the only, factor which delayed the widespread adoption of the wave theory for almost a century from the time Huygens' wavelets were used to show that light travels more slowly in a denser medium to the time Thomas Young demonstrated the interference of light.

8.6.2 *A pattern within a pattern*

Diffraction brings light from the two slits together and allows it to interfere.

The single slit patterns merge into one. Dark bands appear inside the bright peaks. By closing a slit we make the dark bands bright! The actual variation of light intensity across the screen (Figure 8.10), is shown underneath the photograph.

We say that the intensity of the interference pattern is 'modulated' by the diffraction pattern.

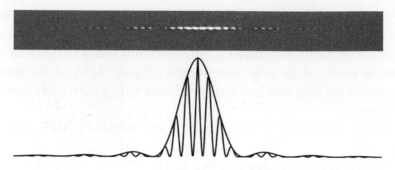

Figure 8.10 Interference between light from a pair of wider slits.

8.7 Interference as a tool

We may make accurate estimates of length using interferometers. In such instruments a single beam of light is generally split into two beams, which then travel along different paths before being recombined. There is invariably *some* difference in the lengths of the paths travelled by the individual beams and an interference pattern may be seen. If the path difference changes, even by some fraction of a wavelength, the positions of individual bright and dark bands (fringes) change. The amount of movement is a measure of the change in the path difference.

8.7.1 *The Michelson interferometer*

The Michelson interferometer was developed in 1880 by Albert Michelson (1852–1931) and Edward Morley (1838–1923) in an attempt to measure the speed of the earth through the ether.

The photograph and corresponding sketch of a typical Michelson interferometer may be seen in Figure 8.11.

A beam of light falls on a semi-silvered glass plate (beam splitter) inclined at 45° to the beam. About half of the light is reflected and travels to a plane mirror, M1. The remainder is transmitted and travels to a second plane mirror, M2. The light beams reflected at the two mirrors retrace their paths back to the semi-silvered mirror. Some of the light from M2 is reflected at the back of the semi-silvered mirror and combines with light from M1 transmitted by that mirror.

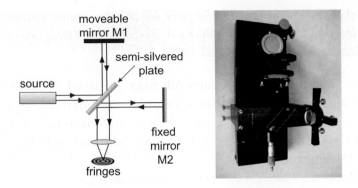

Figure 8.11 Michelson interferometer.

If the position of the moveable mirror M1 is adjusted, the path difference changes and the fringe pattern expands or contracts. In principle, we can measure the distance the mirror is moved by counting the number of fringes which pass a marker in the field of view. In practice the mirror must be moved in individual steps, each so small that any one particular ring remains in the field of view. After each step, the position of the mirror is adjusted to restore the ring to its original position. In this step-wise manner we can count thousands of fringes.

A new standard of length

In the 1880s, one metre was defined to be the distance between two fine scratches on a metal bar held in a museum near Paris. Rulers used in laboratories had in principle been calibrated against the standard metre, albeit indirectly and generally many times removed. This procedure with its potential for gross inaccuracy was very unsatisfactory.

Michelson used his interferometer to define the metre in terms of the wavelength of light. The accuracy he achieved was better than 1/100,000 mm. His method was not only much more accurate, but also had the advantage that direct measurements of length could be made in any reasonably equipped laboratory.

Michelson's work paved the way for a new atomic standard of length, which, although used as early as 1925, was not officially implemented until 1960.

The Michelson interferometer may be used to investigate properties of materials. For example, transparent materials may be inserted into one arm of the instrument, or opaque materials such as metals may be polished and substituted for one of the mirrors.

8.8 Thin films

Thin film interference

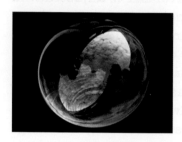

Soap bubble.
Courtesy of Mila Zinkova.

Interference is responsible for the striking displays of colour often given by thin films such as soap bubbles. Light reflected from the upper surface of the film interferes with light reflected from the lower surface to produce interference fringes which generally come in a whole range of colours.

Sound waves stimulate soap films to vibrate.
Courtesy of Stefan Huzler, Trinity College, Dublin.

The photographs on p. 263 show some spectacular patterns in soap films placed horizontally in a vertical resonance tube*. High frequency (above 850 Hz) sound waves in the tube produce thickness variations as they pass through the films. The images (left and right) were caused by sound waves of two different frequencies. The precise form of the pattern at any one frequency depends on the initial thickness of the film and on the amplitude of the wave. The spacing of adjacent fringes reflects the variations in the thickness of the films.

8.8.1 *Newton's rings*

Newton demonstrated interference in the film of air between a convex lens and a flat glass plate. Light reflected at the lower surface of the lens interferes with light reflected at the glass plate, producing a set of alternately bright and dark rings centred on the point of contact between the lens and the glass, as illustrated in Figure 8.12. Newton's measurements showed that the depth of air between the lens and the glass is related to the spacing of the rings. He attempted to explain the phenomenon by using the particle theory of light in conjunction with a convoluted argument.

The interference pattern reflects the circular symmetry of the air film. The rings become progressively closer together towards the edge of the lens, where the thickness of the film

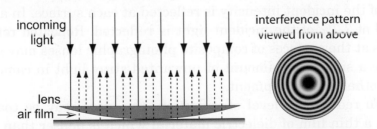

Figure 8.12 Newton's rings.

---

* Visualization of sound waves using regularly spaced soap films. *Eur. J. Phys.* 2: 755–765 (2007).

varies more rapidly. Newton's rings are seen to best advantage in monochromatic light.

In the context of Newton's rings, it is the phase difference of 180° between waves reflected at the two surfaces of the air film at the point of contact that gives rise to a dark spot at

the centre of the pattern. The film is so thin at that point that it does not introduce a significant additional path difference between the waves.

8.8.2 *Non-reflective coatings*

The first recorded reference to the optical effects of thin films was made by Lord Rayleigh in 1886. He noted that the amount of light reflected from glass surfaces decreased as they became slightly tarnished with age. He attributed it to the formation of a superficial layer of different refractive index on the surface. It is now common practice to coat glass surfaces with a dielectric material which has a lower refractive index than the glass to improve aspects of optical performance.

When light passes through a piece of glass, approximately 4% of the incident intensity is reflected at each surface. In all, a total of 7.7% of the incident light is reflected. Repeated reflections at the surfaces of composite photographic lenses may generate a significant amount of unwanted stray light in cameras and other such equipment.

To reduce the level of reflection, the glass may be coated with a thin film of dielectric material which is denser than air, but less dense than glass, as illustrated in Figure 8.13. There is no phase difference due to reflection because *both* waves are reflected at the surface of a denser medium and 'phase-advanced' by 180° (analogous to the inversion of a mechanical wave at a fixed boundary).

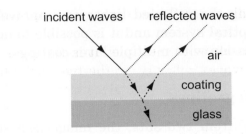

incident waves   reflected waves

air

coating

glass

Figure 8.13 Non-reflective coating (only reflections which make a significant contribution to what is happening at the air surface of the coating are shown).

If the thickness of the film is $\lambda/4$, there is a path difference of $\lambda/2$ between the two reflected waves at normal incidence and destructive interference significantly reduces the intensity of reflected light. A single non-reflective coating can only eliminate reflections for a single value of wavelength, but there will be partial destructive interference for nearby wavelengths. A single coating can reduce total reflection to about 1%[†].

The condition for completely destructive interference between the reflected beams depends on the angle of incidence. In many instances this is not a serious problem as most of the incoming light strikes the surface normally.

Dichroic filters are accurately designed thin film multilayer coatings which selectively transmit a small range of wavelengths. The remaining light is reflected. The picture shows a number of glass discs coated with different dichroic filters.

Dichroic filters.
Courtesy of NASA / JPL.

If the reflected waves cancel, how can they then go forward?

It gets even more mysterious when we read chapter 14 !

[†] See Appendix 8.2.

By minimising the reflected light you improve the transparency of the optical system and it is possible to achieve close to 100% transmission with multiple layer coatings.

8.9 Diffraction gratings

For light coming from two slits, the image is a set of broad bright and dark bands. If the number of slits is increased, the condition for constructive interference, $d \sin \theta = m\lambda$, applies to all pairs of adjacent slits.

The points of maximum intensity are in exactly the same positions on the screen for any number of slits provided that adjacent slits are separated by the same distance and that the experimental arrangement is the same. As the number of slits is increased, more light is transmitted and that light becomes concentrated into narrower bright bands, as illustrated in Figure 8.14.

A *diffraction grating* (which could equally well be called an interference grating) has typically tens of thousands of slits and the maxima form extremely narrow lines. Diffraction gratings *disperse* light because different wavelengths interfere constructively at different angles.

Bright lines are found at all angles where $\sin \theta = m \dfrac{\lambda}{d}$.

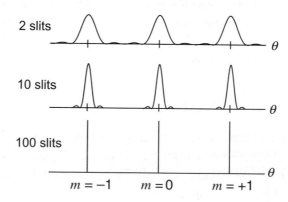

Figure 8.14 Maxima occur in the same place regardless of the number of slits.

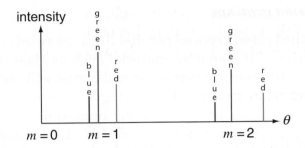

Figure 8.15 Spectrum of an imaginary source with three strong emission lines.

The colour of the zero order bright line reflects the relative intensities of the lines in the spectrum because all the light is in phase there (light from a sodium lamp is orange because of the 'sodium doublet', a pair of strong emission lines at 589.0 nm and 589.6 nm).

Figure 8.15 represents the spectrum of an imaginary source with three strong atomic emission lines. The spectral lines are better separated as m increases but higher order lines are less intense because individual slits have a finite width.

Light is emitted when atomic electrons make transitions between allowed energy levels. Each atom has its own characteristic spectrum — its 'fingerprint'. Heavy atoms have more complex spectra than light atoms; in molecular spectra, some lines may smear into bands. In general, the more tightly bound the atom, the more complex the spectrum. Diffraction gratings enable us to make the high precision measurements of wavelength that are needed for analysis of the properties of what may be very complex materials.

8.9.1 *Practical diffraction gratings*

The basic requirement of a grating is a set of regularly arranged points which act as scattering centres and re-radiate incident light. The precise details are dictated by the requirements of resolving power and spectral sensitivity. A modest research grating might have 600 slits per mm.

Transmission gratings

In one method closely spaced parallel lines are ruled on the surface of a piece of transparent material such as glass. The maxima for the different component wavelengths will form beams at different angles. Thus the grating separates the colours much the same as a prism, with one important difference. In the grating, longer wavelengths are deflected more than shorter wavelengths, while the opposite is true for the prism. To achieve a sharp separation of colours we must use a very large number of very thin slits close together.

The light passes through glass in this type of grating. Glass absorbs light to an extent which depends on the wavelength and this effectively limits the use of transmission gratings to visible wavelengths.

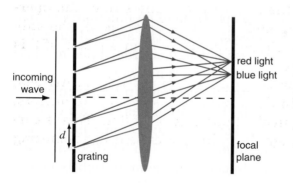

Segment of a transmission grating.

The grooves on a compact disc can act as a grating

Reflection gratings

The grooves are ruled on a highly reflective material such as aluminium sputtered onto a glass surface. When illuminated by the incoming light the edges of the grooves act as coherent light sources in the same way as the slits, but this time the light emerges on the same side as the incoming light. Reflection gratings may be used to examine both visible and ultraviolet light because absorption is negligible. For this reason such gratings are almost always used in spectroscopy research. High quality gratings may have 10,000 grooves per mm.

8.9.2 *Compact discs*

Information is written to a compact disc in the form of tiny pits in a spiral array of tracks on the surface of a clear plastic disc. The pits are approximately 1 μm apart while the track pitch is 1.6 μm corresponding to a track density of 625 grooves per mm. A thin layer of aluminium applied to the surface makes it reflective. When a CD is illuminated with white light there is a symmetrical rainbow effect.

8.10 Other 'lights'

The resolving power of an optical instrument is ultimately limited by diffraction. The angular resolution, a measure of the minimum size of objects that can be resolved, is limited by the wavelength of the radiation and the size of the aperture. For light-emitting and light-reflecting objects, the only parameter that can be changed by more than a factor of 2 is the size of the aperture.

The Large Binocular Telescope at Mount Graham, Arizona, which 'saw first light' in October 2006, has twin 8.4 m mirrors, mounted to give an equivalent aperture of 22.8 m. This is currently the best instrument for looking at distant light-emitting astronomical objects.

8.10.1 *X-ray diffraction*

In certain situations we may use other sorts of 'light' to illuminate objects. X-rays have wavelengths of the order of 1000 times shorter than visible light, making it possible in principle to resolve much smaller objects.

Crystals as diffraction gratings

The German physicist Max Von Laue (1879–1960) was a very versatile scientist whose work and interests ranged over a wide field. In February 1912, his research student P.P. Ewald approached him for advice about crystal optics: '*But during the conversation I was suddenly struck by the obvious question of the behaviour of waves which are short by comparison with the lattice-constants of the space lattice. And it was at that point that my intuition for optics suddenly gave me the answer: lattice spectra would have to ensue. The fact that the lattice constant in a crystal is of an order of 10^{-8} cm was sufficiently known.... The order of X-ray wavelengths was estimated by Wien and Sommerfeld at 10^{-9} cm. Thus the ratio of wavelengths and lattice constants was extremely favourable if X-rays were to be transmitted through a crystal. I immediately told Ewald that I anticipated the occurrence of interference phenomena with X-rays*'.

Von Laue continues his account: '*...acknowledged masters of our science, Wilhem Wien and Arnold Sommerfeld, were sceptical. A certain amount of diplomacy was necessary before Walter Friedrich (1883–1968) and Paul Knipping[‡] (1883–1935) were finally permitted to carry out the experiment according to my plan, using very simple equipment at the outset.*'

The plan was successful beyond all expectations. The two research students observed interference patterns, just as Von Laue had predicted. It was now possible to see crystal structures using X-rays or light of wavelength comparable to the size of the molecules themselves. The 'seeing' is in terms of interference patterns which came directly as a result of reflections from individual atoms and molecules. For conceiving this groundbreaking work in X-ray crystallography, Von Laue was awarded the Nobel Prize for Physics in 1914.

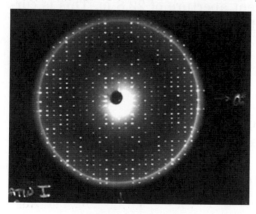

X-ray diffraction pattern
of a protein crystal.
Courtesy of NASA-MSFC.

[‡] W. Friedrich and K. Knipping, research students.

A crystal is a diffraction grating with a difference. It is a three-dimensional arrangement of vast numbers of atoms, unlike the more traditional gratings used to diffract light. Each atom scatters incoming X-rays, giving complex diffraction patterns. Rock salt was the first crystal structure to be fully determined.

A crystal can be visualised as a stack of identical *unit cells*, something like a honeycomb. The basic atomic arrangement is duplicated in each cell and symmetries in the diffraction pattern reflect symmetries in the arrangement of atoms in the cell. No two crystals have exactly the same diffraction pattern; each has its own 'fingerprint' in the same way as a light-emitting atom.

Unit cell.

The Laue pattern of spots was explained theoretically by the English physicist William Henry Bragg (1862–1942). He and his son Lawrence Bragg (1890–1971) jointly received the 1915 Nobel Prize for Physics for the use of X-rays to determine crystal structure. Lawrence, aged 25 at the time, is the youngest-ever Nobel laureate. The Braggs developed X-ray diffraction as a tool to systematically investigate crystal structures. They analysed reflections of X-rays from each one of the numerous atomic surfaces in crystals and were able to determine the internal structure with some precision.

Bragg's law

Bragg's law simplifies the mathematics of crystal diffraction by considering different sets of parallel planes of atoms and writing down the condition for constructive interference between X-rays reflected from them. The arrangement of atoms in each plane in Figure 8.16 looks similar to the arrangement of slits in an optical diffraction grating.

Path difference between the incoming and outgoing rays = $AC + CB = 2d \sin\theta$.

For constructive interference: $2d \sin\theta = n\lambda$ (similar to the condition for light).

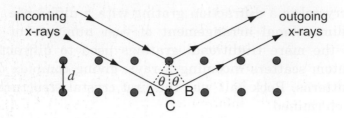

incoming x-rays   outgoing x-rays

d

θ θ

A   B
C

Figure 8.16 Bragg's law.

The investigation of crystal structure using Bragg's law is by no means as straightforward as it might seem at first glance. The condition for constructive interference is simple enough but we have essentially replaced each slit by a unit cell and there may be many sets of parallel planes of atoms at different angles.

X-rays in medical science

X-rays were discovered by the German physicist Wilhelm Conrad Röntgen (1845–1923) in 1896. The very first Nobel Prize for Physics was awarded to Röntgen in 1901 for the discovery of X-rays. He was the first person to use X-rays as a medical imaging tool. The stamp shows an X-ray image of his wife's hand, with a clear impression of a ring on the third finger. X-ray images of this kind are still extensively used for viewing bone structure.

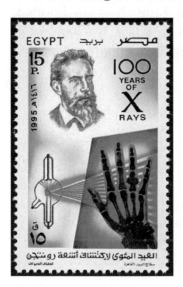

Röntgen Commemoration.
Courtesy of Egyptian Post.

X-ray diffraction techniques were used to disentangle the structure of the complex molecule deoxyribonucleic acid (DNA). The British postgraduate student Francis Harry Compton Crick (1916–2004) and the American postdoctoral fellow in zoology James

Dewey Watson (1928–) believed DNA to be the primary factor in heredity. To establish this they needed to know how the molecule was able to replicate itself, store genetic information and mutate. Maurice Frederick Wilkins (1916–2004), a British biophysicist, was the son of a physician, originally from Dublin. He worked in the biophysics unit at King's College, London. His X-ray diffraction studies of DNA were crucial to the determination of its molecular structure by Watson and Crick. For this work the three scientists were jointly awarded the Nobel Prize for Medicine in 1962.

James Dewey Watson.
*Courtesy of US National
Library of Mecicine.*

Rosalind Elsie Franklin (1920–1958), a molecular biologist, made a vital contribution to the identification of DNA. She and Wilkins worked in the biophysics unit at King's College, London, and her X-ray diffraction photographs were key data. She died of cancer four years before the award of the Nobel Prize.

8.10.2 *Electron diffraction*

Optical microscopes are severely limited by diffraction and can only magnify up to about 2,000 times, at best. Electron with energies of 120 keV have a wavelength about 100,000 times smaller than the wavelength of red light so an electron microscope has much higher magnification than its optical counterpart.

Electron microscopes

To prevent the scattering of the electrons by air molecules, the microscope is enclosed in a metal cylinder maintained at a high vacum. The electron beam is focused by magnetic, rather than

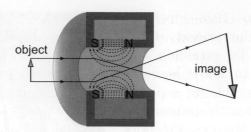

Figure 8.17 Magnetic lens.

optical, lenses. A magnetic lens consists of a coil of copper wires mounted inside iron pole pieces. A current passing through the wires creates a magnetic field between the pole pieces (see Chapter 10).

The magnetic field is weakest along the axis as illustrated in Figure 8.17 so electrons close to the axis are less strongly deflected than those far from the axis. The overall effect is that a beam of parallel electrons is focused into a spot. The electrons in magnetic lenses are much less well behaved than light in optical lenses; they spiral in the magnetic field, so the image is rotated with respect to the object.

Transmission microscopes. Electrons penetrate even transparent materials to a very limited extent. Specimens used in transmission electron microscopy must be thin (between 10 and 200 nm) to be transparent to electrons and carefully prepared to withstand the high vaccum.

Reflection microscopes. The surface of a solid can be used to diffract electrons; it behaves rather like an optical reflection grating. The energies and intensities of the diffracted electrons reflect the atomic arrangement on the surface.

Electron microscope were first used in Material Science to examine structures at almost atomic levels. They are used extensively in Biological and Medical Sciences for the study of cells.

8.11 Coherence

Light waves, unlike mechanical waves, do not come from oscillators which emit waves continuously and at a fixed frequency. Take the example of a neon lamp which emits light when an electric current is passed through the gas. Some of the electrical energy is transferred to individual atoms and they become excited. There is no way of knowing which atom was the first to react to the stimulus or exactly how many atoms are emitting light at any one time, with what phase, or in what direction. The atoms are 'a law unto themselves'. If one atom emits light, its neighbours are unaffected. We say *the atoms do not act co-operatively* (or *coherently*).

It is like watching a large fireworks display and being enthralled by the completely random, multicoloured bursts of light, without knowing where and when you will see the next one, what colour it will be or how it will behave.

Dancers in a sequence from *Riverdance* often create the same repetitive sound (they are a *coherent* set of dancers) for long periods of time. In many sequences, the line of dancers rotates about one particular person who dances almost on the same spot. All the dancers who move in circles about that person have to increase their step length to keep in time. They may be physically independent of one another, yet they are stimulated to keep in time by one another.

Stimulation of atoms is crucial to the creation of a coherent source of light.

8.11.1 *The question of phase*

We might expect from past experience that Young would have used two lamps in his demonstration of interference, in the same way that two oscillators may be used to demonstrate the interference of mechanical waves. He would not have been successful because of the way light is produced.

The basic condition for interference between two waves of the same frequency is that the phase difference between them remains constant. The phase difference between sthe light waves from two lamps varies in a completely arbitrary way.

Why was Young's experiment not done much earlier?

They didn't know about coherence.

The light from a lamp consists of an enormous number of individual wave trains which have no fixed phase relationship between them (like crowds streaming into a soccer match) and the phase varies in a completely random way. We cannot use the light from two lamps to demonstrate interference. In interference experiments such as Young's slits, the two wave trains originate from the same source and are therefore coherent. Any random phase changes occur simultaneously in the light from both slits.

8.12 Polarisation

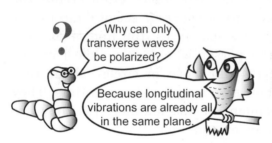

The word 'polarisation' comes from 'polar' — 'can have an orientation'.

All electromagnetic waves are transverse waves and therefore can be polarised.

8.12.1 *Polarisation of electromagnetic waves*

Electromagnetic waves are transverse and can be polarised.

Radio waves. In a simple radio transmitter, electrons vibrate up and down in a wire beside the aerial and this induces the electrons in the aerial to do the same (just as if they were bonded together). Since the electrons vibrate in one direction, so does the electric field and the waves are polarised.

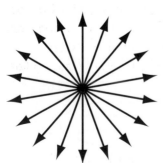

Light waves. Light waves may be unpolarised, partially polarised or completely polarised. In *unpolarised* light the direction of the electric field varies randomly in the plane perpendicular to the direction of propagation.

The electric field of *plane polarised* light vibrates in a single direction in that plane.

8.12.2 *What happens to light as it passes through a polaroid?*

In this simple model we represent the polariser by a thin slit. The incoming light has its electric field vectors scattered all over the plane perpendicular to the direction of propagation. The vectors which are parallel to the slit get through unscathed. For the remainder, as illustrated in Figure 8.21, only the component parallel to the slit emerges. These vibrations emerge polarised in the direction of the slit and reduced in size.

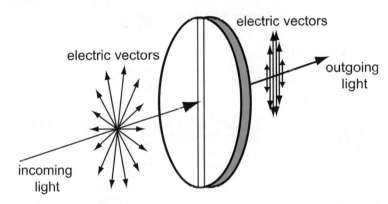

Figure 8.21 A model of the process of polarisation.

Changing the plane of polarisation

Let us consider a light wave which is already polarised when it enters a polaroid. The fraction of the intensity transmitted will depend on the relative direction of polarisation. A law named after Etienne-Louis Malus (1775–1812) follows directly from Figure 8.22. If the electric vector of the incident light has an amplitude E_0 and is at an angle θ to the slit, only the component in the direction of the slit ($E_0 \cos \theta$) will be allowed through.

The intensity of the transmitted wave is proportional to the square of the amplitude:

$$I \propto (E_0 \cos \theta)^2$$

Intensity of light transmitted $I \propto \cos^2 \theta$ (Malus's law)

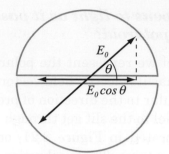

E_0 = incident wave amplitude
$E_0 \cos \theta$ = transmitted wave amplitude

Figure 8.22 Malus's law.

'Crossed' polaroids

Let us next consider two polaroids arranged so that their axes of polarisation are perpendicular. The light transmitted by the first polaroid is polarised at an angle, $\theta = 90°$, relative to the second slit and no light will get through.

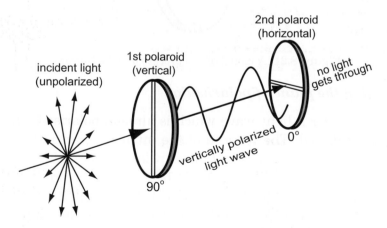

Figure 8.23 'Crossed' polaroids.

Polarising materials

Transparent materials that transform unpolarised light to plane polarised light have a characteristic molecular structure.

Typical polarising material: The molecules tend to be aligned in one direction and look somewhat 'stretched'.

Typical non-polarising material: The atomic arrangement is too symmetrical to let the light squeeze through with its electrical energy absorbed in all but one single direction.

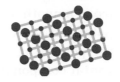

Polarisation and spiders

The *Drassodes cupreus* spider has a pair of optical sensors behind its principal eyes. These are not 'normal' eyes, but filters which sense the direction of polarisation of sunlight, which varies as the sun changes position. Research indicates that the 'spider-compass' works best at dawn and dusk and when the spider is out and about searching for food.

8.12.3 *Polarisation by reflection*

When white light illuminates the surface of a transparent material, the electric fields of the waves interact with the surface molecules, giving rise to partial polarisation of the light.

The reflected and refracted waves come from a single beam of unpolarised light, so if the reflected wave is partially polarised in one direction, the refracted wave is also partially polarised in a way

which complements it. (If the two were to be rejoined, the resulting light would be unpolarised.)

Sunglasses reduce glare

Polarising filters are used in sunglasses. Sunlight reflected from smooth horizontal surfaces such as water is strongly polarised in the horizontal plane. Sunglasses are worn with the polarising axis of the lenses vertical and much of the horizontal component of the light is absorbed, significantly reducing glare.

A historical interlude: Thomas Young (1773–1829)

Thomas Young (1773-1829)

Thomas Young was born in Somerset, England in 1773, into an affluent Quaker family. He was an infant prodigy, able to read fluently at the age of two. By the time he was six years old, he had read the Bible twice. A year later, he was sent to boarding school, where he studied mathematics and languages. He developed a good reading knowledge of French, Italian, Greek and Latin. He independently studied physics and natural history and learned to make telescopes and microscopes. After leaving school, he began to study languages such as Arabic and Turkish, and by the time he entered university he was an accomplished Greek and Latin scholar and had read the works of Isaac Newton. Towards the end of his life he stated, apparently with great satisfaction, that he had never spent an idle day in his life.

Young studied medicine at the universities of London, Edinburgh, Göttingen and Cambridge. During his university years, he drifted away from the Quakerism of his youth and enjoyed social pursuits such as music, dancing and the theatre. In effect he had all the external attributes of a 'gentleman scholar'. Young inherited

(Continued)

£10,000 and a house in London on the death of his uncle Sir Richard Brocklesby in 1797, a year after his graduation from Göttingen. Three years later Young moved to London and began to practise medicine. He failed to establish a really successful practice, owing to an apparent inability to gain the confidence of his patients.

Being a man of independent means, Young was able to pursue other interests and regularly attended meetings of the Royal Society of London, where he came to the attention of Joseph Banks, president of the Royal Society, and Benjamin Thompson, founder of the Royal Institution. On their recommendation he was appointed 'Professor of Natural Philosophy, Editor of the Journals and Superintendent of the House' in 1801. He lectured on a wide range of topics, such as optics, animal life, vegetation and techniques of measurement. Young was required to give 'popular' lectures to a more general audience. These were often difficult to understand and had limited appeal, in sharp contrast to the popular lectures in chemistry given by Humphry Davy. A year later, Rumford left England and Young subsequently resigned and returned to the practice of medicine.

Young had a passion for languages and in 1813 he began to study hieroglyphic script on the Rosetta stone, a black basalt slab discovered near the Egyptian town of Rosetta at the mouth of the river Nile. The stone was unearthed by a group of Napoleonic

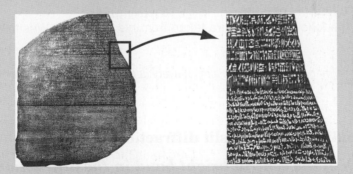

Rosetta stone. *Courtesy of Bibliotheca Alexandria, Centre for Documentation of Cultural and Natural Heritage.*

(Continued)

soldiers digging to build a fort. Young's work, published anonymously in a supplement to the *Encyclopaedia Britannica*, laid the foundation for later, more comprehensive, work on deciphering hieroglyphics. It was established that the Rosetta stone was set up to commemorate the first anniversary of the Egyptian king Ptolemy V in 196 BC.

The name of Thomas Young is synonymous with the experiment which unambiguously established the wave nature of light, but he could also be considered to be the founder of physiological optics. Young explained how the eye focuses objects at different distances by changing the curvature of the lens, he described and measured astigmatism, and he presented a theory of colour vision in terms of three different colour receptors on the basis of experiments with different mixes of primary colours. His scientific work gained international recognition in 1827, when he was elected a foreign associate of the French Academy of Sciences.

Young was physician to the Royal College of Surgeons from 1811 to his death. He was, at various times, foreign secretary of the Royal Society, secretary to the Board of Longitude, editor of the *Nautical Almanac*, and inspector of calculations and physician to the Royal Palladium Insurance Company.

Young had such a wide range of interests that he was probably not sufficiently focused to make a consistent contribution in any one area. He was a true scholar, with a love for knowledge of any subject, no matter how obscure.

The following words are taken from his epitaph in Westminster Abbey:

'a man alike eminent in almost every department of human learning.'

Appendix 8.1 Single slit diffraction

Width of the central bright band in the diffraction pattern of a single slit

The widths and separations of the bright bands on the screen behind the slit depend on the distance between the slit and the screen. For this reason, the distribution of light intensity on a

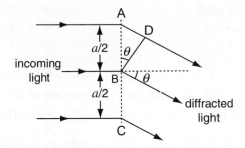

Figure 8.24 Calculating the position of the first minimum.

screen behind the slit is usually described not as a function of position on the screen but in terms of angular position (the angle at which light leaves the slit).

The angular positions of the minima on either side of the central bright band can be calculated by mentally dividing the slit into two and calculating the angle at which wavelets from pairs of points such as A and B in Figure 8.24 interfere destructively.

The path difference between the rays from points A and B is AD.

Triangle ADB is right angled:

$$\sin\theta = \frac{AD}{AB} \quad \text{and} \quad AD = \frac{a}{2}\sin\theta$$

The rays are completely out of phase if

$$AD = \frac{\lambda}{2}$$

The corresponding value of θ is given by $\sin\theta = \dfrac{\lambda/2}{a/2} = \dfrac{\lambda}{a}$

$$\theta = \sin^{-1}\frac{\lambda}{a}$$

For small values of θ,

$$\theta \approx \frac{\lambda}{a}$$

For every point such as A in the upper half of the slit there is a point such as B in the lower half. Light travelling at an angle $\theta \approx \frac{\lambda}{a}$ from all such pairs of points interferes destructively, giving rise to the first minimum. Corresponding rays directed upwards at the same angle also interfere destructively. Higher order minima occur at angles where the path difference between the two beams is any odd multiple of $\lambda/2$.

Variation of intensity with angular position

To find an expression for the intensity in terms of angular position, we need to calculate the total amplitude of Huygens' wavelets arriving at different points on the screen. The wavelets are assumed to leave the slit in phase, so phase differences at the screen are related solely to path differences.

A path difference of $\lambda/2$ corresponds to a phase difference of π.

$$\text{In general, phase difference} = \left(\frac{2\pi}{\lambda}\right) \times (\text{path difference})$$

For light travelling at an angle θ from two adjacent sources separated by Δx, the path difference $= \Delta x \sin\theta$.

Phase difference between adjacent wavelets $\Delta\varphi = \frac{2\pi}{\lambda}\Delta x \sin\theta$.

We use a vector technique called the addition of phasors, in which the slit is treated as a very large number of identical and equally spaced sources of Huygens' wavelets. Each wavelet is represented by a phasor, whose length represents the amplitude of the electric field. For identical sources, all phasors have the same length. The phase difference between wavelets is represented by the angle between the corresponding phasors.

To sum over all the wavelets arriving at any point on the screen, we arrange the phasors from adjacent sources head to tail in vector fashion, starting at the top of the slit. The total amplitude is given by the line joining the two ends of the 'phasor chain', as illustrated in Figure 8.25.

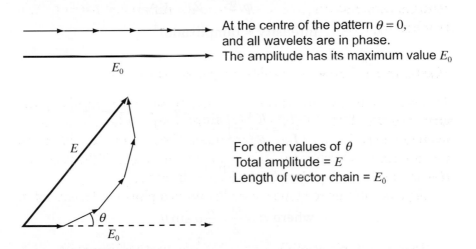

At the centre of the pattern $\theta = 0$, and all wavelets are in phase. The amplitude has its maximum value E_0

For other values of θ
Total amplitude = E
Length of vector chain = E_0

Figure 8.25 Phasor diagrams for five wavelets.

For very large numbers of sources, the individual amplitudes are so small that the phasor chain forms a smooth circular arc. We can express E in terms of E_0 using Figure 8.26.

$$\text{Arc AB} = R\varphi = E_0$$

$$\sin\frac{\varphi}{2} = \frac{\text{AB}}{\text{OA}} = \frac{E}{2R}$$

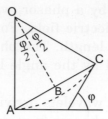

Figure 8.26 Large numbers of phasors.

$$E = \frac{E_0}{\varphi/2}\sin\frac{\varphi}{2}$$

Intensity $\propto$ (amplitude)^2

Let $\alpha = \dfrac{\varphi}{2}$:

$$\frac{I}{I_0} = \frac{E^2}{E_0^2} = \left(\frac{\sin\alpha}{\alpha}\right)^2$$

$$I = I_0\frac{\sin^2\alpha}{\alpha^2}$$

where $\alpha = \dfrac{\varphi}{2} = \dfrac{\pi}{\lambda}a\sin\theta$

Figure 8.27 shows how the intensity varies across the screen as a function of the angle and includes the angular positions of minima.

The angular separation of the minima on either side of the central peak is often referred to as the angular width of the main maximum.

$$\text{Angular width of the main maximum} = \frac{2\lambda}{a}$$

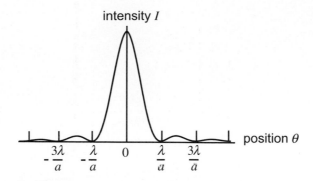

Figure 8.27 Graph of intensity as a function of angular position.

Appendix 8.2   Reflectance of thin films

The percentage of light reflected at an interface between two materials (reflectance) depends on the refractive index and the angle of incidence.

For an uncoated piece of crown glass (refractive index = 1.52),

$$\text{Reflectance } R = \left\{ \frac{1 - n_{\text{glass}}}{1 + n_{\text{glass}}} \right\}^2 = 4.3\%$$

For the same glass coated with a material of index n_c,

$$R = 100 \times \left\{ \frac{n_{\text{glass}} - n_c^2}{n_{\text{glass}} + n_c^2} \right\}^2$$

This gives the interesting result that $R = 0$ when $n_c = \sqrt{n_{\text{glass}}} = 1.23$.

There are no suitable materials with this refractive index. Magnesium fluoride ($n_c = 1.38$) is cheap and durable and widely used. For a single coating of MgF_2, $R = 1.3\%$

Figure 8.27 Graph of intensity as a function of angular position θ.

Appendix 8.2 Reflectance of thin films at normal incidence

The percentage of light reflected at an interface between two materials (reflectance) depends on the refractive index and the angle of incidence.

For an uncoated piece of crown glass (refractive index $n = 1.5$) in air, the reflectance for a beam of light arriving at normal incidence is given by

$$\text{Reflectance } R = \left(\frac{n-1}{n+1}\right)^2$$

For the case of a material with a material of index n,

$$R = 100 \times \left(\frac{n-1}{n+1}\right)^2$$

This gives the interesting result that $R = 0$ when $n = 1$.

There are no suitable materials with this refractive index. Magnesium fluoride ($n = 1.38$) is cheap and durable and widely used. For a single coating of MgF_2, $N = 1.38$.

Chapter 9
Making Images

Now, we come to the making of images. We follow developments from the days of the early Greeks through to modern imaging. A photograph, although it may be truly a work of art, is just a two-dimensional image; there is no change in perspective when we look at it from different angles.

Holography, or three-dimensional imaging, was invented by the Hungarian physicist Dennis Gabor. We discuss what information the photograph lacks, its third dimension, and how we can encode that information to make a complete picture. We see that holography is a technique which has been rapidly incorporated as an art form and a scientific tool with practical and commercial applications.

9.1 Creating images

The optical principle of imaging was already known in the fourth century BC, when Aristotle described a method of watching a solar eclipse without damaging the eyes. He passed light from the sun through small holes punched in a metal plate, forming an image on the ground underneath the plate. Changes in the shape or position of the sun were reproduced in the image, which could be viewed without risk to the eye.

9.1.1 *Photography*

The optical principle of the camera obscura, or 'dark chamber', dates back to Aristotle but the earliest record of its use is found in the manuscripts of Leonardo da Vinci (1442–1519). It can be as simple as a box or room with a small hole in one wall. Light entering the box is reflected from a mirror on the opposite wall and the image is seen on a piece of ground glass set into the top of the box. The camera projects visual images in a two-dimensional form. In the 17th century, the camera obscura was used by both scientists and artists.

Johannes Kepler made his solar observations in the early 1600s using one of these instruments and it is said that he was the first to give the instrument its name. The design of his portable camera obscura was very ingenious. The structure was tent-like, serving both as the dark chamber and as a 'laboratory'.

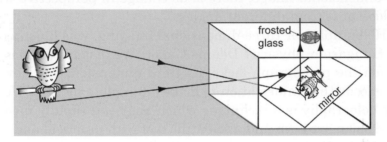

Camera obscura.

The camera obscura was used for making drawings. The image was projected onto a piece of paper and the details were traced by the artist, usually as the basis for a painting. (It is much easier to draw from a two-dimensional image than from the real three-dimensional subject.) Johannes Vermeer, the famous Dutch 17th century artist, may have used a camera obscura. (There is no real documentary evidence of this, but studies of his style have led to the belief that he did.)

The camera obscura came in a variety of shapes and sizes. Some of them could accommodate more than ten people. By the 16th century, the design had been modified by putting a lens at the entrance to the aperture, improving the quality of the image.

In 1685, Johann Zahn (1631–1707), a German monk, further refined the design by using a system of lenses. His camera was the prototype for the modern camera, with a design similar to today's single lens reflex camera. All it lacked was a mechanical shutter and a way to record a permanent image of the object.

9.1.2 *History of the photograph*

Niépce's heliograph

In 1727, the German physician, Johann Heinrich Schulze (1687–744) discovered that light darkens silver salts, the key to creating permanent photographic images, but he did not realize the significance of his discovery. The first permanent photograph from nature was taken by the Frenchman Joseph Nicéphore Niépce (1765–1833). In summer 1826, using a camera obscura, he projected an image of the view from his country house in Le Gras onto a pewter plate coated with a light-sensitive solution of bitumen of Judea. Bitumen gradually hardens on exposure to light and, after about 8 hours, the unexposed

Hard to keep still for 8 hours.

parts can be washed away to reveal an image. Niépce brought his photograph to London and presented an account of his process, which he called *Heliography*, to the Royal Society, but his invention failed to produce any interest.

Disappointed, Niépce returned to France leaving his heliograph in the house of his host Francis Bauer, where it laid, stored away and forgotten, until 1952, when it was discovered by a photographic historian Helmut Gemsheim (1913–1995). Gernsheim verified its authenticity and obtained it for his collection which is now exhibited in the Harry Ransom Center at the University of Texas. It is kept sealed in an atmosphere of inert gas, and must be viewed in a darkened environment under controlled lighting.

The picture on the upper left hand side of p. 292 shows a gelatine silver print of Niépce's photograph made in 1952 at the Kodak Research Laboratory in Harrow. It is a precise copy of the original, to the extent that the texture and the unevenness

Reproduction of Joseph Nicéphore Niépce's 'View from the window at Le Gras'.

The same reproduction touched up with watercolours by Helmut Gersheim.

Images courtesy of Harry Ransom Centre.

of the pewter plate is clearly visible. Gersheim was unhappy with it and did not allow its reproduction until 1977. The picture on the right hand side shows the Kodak print touched up by Gernsheim himself using watercolours. On the left of this print is the upper left of the family home; beside it a pear tree with a patch of sky showing through an opening in the branches. At the centre we see the slanting roof of a barn with another wing of the family house on the right.

Daguerreotypes

Louis Daguerre

Louis Daguerre (1787–1851), a French painter who had also been experimenting with photography, worked in partnership with Niépce from 1929 until his death in 1833. Daguerre went on to discover that the exposure of a copper plate coated with silver iodide creates a latent image, which can be developed by exposure to mercury vapour, and then fixed using a solution of common salt. The result is a negative from which copies may be made. This was the birth of practical photography. In 1839, 'daguerreotypes', as they were known, were shown at a joint meeting of the Académie des Sciences and the Académic des-Beaux Arts in Paris by the distinguished astronomer and physicist, François Jean Dominique Arago (1786–1853).

L'atelier de l'artiste. Daguerre 1837.

Over the next forty years, the quality of photographs improved very significantly. Astronomers rapidly took advantage of the technique. The first daguerreotype of the moon was made in 1840 and of the star Vega in 1850. Astronomical exposure times were long because of the low light levels but the development of more sensitive plates did a great deal to improve the situation.

In 1879, George Eastman (1854–1932), a junior bank clerk and dedicated amateur photographer, was awarded his first patent for coating dry plates with photographic emulsion. The following year he went into the commercial production of dry plates, and the Eastman Dry Plate Company, based in Rochester, New York, was founded in January 1881, with financial help from businessman Henry A. Strong. A major step forward came with the development and patenting (in 1885) of the rolled paper holder by Eastman and William Walker. This was a cassette containing rolled paper film, which could be attached to the back of almost all commercially available plate cameras, in place of the plate holder. Eastman's dry emulsion plates were extensively used in astronomy; in 1880, the American physician and amateur astronomer, Henry Draper (1837–1882) photographed the Orion nebula, a diffuse patch of light in the sword of the Hunter in the constellation Orion. The nebula is just visible to the unaided eye.

Kodak camera

The launch of the Kodak camera in 1888 opened photography to the general public. The camera, containing a 100-exposure film, was priced at $25. After use, the entire camera was returned to the factory, where the film was developed. Photographs and camera (complete with new film) were dispatched to the client, at a cost of $10. The slogan was: *"You press the button, we do the rest"*. Transparent film, made from cellulite, replaced paper film in 1889.

Box Brownie.
*Courtesy of
Lynn Mooney.*

The *'Box 'Brownie'*. In 1900, Eastman marketed a cheap portable camera. The body was a simple box and the film cassette could be removed for processing. The camera, named the Brownie, sold for $1 and the film for 15 cents.

The Kodak camera and the Brownie established the name of Kodak in the annals of photographic history.

Colour photography

James Clerk Maxwell (1831–1879) did not confine himself to the theory of light and electromagnetic radiation to be described in

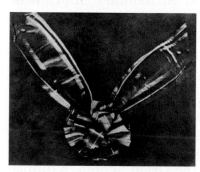

Image of tartan ribbon by
James Clerk Maxwell.

the next chapter. He had a practical interest in the formation of colours and demonstrated a system of colour photography in the 1860s. He made three separate black and white images of a tartan ribbon, using blue, green and red filters. The colours of the object were re-created by illuminating each black and white image with light passed through the appropriate filter and projected onto a screen. This was the first example of what is referred to as additive synthesis, building up a coloured image by adding appropriate amounts of coloured light.

Practical colour photography first made its appearance in France in 1904 when the Lumière brothers, Auguste and Louis, presented their Autochrome process to the French Académie des Sciences. It was launched onto the market in 1907. Basically, it combined filters and light sensitive emulsion on a single plate. The filters were tiny grains of starch, dyed and spread over a glass plate, which was then coated with transparent emulsion. After exposure, the plates were processed to produce a transparent image (slide), which was viewed in transmitted light. As with Maxwell's more rudimentary process, the photographic image was created by addition, and could be viewed only in projection. Autochrome plates were popular as they could be used in existing cameras, although exposure times were long (at least a second, even in bright sunlight compared to less than 0.1 seconds for black and white film).

It was the Eastman-Kodak Company, once again, which brought colour photography to the general public in 1935, when it launched Kodachrome, the first colour negative film. These films are layered with three different emulsions. Each one is sensitive to a different primary colour and a 'colour coupler' is introduced, during development, to deposit an appropriate dye in areas where silver forms. This film may be used for prints or projection (slides).

Digital photography

Digital imaging has almost completely replaced imaging based on chemical changes in light sensitive films. In digital cameras, images are recorded using electronic image sensors (pixels) arranged in a grid, rather like the retina of the eye. Light striking individual sensors during an exposure is converted to electrons (photoelectric effect, see Chapter 13) which are registered as electrical signals. After the exposure, the signals are stored in a memory chip and the camera is ready for the next exposure.

9.1.3 *Nuclear photographic emulsion*

In photographic emulsion, crystalline grains of silver bromide (with a small proportion of silver iodide) are suspended in gelatine. Exposure to light creates the 'latent' image by modifying some of the grains. The chemical process of development turns modified grains into blackened grains of silver. Unmodified grains are washed out using a 'fixing' solution.

In 1945 it was discovered that photographic emulsion could be used for an entirely different purpose. When fast charged particles of nuclear size cross the emulsion, they produce the same effect as light. They leave a trail of modified grains of silver which upon development appears as a track somewhat like a condensation trail of a high flying jet.

The Ilford company, to their credit, put considerable effort into the development of non profit-making photographic emulsion for fundamental particle research. By producing emulsion which contained about eight times the normal ratio of silver to gelatine, they succeeded in making it so sensitive that the ionisation produced by the passage of a single proton or electron was sufficient to make a track clearly visible under a high power microscope. As we shall see in Chapter 17, the technique played a key role in a number of pioneering experiments in particle physics.

9.1.4 *Interpretation of photographic images*

The worm can put a number of interpretations on what it sees in Figure 9.2, but the addition of a cone as shown in Figure 9.3 may remind the worm of something it has seen before.

This an example of what generally happens when we look at a photograph. Our brain uses its enormous memory bank of similar images to interpret two-dimensional images in terms of

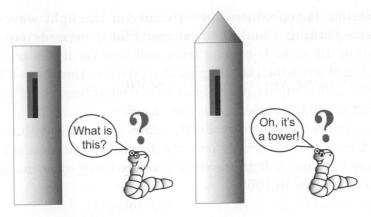

Figure 9.2 The question answered?

three-dimensional objects. The interpretation may not be correct because there is a limited amount of information available.

Just as the worm cannot see around the sides of the tower in Figure 9.2, so we are unable to look around an object in a photograph. We do not see anything new no matter how we look at the photograph (from above, from below, from the left or from the right).

9.2 Holography

9.2.1 *The inventor*

The word 'hologram' is derived from the Greek words *holos* and *gramma*, which translate as 'whole writing'.

Holography was invented in 1947 by the Hungarian scientist Dennis Gabor (1900–1979), while trying to develop a method of producing 'lensless' images to improve the poor resolving power of electron microscopes. Gabor was awarded the Nobel Prize in 1971 for his work on holography.

In Gabor's Nobel acceptance speech he explained the difference between photography and holography in the following terms: 'In an ordinary photographic image important information is missing. It reproduces the intensity of the light waves but tells us nothing about their phase. Phase depends on what direction the wave is coming from and how far it has travelled from the object which is being imaged. We lose the phase if there is nothing to compare it with! Let us see what happens if we add a standard to it, a *coherent background'*.

There are some practical difficulties associated with obtaining interference patterns for light from conventional sources, and high quality holograms were produced only after the invention of the laser in 1960.

9.2.2 *The principle*

The principle of holography is illustrated in Figure 9.3 by considering the simplest case of a single point object. Let us assume that the object is at the point O. We illuminate the object with light from a reference source S (preferably a laser). The diagram shows light waves from the reference source spreading directly towards the right. Waves from S also illuminate the object at O, are reflected from O, and then overlap the waves coming directly from the reference source. The two waves combine and interfere constructively and destructively at stationary points in space.

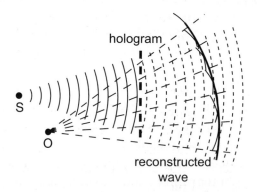

Figure 9.3 The principle of holography.

The hologram is made by placing photographic film in the area of overlap. The brightest light will be in places where the phases of the two waves are identical and there is constructive interference. When the film is developed we can make a hard positive record, so that it transmits only at these maxima. This is now our hologram. Let us now put the hologram back into the place where the film was originally.

We can now use Huygens' construction for the waves leaving the hologram. The wavelets originating at the points of transmission all start out in phase independent of their previous 'history' before they reached the hologram. We can take away the object, and illuminate it with the reference source alone. To quote once more from Gabor's speech: 'They (the phases) are right for the reference source S, but as at the slits the phases are identical, they must be right also for O, therefore the wave from O must appear reconstructed.'

9.2.3 *Making a hologram*

Light from a laser passes through a beam splitter. One half (*the reference beam*) is reflected through 90°, passes through a diverging lens and then continues on to illuminate holographic film at the right in Figure 9.4. The other half (*the object beam*) goes straight through the beam splitter and is then reflected back to another diverging lens to illuminate the object (e.g. chess pieces). The beam is reflected from the object to merge with the reference beam at the holographic plate.

In Figure 9.5 the photographic plate has been developed and the hologram is held approximately in the same place as before. The chess board and chess pieces are gone, and so is the beam splitter, leaving only the reference beam. Looking through the hologram we still see the chess pieces as a three-dimensional image. The right hand picture is a photograph of what we see in the hologram — the 'ghosts of the chess game'.

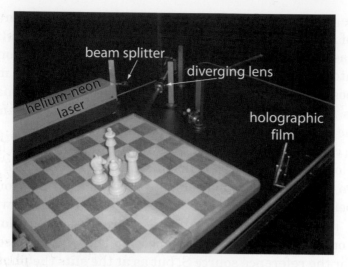

Figure 9.4 Making a hologram.

Figure 9.5 The finished product. *Courtesy of James Ellis, School of Physics, UCD, Dublin.*

The developed holographic film bears no resemblance to a 'normal' photograph. It consists of a pattern of irregular black and white fringes. These are interference fringes, which are very different from, for example, those in Young's slits experiment, because the wave fronts of the light reflected from the object are irregular.

The brightness at any place on the film is not only determined by the intensity of the reflected light but also by the path differences between the object and reference beams. The distances travelled by light from different parts of the object depend on its shape, so the interference pattern is an encoded map of the surface of the object. The image is virtual.

9.2.4 *Why does a holographic image look so real?*

As we look at the image from different angles we see different sides of the object — a complete three-dimensional picture! This is of course completely different from a photograph,

where you see only part of the image on any one piece of the film. (In Figure 9.6, the angular separation between view 1 and view 2 is about 45°; in practice the separation is considerably smaller.)

The holograms we have described are transmission holograms, made by passing the object beam and the reflected beam through the film in the same direction. Reflection holograms are made by passing the object beam and reference beams through the film in opposite directions. They are viewed by reflecting light from the surface of the hologram.

The first 'white light' hologram, a reflection hologram, was produced in 1947, and the first white light transmission hologram was produced in 1968. These developments brought holography to the attention of the general public.

9.2.5 *Applications of holography*

Holographic techniques are very widely used. Well-known applications are the lasers scanners used at supermarket checkout counters and the embossed images used on credit cards. Medical applications include holographic imaging of the interior of inside

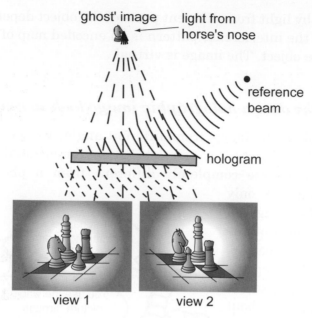

Figure 9.6 Viewing the image from different sides.

live organs using optical fibres and three-dimensional CT (computed tomography) scanning. Recent years have seen the development of materials for holographic storage of computerised data.

The Lindow man

Holograms may be used to store information about very fragile objects, such as museum artifacts. In 1984, the remains of an Iron Age man were found in Lindow Moss (a peat bog in Cheshire, England). Holographic images were used to construct a model of his original features.

Holograms from other waves

Holographic imaging is not restricted to light waves. Any electromagnetic radiation and even sound waves can also be used to make holograms. Ultraviolet and X-ray holograms have higher resolution than visual holograms.

Chapter 10

There Was Electricity, There Was Magnetism, and Then There Was Light...

In our universe there exist certain phenomena which involve a mysterious 'action at a distance'. The best-known of these is gravitation, which Isaac Newton showed to be governed by one universal law of attraction.

Electric charges also exert forces on one another, but these can be both attractive and repulsive. If the charges are moving relative to one another, a new force appears which is governed by a more complicated set of laws.

It took the genius of James Clerk Maxwell to put these laws together and to show that they lead to the propagation of *electromagnetic waves,* which we now know criss-cross the universe at a certain fixed speed, precisely as Maxwell had predicted.

10.1 The mystery of 'action at a distance'

10.1.1 *The gravitational force*

We begin with a topic which we have already dealt with in Chapter 5, apparently far removed from electricity, the relevance of which will become clear shortly. Objects fall because a mysterious force called gravity attracts them to the earth. As a child learns to walk it learns the art of balancing to combat this force. Familiarity may have taken away the sense of mystery, but the puzzle remains: How is it possible for things to be attracted to the earth without a visible link or connection? It is the first example we meet of a number of curious phenomena in nature of *action at a distance*.

Isaac Newton recognised gravity as a fundamental law of nature. A logical argument further led him to assume that as the separation of the masses increased, their mutual attraction would decrease in proportion to the *square* of the distance. This led him to the formula, to be confirmed by experiment, for the force of attraction between two masses m_1 and m_2 separated by a distance d:

$$F = \frac{Gm_1m_2}{d^2} \quad \text{(Newton's universal law of gravitation)}$$

where G is a constant which has the *same value* everywhere in the universe.

'*Hypotheses non fingo*' — Newton did not wish to speculate on any philosophical implications of such 'action at a distance', In Book III of his *Principia* he acknowledges:

> '*But hitherto I have not been able to discover the cause of these properties of gravity...I frame no hypotheses...it is enough that gravity really exists and acts according to the laws we have described.*'

As might be expected, Newton had no shortage of critics. In a letter to Gottfried von Leibnitz (1646–1716), Christiaan Huygens wrote: '*I am far from satisfied nor do I feel happy about theories built on his (Newton's) principle of attraction, which to me seems absurd.*' The French philosopher René Descartes (1596–1650) expressed his reservations poetically: '*We must assume that...these souls of material particles are endowed with knowledge of a truly divine sort, so that they may know without any medium what takes place at very great distances, and act accordingly.*'

Leaving aside philosophical reservations, the fact remains that every piece of matter, merely by its existence, exerts an influence which permeates the space around it and attracts other matter. This influence decreases gradually with distance, but never completely dies away. We call it *gravitation*.

10.1.2 *The electrostatic force*

The ancient Greeks discovered that amber rubbed with fur attracts little pieces of straw. This phenomenon we now call *electrification*, derived from ελεκτρον, the Greek word for

'amber'. The Greeks could hardly have realised what they were seeing was a tiny clue to a hidden force of enormous dimensions. The source of this force we call *electric charge*. This force also acts at a distance. It is much stronger than gravitation — in fact, stronger by a factor of about 10^{36}!*

A major difference between the electrical force and the force of gravitation is that while there is only one kind of matter, and the force of gravitation is always attractive, there are two kinds of electrical charge. We call them *positive* and *negative*. (These names are purely conventional; we could equally have called them 'black' and 'white', or any two names which denote opposites.) Opposite charges attract while like charges repel. This is the force which holds atoms together as the electrons which carry negative charge are attracted to the positively charged protons in the nucleus. The same force is also responsible for the chemical binding of molecules, with the result that matter is a

---

* The strength of interaction can be represented by a number called *the dimensionless coupling constant*, which has the value $1/137 = 7.3 \times 10^{-3}$ for the *interaction* between charges, compared to 5.3×10^{-39} for the gravitational interaction.

tightly bound mixture held together by electrical forces. Electric forces are in action 'behind the scenes' in all chemical and bio-chemical processes.

The balance of positive and negative charges is so perfect that we cannot see the huge electrical tensions within a piece of matter. The positive charge of the nucleus of an atom is electrically neutralised by negatively charged electrons. 'Electrified' materials have a tiny surplus of electrons which results in net negative or positive charge (less than one charge in 10^{10}), and as a result the electric forces 'are brought out into the open'.

Quantitative electrical experiments began in the 18th century, and various theories on the nature of the phenomena were put forward. In America, Benjamin Franklin (1706–1790) carried out spectacular and dangerous experiments during thunderstorms, which led him to recognise lightning as an electrical effect. Joseph Priestley (1733–1804), an English clergyman and schoolteacher, had met Franklin when he was in England, and they became great friends. Priestley carried out more precise if less adventurous experiments using a small pith ball and an electrified metal cup. He found that when the pith ball was suspended close to the outside of the cup, it experienced a force of attraction, but it experienced no net force when it was suspended anywhere inside the cup. This reminded him of Newton's calculations in relation to the force of gravity, and led him to infer that there was a similarity between electrical and gravitational phenomena. He concluded that *the attraction of electricity is subject to the same laws as gravitation, and therefore according to the inverse squares of the distance'*. Henry Cavendish (1731–1810) confirmed Priestley's hypothesis with a similar but more precise experiment, but he was an extremely shy man and did not publish his results, which only became known many years after his death.

10.1.3 *Coulomb's law*

Charles Coulomb (1736–1806) is credited with the definitive verification of the hypothesis of Priestley and Cavendish, and showed that the force between two electrically charged spheres varies inversely as the square of the distance separating them. He designed a very sensitive torsion balance which could measure forces as small as 'a ten thousandth part of a grain'.[†] He established the expression, which bears his name, for the law of force between two stationary electric charges of magnitude q_1 and q_2:

$$F = \frac{kq_1q_2}{d^2} \quad \text{(Coulomb's law)}$$

where k is another universal constant, and q_1 and q_2 are the two electric charges.

The numerical value of k depends on the units used. If we use the standard units for force and distance, and express the charges q_1 and q_2 in coulombs, a unit which we will define later, then k has the value

$$k = \frac{1}{4\pi\varepsilon_0} = 8.99 \times 10^9 \text{ N} \cdot \text{m}^2/\text{C}^2$$

The constant $\varepsilon_0 = 8.85 \times 10^{-12} \text{ C}^2/\text{N} \cdot \text{m}^2$ is known as the *permittivity of free space*.

Electric charges exert forces on one another across empty space. We have 'action at a distance', which, just as in the case of gravitation, presents a major puzzle. Again, in the words of Descartes, one might ascribe to electric charges 'knowledge of a truly divine sort of what is taking place at a distance'.

As the wise owl observes, Coulomb's law does not tell the whole story of electric forces. The law applies only when charges are stationary. The set of phenomena associated with charges at

---

[†] The *grain* was conceptually the mass of a grain of wheat, originally defined in France as just over 50 milligrams.

rest forms the subject of *electrostatics*. A further complication comes into play when charges are moving. This complication turns out to be very important — without it there would be no light, no universe, and we would not exist! It is a central feature of the subject of *electrodynamics* (Section 10.4).

10.2 'Fields of force'

10.2.1 *Vector fields*

A representation of gravitational and electrical forces which circumvents some of the conceptual difficulties associated with action at a distance is that each particle of matter and each electrical charge influences the surrounding space, setting up a *'field of force'*. Matter, simply by its presence, creates a *gravitational field*, while an electric charge similarly sets up an *electric field*. These fields spread out continuously from the source in all directions. If a second particle of matter is located in the gravitational field created by the first particle, it experiences a force, and the same is true for a second charge in an electric field. At the same time, gravitational fields have no effect on electric charges, nor are uncharged particles of matter sensitive to electric fields. It is assumed that such fields are present whether or not there is anything there to experience them.

We can define the *strength of an electric field* $(\vec{E})$ as being proportional to the force exerted by the field on a unit positive charge at a given point:

$$\vec{F} = q\vec{E}$$

Force is a vector quantity with direction as well as magnitude. Hence gravitational and electric fields are *vector fields*.

To help us visualise the field, we might draw the field vector at a sample number of points, where the strength of the field could be represented by the length and thickness of the vector. Such a picture is easy to interpret and is often used to describe winds and currents on weather charts.

10.2.2 *A picture to represent a physical law*

There would be an extra advantage if one could describe diagramatically the physical laws which govern a field. Figure 10.1 may describe which way the wind is blowing or map out the strength of currents in a body of water. It does not give information on the gas laws or on the laws of hydrodynamics. What we need is a picture which not only describes the features of the field but also tells about the physical laws which govern the field and can perhaps be used to predict consequences of these laws.

A representation of electric fields which incorporates the conditions imposed by Coulomb's law is generally credited to Michael Faraday (1791–1867). His idea was slightly different from the representation shown in Figure 10.1, in that he drew

Figure 10.1 A vector field.

hypothetical 'lines of force' instead of vectors. The tangent to the line of force at any point indicated the *direction* of the field at that point. The *density* of the lines indicated the strength of the field.

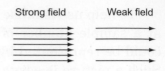

Strong field   Weak field

Faraday's convention represents the strength of the field by the density of the lines of force.

The simplest electric field, that due to a single positive electric charge, is illustrated in Figure 10.2. The lines of force spread out symmetrically. All lines start at the charge, and once we move away from the charge, no lines are created or destroyed. It follows that if we draw an imaginary sphere around the charge, every line of force will cross that sphere. Surprisingly, this simple and obvious observation leads to important physical consequences.

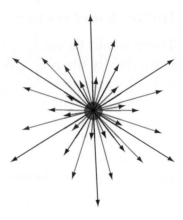

Figure 10.2 Electric field due to a single positive point charge.

Michael Faraday

We make the rule that the number of lines leaving the source charge is proportional to the magnitude of the charge. By convention the number of lines emerging from a charge q is chosen to be q/ε_0. This means that 1.13×10^{11} lines leave a charge of one unit charge (1 coulomb).

There is an advantage which arises quite naturally in Faraday's method. The surface area of our imaginary sphere *increases* in proportion to the square of the radius as the sphere gets bigger.* Therefore, as we move away from the source, the density of the lines of force crossing the spherical surface *decreases* in proportion to $1/(radius)^2$. If no new

lines appear, and no lines disappear in space, the picture tells us that the force field obeys the *inverse square law*! The above argument is not limited to the choice of a spherical surface. If we surround the charge by *any closed surface* all lines will cross it sooner or later (some might cross and re-enter, but eventually will leave again).

10.2.3 *Gauss's theorem*

To develop the diagrammatic representation of the field it is convenient to use a mathematical theorem due to Karl Friedrich Gauss (1777–1855), we define a quantity called *electric flux* φ_E through a surface as the product of the area A by the normal component of the electric field E.

When the E field is perpendicular to the surface ($\theta = 0°$), $\cos \theta = 1$, and the flux through the surface has its maximum value; similarly, the flux is zero when the field is parallel to the surface. ($\theta = 90°$ and $\cos \theta = 0$). The number of field lines

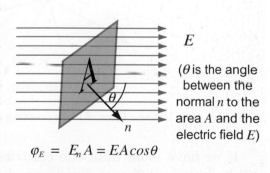

(θ is the angle between the normal n to the area A and the electric field E)

$$\varphi_E = E_n A = EA \cos\theta$$

---

* Compare Newton's reasoning p. 126 and Figure 5.3.

passing through the surface also depends on the inclination of the surface to the field. It is easy to see that it is also proportional to cos θ and therefore proportional to the flux. In the diagrammatic representation it is simple and convenient to define the electric flux φ_E as the *number of field lines through an area*, irrespective of the angle at which the field lines cross.

Gaussian surface

Imagine that a surface (of any shape or size) encloses a volume containing one or more electric charges. The *total outward flux* is equal to the number of lines crossing the surface (lines out count as positive, and lines in as negative). The angle at which the lines cross at any particular point is not relevant as far as total flux is concerned.

Such an imaginary surface is known as a *Gaussian surface*.

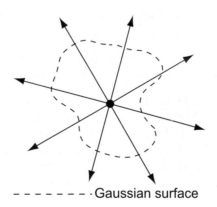

------- ·Gaussian surface

Figure 10.3 Electric lines of force crossing an arbitrary closed surface.

If we have more than one electric charge the lines of force will be bent in different directions, depending on the positions and signs of the other charges. If there are just two charges the diagram is relatively simple to draw, as in Figure 10.4.

It is not difficult to see that, if only one or two charges are involved, and if no new lines are created or destroyed, the net number of lines crossing the surrounding surface is proportional to

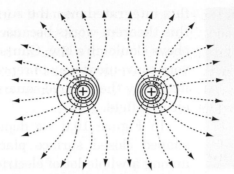

Figure 10.4 Electric field due to two equal positive electric charges. The total flux crossing any surface which surrounds both charges will be twice the flux due to either charge.

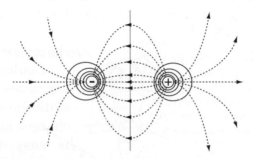

Figure 10.5 Electric field due to two equal but opposite electric charges. All field lines begin on the positive charge and end on the negative charge, including those which appear to be going towards infinity. Such lines eventually come round and finish on the negative charge. Some lines never reach an arbitrary Gaussian surface which encloses both charges, but any line which exits through the surface will eventually come back. The total flux through such a surface is zero.

the algebraic sum of the charges inside. It is not quite so obvious if we have many charges scattered randomly, some positive, some negative. It would not be practical to draw the field lines — they would look like a pile of spaghetti! Gauss showed that if we surround *any number of charges, distributed in any fashion* in space, *by any closed surface*, the total net flux through that surface is proportional to the net total charge enclosed. Stated as an equation, $\int E_n dA = q/\varepsilon_0$, where the left-hand side expresses the

Karl Friedrich Gauss
(1777–1855)

flux integrated over the surface. Note that this theorem works because the force field obeys Coulomb's law. Gauss's theorem is a very convenient way to represent diagrammatically the inverse square nature of the electric field.

In Figure 10.6 we depict a randomly shaped closed surface placed at random among a whole lot of electric charges, both positive and negative. Some of these charges are outside and some are inside the surface. The theorem still holds; the net flux outwards through the surface is proportional to the net total charge inside.

Coulomb's law expresses the electrostatic force between two charges in a way which is easy to visualise.

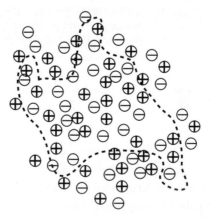

Figure 10.6 Gauss's Law: The net flux through any imaginary closed surface is directly proportional to the net charge enclosed by that surface. A physical law expressed as a diagram.

Gauss's theorem expresses the same law without reference to the distribution of charges and applies to electric fields created by any number of charges distributed in space in any random manner. It expresses an intrinsic property of electric fields without reference to any particular situation and can easily be combined mathematically with other laws governing electricity and magnetism.

10.2.4 *The energy in an electric field*

In general, if we wish to assemble a system consisting of particles which are subject to forces, we have to do work. Where there is work there is energy. For example, a mass which is raised away from the earth's surface acquires 'potential energy'. This energy does not belong to that mass alone, but to the earth–mass system, and the same applies to any systems consisting of any number of particles under gravitational interaction.

Similarly, a system of electric charges possesses electrical potential energy. An example is an electrical capacitor such as the parallel plate capacitor illustrated in Figure 10.7.

The positive and negative charges on opposite plates of the capacitor attract one another but are held apart since they cannot cross the medium between the plates. The system resembles a stretched spring and, like a spring, possesses potential energy.

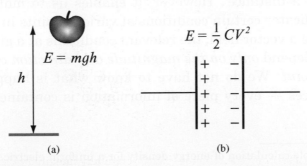

$$E = \frac{1}{2} CV^2$$

$$E = mgh$$

h

(a)   (b)

Figure 10.7 The energy of (a) an apple above the earth's surface, and (b) a charged electrical capacitor.

The energy of system (a) is due to the position of the apple relative to the earth's surface and belongs to the apple–earth system. In case (b) the energy is due to the position of the electric charges, but can we say where it is located? This question may appear to be quite trivial, but must be answered if we wish to extend the principle of conservation of energy so that energy is conserved not only over a whole system, but also *locally*, i.e. at every point in space.

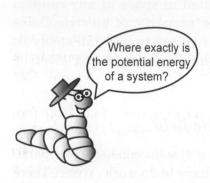

Local conservation demands that if energy disappears from one region, it must *flow* across the boundaries of that region, and also that the rate of change of energy in a certain volume be equal to the total net flow of energy across the boundaries of that volume. If energy is not located at any definite point, local conservation does not make sense. We can achieve local conservation by assuming that energy is *a property of the field* rather than of the particles in the field. There is *a certain energy density* at every point in the field. This concept fits nicely into the model of both gravitational and electric fields.[‡]

Of course, the idea of a field does not solve the mystery of 'action at a distance'. However, it enables us to make a map which indicates certain conditions at various points in space. In the case of a vector field, the relevant conditions at a given point in space depend *only on the magnitude and direction of a vector at that point*. We do not have to know what is happening at other places — every piece of information is contained in that vector.

[‡] For a simple calculation of energy density for a uniform electric field E, see Appendix 10.1.

10.3 Magnetism

10.3.1 *Magnetic materials*

There exists another mysterious force which has been known since the time of ancient civilisations. Pieces of iron ore, or *lodestone*, found near the city of *Magnesia*, had the property of exerting forces on one another without touching. We call this property *magnetisation*.

When a piece of lodestone is suspended so that it is free to rotate, it tends to align itself in a north–south direction. As such, lodestone was used for navigation as early as the 13th century. Petrus Perigrinus de Maricourt, in a letter to a friend in 1269, described a magnetic compass as consisting of a north-seeking and a south-seeking *pole*. He also observed that like poles repel and unlike poles attract, and was probably the first to realise that the earth itself is a magnet with its south pole located at the north geographic pole, and vice versa.

In his autobiography, Albert Einstein refers to his fascination, at the age of four, with the magnetic compass: *'There was this needle completely isolated, with an uncontrollable urge to point north!'* Clearly, the force of gravity which gives everything an 'uncontrollable urge' to fall downwards was too commonplace to young Einstein — but he was to return to gravity as an object of wonder some years later!

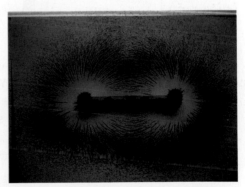

Figure 10.8 Iron filings around a magnet. *Courtesy of James Ellis, UCD School of Physics, Dublin.*

When a piece of loadstone is brought near to a piece of ordinary iron, the iron becomes magnetised. This can be demonstrated by sprinkling the surrounding area with iron filings.

The filings tend to arrange themselves 'head to toe' in a pattern of lines, as shown in Figure 10.8.

The pattern of iron filings gives a map of the magnetic field. The field can also be detected using a small compass needle. The lines of force do not have a real existence, any more than the lines of an electric or gravitational field. If we tap the surface, the filings will rearrange themselves in a similar pattern but not necessarily precisely along the same lines.

A feature of magnetism is that it is not possible to isolate a single magnetic pole. If we cut a magnet in two we are left with

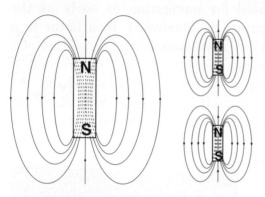

two magnets, each with a north and a south pole. Magnetic lines of force always form closed loops; unlike electric or gravitational lines, they never originate from a point source. We can imagine that each line continues on inside a bar magnet to complete a loop.

Figure 10.9 Magnetic loops.

Magnetic poles always come in pairs. We cannot have one without the other! Mathematically, we can express this property by adapting Gauss's theorem to magnetism, stating that no surface can enclose a single magnetic pole.

> Gauss's law for magnetism: The net magnetic flux through any imaginary closed surface is always zero.

10.4 Electrodynamics

10.4.1 *Electric currents*

So far we have discussed only electric charges at rest. However, if an unbound charge is introduced into an electric field it will move.

$$\longleftarrow E$$

'checkpoint'

Figure 10.10 The magnitude of an electric current is the total net charge passing a point per unit time.

The figure below depicts schematically an electric current as the net movement of electric charge in a given direction. It can be positive charges moving in the direction of the electric field, or negative charges in the opposite direction, or a mixture of the two.

10.4.2 *Ampère's discovery*

André Marie Ampère
(1775–1836)

In 1820, André Marie Ampère made the surprising discovery that moving charges produce a new force which is not there when the charges are stationary. This force manifested itself in a very simple experiment where two electric currents were running parallel to each other. Ampère noticed that two current-carrying wires attract each other when the currents are running in the same direction and repel each other when their directions are opposite! There is no force before the currents are switched on, or if current flows in one wire only.

The two wires are electrically neutral whether the current flows or does not flow. Ampère could see no reason why positive and negative charges should not remain delicately balanced at all times, whether moving or stationary. Something happens when charges move — nature has given us yet another law!

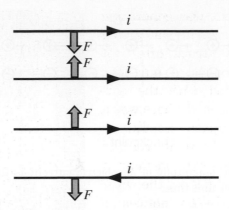

Figure 10.11 Parallel currents attract and anti-parallel currents repel.

10.4.3 *Definition of electrical units*

The SI unit of current, the ampere, was defined in 1946 on the basis of Ampère's experiment. The ampere is *the constant current in two straight parallel conductors of infinite length and negligible cross-section, placed 1 m apart in a vacuum, which will produce between them a force of 2×10^{-7} newtons per unit length*. All other electrical units are derived from this definition; for example, the coulomb is the amount of charge which passes a point per second when a current of 1 ampere is flowing (see Appendix 10.2).

> Nature's elementary charge carrier, the electron, has a negative charge: $e = 1.6 \times 10^{-19}$ coulombs.

10.4.4 *Oersted's discovery*

At about the same time as Coulomb was making his discoveries, a Danish scientist, Hans Christian Oersted (1777–1851), was also experimenting with electrical currents. Oersted gave a public lecture in April 1820 in which he demonstrated that an electric current could be produced in a wire by means of a battery. The wire began to glow as its temperature increased, showing that the current was generating heat and light. As he was giving the lecture, it struck him that there might be some magnetic effect associated with the current. As it happened, there was a magnetic compass needle to hand, which he brought over and placed under the wire. Perhaps it would align itself along the wire? Apart from a barely noticeable sideways twitch, nothing much seemed to happen. The audience became a little restive, so Oersted put aside his equipment and con-

Hans Christian Oersted

tinued with the lecture. He had broken the cardinal rule that one should never try a public demonstration of anything which has not been tried beforehand!

Later, in the privacy of his laboratory, he tried again. This time he used a stronger current and the effect on the needle became more visible. It turned in a direction *perpendicular* to the current. Oersted could only come to one conclusion: there was a connection between electricity and magnetism. It became

clear that a fundamental discovery had been made, probably for the first and last time in the history of science, in front of a live audience!

In the months after Oersted's discovery there followed great activity in the study of the magnetic effect of electric current. Ampère refined Oersted's experiments and found that the magnetic lines of

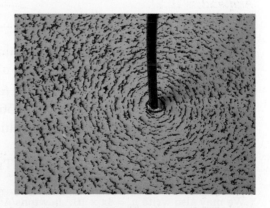

Iron filings around a current-carrying wire. *Courtesy of James Ellis, UCD School of Physics.*

force formed circles around the current-carrying wire in a plane perpendicular to the wire.

Two French physicists, Jean-Baptiste Biot (1774–1862) and Felix Savart (1791–1841), devised a mathematical formula for the magnitude and direction of this magnetic field in the vicinity of a current-carrying wire. According to the Biot–Savart law, at the point P, the magnitude of the contribution to the magnetic field dB due to the section dl of wire carrying a current i is

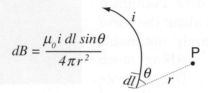

$$dB = \frac{\mu_0 i \, dl \, \sin\theta}{4\pi r^2}$$

Figure 10.12 The Biot–Savart law.

Applying this formula to an 'infinitely long' straight wire shows that a current i in such a wire gives rise to a magnetic field which can be represented by a series of rings which encircle the wire, as illustrated below. The magnitude of the magnetic field B at a perpendicular distance r from the wire is

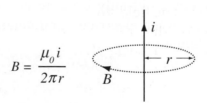

$$B = \frac{\mu_0 i}{2\pi r}$$

The magnetic field is measured in *teslas,* (T) a unit named after the Serbian scientist Nikola Tesla (1856–1943). μ_0 is called the *magnetic constant,* or *permeability.* Its value in SI units is by definition

$$\mu_0 = 4\pi \times 10^{-7} \; \text{Tm/A} \; *$$

---

* We may also write $\mu_0 = 4\pi \times 10^{-7}$ newtons/A²; a summary of the logical order in which the units of current, charge and magnetic field are defined is given in Appendix 10.2.

It is interesting to note that, while the contribution of each segment of current decreases as the *square* of the distance to the point P, the magnetic field B produced by the whole infinite wire decreases as $1/R$. The direction of B is at all times in a plane perpendicular to the current. This somewhat complicated relation between cause and effect results in magnetic field lines which form the aforementioned concentric circles around the wire, just as the iron filings would suggest.

10.4.5 *Ampère's law*

When Ampère heard of Oersted's discovery, he immediately returned to his work on electric currents, and devoted his attention to developing a mathematical theory of magnetic phenomena. Magnetic field lines surrounding a current are quite different from electric field lines, in that magnetic lines form a closed loop with no beginning and no end, whereas electric lines always begin and end on electric charges. As we saw above, there was no evidence for single isolated magnetic poles acting as sources of magnetic field lines, unlike electric charges which act as sources of electric field lines.

Ampère continued the argument in that, if isolated magnetic poles existed, one such pole, placed near a current-carrying wire, would experience a force which would drive it around the magnetic loop for ever. What is more, the force would be tangential, causing a continuous increase in angular velocity, which would continue for ever and be difficult to reconcile with the principle of conservation of energy!

A bar magnet, which of course does exist, will experience different forces on its north and south poles, depending on its orientation. Again, from energy conservation, one could form the hypothesis that no matter what its orientation, *no mechanical work is done in bringing a magnet along any path completely around the current*. The work done on the individual north and south poles must be equal and opposite.

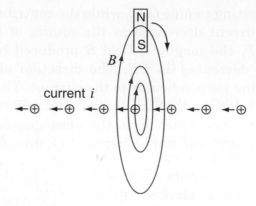

Figure 10.13 Bringing a bar magnet completely around a current.

Since for any circular path $B \propto 1/R$ and the path length $\propto R$, the work done in bringing the north end of the magnet around the large circle is exactly cancelled by the work done by the south end around the smaller concentric path.

The work done for a complete lap around any circle is the same. Since the hypothesis is not confined to a circular path, it can be generalised by stating that the work done in bringing a (imaginary) unit magnetic pole *around any closed loop* is the same and is directly proportional to the current i. The proportionality constant is the magnetic constant μ_0, which we have just met in the Biot–Savart law.

Numerous experimental tests have confirmed this hypothesis and can be stated now as a law written as follows:

$$\int \vec{B} \cdot \vec{ds} = \mu_0 i \quad \text{(Ampère's law)}$$

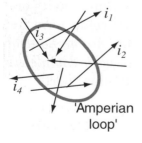

The law can be extended to apply to a magnetic field created by *any* number of electric currents going in random directions. Imagine current-carrying wires criss-crossing through space in all directions. If $i_{\text{total}} = i_1 + i_2 + i_3 + i_4 + \cdots$ is the algebraic sum of these currents passing through the loop, then

$$\int B \cdot ds = \mu_0 (i_{\text{total}})$$

(Any currents which pass outside the loop do not contribute to the integral.)

This formula is similar to Gauss's law in electrostatics except that, instead of an arbitrary *Gaussian surface* surrounding electric charge, we draw an arbitrary *Amperian loop,* and instead of *total enclosed charge inside the Gaussian surface,* we have *total current through the Amperian loop.*

André Ampère.
Courtesy of La Poste, Monaco.

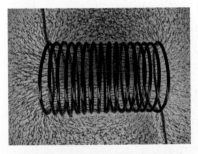

In practice the most efficient way to create a magnetic field is to send a current i through a coil of many turns. The magnetic field thus created looks just like the field produced by a bar magnet, as can be seen from the pattern of iron filings in the diagram on the left.

Ampère's law was given a stronger footing by a mathematical theorem due to George Stokes, an Irish mathematician who held the post of Lucasian Professor of Mathematics at Cambridge from 1849. Stokes' theorem applies to all vector fields and equates a path integral called *the circulation* around any closed loop to an area integral of *vector flux* over any surface bounded by that loop. Applying Stokes' theorem to electromagnetism we get a relationship identical to Ampère's law.

George Stokes
(1819–1903)

10.4.6 *The effect of a magnetic field on an electric charge*

The fact that electric currents create magnetic fields establishes a link between electricity and magnetism. It could perhaps be expected that magnetic fields in turn should exert a force on electric charges. But the experimental evidence does not support this idea:

> A stationary electric charge is not affected by a magnetic field.

This is by no means the whole story. Such immunity of charge to magnetism is not complete, and applies only if the charge is 'sitting still' in the magnetic field!*

Ampère's discovery that parallel currents repel one another and anti-parallel currents attract one another provided the first clue that something new happens when electric charges *move*. The mechanism for the force between currents can be formulated as follows:

The moving charges in one conductor create a magnetic field represented by circular magnetic lines of force around that conductor. The charges in the second conductor are moving in that magnetic field, which in this case is perpendicular to their direction of motion, and causes them to be subjected to a force. Similarly, the moving charges in the first conductor are affected by the magnetic field from the current in the second conductor. In each case the charges experience a force which is at right angles to their direction of motion, as illustrated in Figure 10.14.

We can now formulate a law: To experience a force from a magnetic field a charge must be *moving* in a direction *at an angle* to the magnetic lines of force. This force is perpendicular

---

* We will not try to define the meaning of 'stationary' at this point.

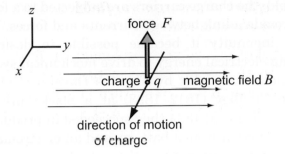

Figure 10.14 The force on an electric charge moving through a magnetic field.

to the direction of the magnetic field and also perpendicular to its direction of motion.

Figure 10.14 illustrates a *positive* charge q moving *outwards* in a direction perpendicular to the page (the x direction). The magnetic field is in the y direction. The resulting force then acts in the z direction.

10.4.7 *Electromagnetism*

The year 1820 marked a milestone in the unification of electricity and magnetism, which up to then had been considered as two independent subjects. The discoveries of Oersted and Coulomb provided the first clues that not only are electricity and magnetism interlinked in a complex and fascinating way, but they are basically two manifestations of the same phenomenon of nature. The laws of *electromagnetism* are subtle and complicated, and the representation of electric and magnetic lines of force paints a picture which illustrates the phenomenon.

Invention of the electric motor

On the practical side, it soon became apparent that the newly discovered laws point to a method for turning electrical energy into mechanical energy of motion. Electric currents make magnetic fields. When an electric current flows through a magnetic field, the charge carriers are subjected to a force. These two facts provide a link between currents and forces. With some mechanical ingenuity it became possible to design electric motors, using electrical energy to drive mechanical systems. The first such motor was built by Thomas Davenport (1802–1851) and Orange Smalley (1812–1893) at a blacksmith's shop in Vermont, New York. In 1834 they succeeded in producing rotary motion using current from a battery, and an electromagnet powered by the same battery. In other words, the same current which produced the magnetic field was subsequently acted upon by its own magnetic field!

At that time the generation for mechanical power from electricity did not seem important. This world-changing invention was not yet a commercial success.

10.4.8 *The interaction between moving charges*

When electric charges move, they exert forces on one another over and above the electrostatic force. The directions and magnitudes of these forces depend on the speed and direction of motion of each charge and on their relative positions at a given instant.

The concept of a magnetic field makes it much easier to describe these forces, and magnetic lines of force are particularly

Fig 10.16   Charges in motion.

useful in visualising the laws which govern the interaction of moving charges. One must remember that the lines of force and indeed the whole concept of magnetism are only a tool to help the description. The basic entities are electric charges at rest and in motion. There are no such entities as 'magnetically charged particles'. Magnetism has no reality of its own without electricity.

10.5 Getting electric charges to move with the help of magnetism

10.5.1 *Faraday's discovery*

The discoveries of the 1820s prompted speculation that since electric currents created magnetic fields, could magnets in turn be used to make electric fields? If so, the electric field created by a magnet could exert a force on electric charges, causing them to move and creating an electric current.

The simplest experiment was to place a magnet beside a piece of wire — perhaps it would produce a current in the wire? Experiments using the strongest available magnets failed to produce the slightest hint of a current.

In subsequent attempts the magnet was replaced by a wire carrying the strongest possible current. Would the current in the first wire somehow 'drag' even the smallest current parallel to it in the second

strongest possible
steady current

no charges 'dragged along'
in neighbouring conductor

wire? Again, no matter how large the current in the first wire was, and no matter how close they were together, nobody could detect the slightest movement of charge in the wire.

In 1840 Michael Faraday (1791–1867) realised that what had been missed was an essential feature. A magnetic field does not, simply by its presence, create an accompanying electric field. Electric effects appear only if a magnetic field is changing. When, rather than using a steady current, he decided to change the current in the first wire, a surprising thing happened — a current

suddenly appeared in the second wire! We now call this phenomenon *electromagnetic induction*, where the current in the second wire is an *induced current*. By stopping and starting an electric current Faraday produced a current in the neighbouring wire, which was also stopping and starting.

To make a quantitative statement about electromagnetic induction it is convenient to define *magnetic flux* φ_B in a similar manner to the way we defined *electric flux* φ_E. Again the name *'flux'* does not imply actual physical movement. *Magnetic flux* is a measure of the number of magnetic field lines crossing an area.

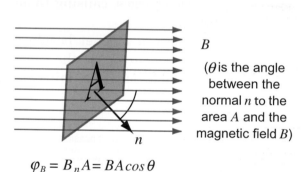

B

(θ is the angle between the normal n to the area A and the magnetic field B)

$$\varphi_B = B_n A = BA \cos \theta$$

There are many ways of inducing current. We could change the magnetic field by moving a magnet towards a loop of wire (Figures 10.15 and 10.16). As we move the magnet, there is an increase in magnetic flux through the loop, and a current appears in the wire. This current is present as the flux is increasing, but stops when the flux reaches its maximum value. If the magnet is then moved away from the loop, the current flows in the opposite direction. The direction of the current depends on whether the flux through the loop is increasing or decreasing. The magnitude of the current depends on the speed at which the magnet is moved.

Rotating a coil in a magnetic field is another, more practical, way to produce induced current (Figure 10.17). The current alternates at the rate of two cycles for every complete revolution of the loop. When the coil stops rotating the current stops.

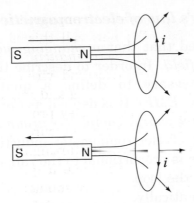

Figure 10.15 Inducing a current by changing the magnetic flux through a loop.

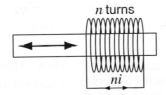

Figure 10.16 The current is increased n times.

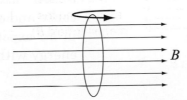

Figure 10.17 The principle of the dynamo.

To create larger currents a rotating coil can be used instead of a single loop. Such a system is known as a *dynamo*.

It is interesting to note that in the dynamo the role of magnetism is passive. Its presence, however, makes it possible to change the mechanical energy of the rotating coil into the energy of an electric current. Work has to be done to keep the coil rotating — the principle of conservation of energy ensures that we cannot get something for nothing!

10.5.2 *Faraday's law of electromagnetic induction*

Faraday discovered that that *a changing magnetic field produces an electric field*. In order to describe this fact quantitatively, it is necessary to define a quantity called the *electromotive force (EMF)*. It is defined as *the work required to bring a unit positive charge completely around the loop*.

Faraday's law of electromagnetic induction: The induced electromotive force is proportional to the rate of change of magnetic flux through the loop.

Stated mathematically,

$$\text{EMF} = \int \vec{E} \cdot \vec{ds} = \frac{-d\varphi_B}{dt}$$

That little minus sign saves the universe from self-destruction!

(The negative sign indicates that the direction of the induced E field is such as to oppose the change in magnetic flux. If this were not so, a self-sustaining chain of cause and effect would result in an exponential growth of energy without limit.)

10.6 Maxwell's synthesis

10.6.1 *Putting facts together*

It is hardly surprising that, by the middle of the 19th century, the experimental results involving electrical and magnetic phenomena presented a complex and apparently disjointed picture. There seemed to be a lot of independent laws involving electric fields and currents, magnetic fields and induced currents. These were often quite complicated, with no logical connection between one law and another. In the year 1864, James Clerk Maxwell (1831–1879) put all the laws together and showed that *in combination* they form a beautiful unified picture. Maxwell realised that in essence there are only four laws in total: *Gauss's theorem* for electricity, *Gauss's theorem* for magnetism, *Ampère's law* and *Faraday's law*.

At that stage all four laws were very well established and had been confirmed by experiment. Maxwell expressed them as simultaneous equations — mathematical statements of facts which are all true at the same time — and could now use the power of mathematics to explore the consequences which follow from the four laws in combination.

The consequences were as surprising as they were dramatic. The solution of the equations showed how the interplay of electricity and magnetism gives rise to an electromagnetic signal which propagates at an enormous speed and provides a means of communicating information and transporting energy even across billions of miles of empty space. *A new chapter had opened in the history of physics, with practical consequences of unimagined dimensions.*

10.6.2 *An important extension to Ampère's law*

Maxwell soon realised that one of the laws, namely Ampère's law, as given by equation (10.1), is incomplete. This equation correctly quantifies the magnetic field produced by an electric current, i.e. by moving electric charges. Maxwell pointed out that it is also possible to have a magnetic field without physically moving electric charges. We can most easily understand why this is so by looking at what happens during the process of charging an electrical capacitor:

A capacitor is a device which stores charge. In its simplest form it consists of two parallel plates separated by a certain distance. Figure 10.18 depicts an electric circuit consisting of a battery, a switch and a capacitor. When the switch is closed a current flows, but the electric charges cannot cross the gap between the capacitor plates. Positive and negative charge build

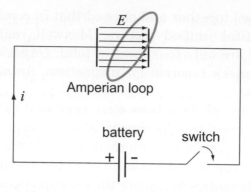

Figure 10.18 As a capacitor is being charged an increasing electric field E appears between the plates.

up on opposite plates of the capacitor until the capacitor is fully charged.

While the current is flowing it produces a magnetic field, with magnetic lines of force encircling the wire according to Ampère's law. But are there any magnetic effects around the region where there is no actual flow of charge, i.e. around the central part of the capacitor? We can investigate this experimentally with a compass needle, but we can also make a logical deduction. The argument which forms the basis of Ampère's law does not depend on the size or shape of the loop. If an *Amperian loop* is drawn arbitrarily around the circuit, the integral of the magnetic field around the loop (recall $\int B \cdot ds = \mu_0 i$) is equal to the current passing through *any surface* bounded by the loop. We could choose a surface which intersects the current-carrying wire, or a surface within the capacitor, where there is no current. It follows that whatever is happening within the capacitor must equally contribute to the making of the magnetic field.

Maxwell noted that while there is no current through the capacitor, there is a *changing electric field E*. This results in a changing electric flux φ_E through any *Amperian loop* around the capacitor. He proposed that this changing electric flux had the same magnetic effect as the flow of electric charge, and generalised Ampère's law by adding a second term of the form $\varepsilon_0 d\varphi_E/dt$ to the right-hand side of the equation. He called this term the *displacement current* (i_d). This is not a true current in the sense that there is no flow of charge, but has the same effect (and the same dimensions) as a physical current. The generalised form of Ampère's law now becomes:

$$\int \vec{B} \cdot \vec{ds} = \mu_0(i + i_d) = \mu_0 i + \mu_0 \varepsilon_0 d\varphi_E/dt$$

The value of the integral on the LHS is independent of the location of the *Amperian loop* and equals μ_0 times the sum of the charge current plus displacement current threading the loop at that instant.

Maxwell's extension does much more than add an extra term to a certain equation. It states that a magnetic field can be produced in the absence of moving charges. It can also be created by an electric field which is changing with time. The charges which originally created the electric field no longer play a role in the process, which continues without them. The fields take on a new reality.

A changing electric field is all we need.

Maxwell's fundamental work *Treatise on Electricity and Magnetism* was first published in 1873 and can be ranked alongside Newton's *Principia* as one of the cornerstones of a new scientific era.

10.6.3 *The four laws*

1. *Coulomb's law/Gauss's theorem*

$$\int E_n dA = \frac{q}{\varepsilon_0}$$

The electric flux through any closed surface = net charge inside.

2. *Magnetic lines of force always form closed loops*

There are no single magnetic poles. Expressed in Gaussian form

$$\int B_n dA = 0$$

Magnetic flux through any closed surface = zero.

3. *Ampère's law (plus Maxwell's extension)*

$$\int \vec{B} \cdot \vec{ds} = \mu_0 i + \mu_0 \varepsilon_0 \frac{d\varphi_E}{dt}$$

An electric current or a changing electric field produces a magnetic field.

4. *Faraday's law*

A changing magnetic flux produces an electric field:

$$\int \vec{E} \cdot \vec{ds} = \frac{-d\varphi_B}{dt}$$

10.6.4 *As we turn on a current ...*

Maxwell's equations presented the laws of electromagnetism in quantitative form, and made it possible to use the powerful techniques of mathematics to draw logical conclusions based on the combination of diverse experimental data. Maxwell showed that the solution of the equations predicted that an electromagnetic signal, once initiated, will propagate at a certain fixed velocity.

Rather than solving the equations formally, it is perhaps more instructive to consider each equation in turn, and show that a remarkable chain of events is set in motion by a simple act such as switching on an electric current.

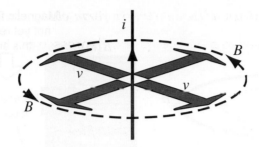

Figure 10.19 Magnetic field around a current-carrying wire.

As soon as we turn on the current a magnetic field (B) appears which encircles the wire, anti-clockwise as we look at it from above. The diagram depicts the situation where we have a steady current i flowing in the wire. If we go back to the moment when the current was switched on, there was at first no current and no magnetic field. Then, as the current started to flow, there was still no magnetic field at points removed from the wire. We can deduce from Faraday's law that *the magnetic field cannot appear instantly.* If it did, there would be an instantaneous change of the magnetic field from zero to its final value everywhere, resulting in an infinitely large rate of change of flux, $d\varphi_B/dt$, in all areas of the surrounding space, which in turn would induce an infinitely large electric field E. We have no option but to conclude that the creation of the B field can only *spread out from the wire at a finite speed (v).*

To make the discussion simpler, let us assume that the magnetic field created by the current is uniform. (Such a uniform B field would result if we had an infinitely large moving sheet of charge instead of the electric wire.) The rate of propagation of the magnetic field is the same in the two cases.

10.6.5 *The propagating magnetic field*

The moment the electric current is switched on, the resulting magnetic field spreads out in all directions like a tidal wave. In Figure 10.20 a segment of the leading edge of the magnetic field, moving from left to right, has reached a certain point.

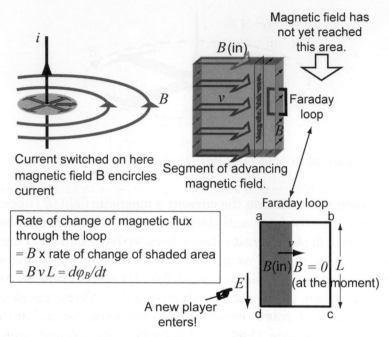

Figure 10.20 Tidal wave of magnetism.

To the left there is now a magnetic field in the direction perpendicular to the page; to the right there is none. The 'tidal wave' has not yet reached beyond that point.

Let us place a *Faraday loop* at some point on the advancing edge of the 'magnetic tidal wave'. (Lower half of Figure 10.20) We can make the loop as small as we like so that we do not have to worry whether B is constant across the loop.

Faraday's law states that as the magnetic flux through the loop changes it will induce an electric field E such that

$$\int E \cdot ds = \frac{-d\varphi_B}{dt}$$

But $\dfrac{-d\varphi_B}{dt} = BvL$ *   and $\int E \cdot ds = EL$

$\Rightarrow$ $BvL = EL$ and $v = \dfrac{E}{B}$

The electric field E is zero along the side bc of the loop and is perpendicular to the sides ab and cd, and therefore contributes to the integral only along ad.

---

* See box, Figure 10.20.

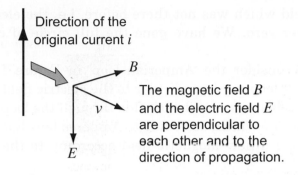

Direction of the original current i

B

v

E

The magnetic field B and the electric field E are perpendicular to each other and to the direction of propagation.

Figure 10.21 Electric field induced at right angles to the magnetic field.

We have deduced that the incoming magnetic field B generates an electric field E which is perpendicular to the magnetic field and in the direction opposite to that of the original current. In addition, by solving the simultaneous equations relating to Figure 10.20 we have obtained an expression for v. The two fields will propagate together at a certain fixed speed which is equal to the ratio of the magnitudes of the two fields:

$$v = \frac{E}{B}$$

The next step is described in Figure 10.22. A *new player* has now appeared and is contributing to the action. It is an

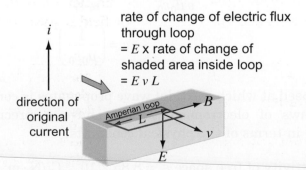

i

rate of change of electric flux through loop
$= E$ x rate of change of shaded area inside loop
$= E v L$

direction of original current

Amperian loop

L

B

v

E

Figure 10.22 Changing electric flux through an Amperian loop.

electric field which was not there before, i.e. the electric field is no longer zero. We have gone the full circle of cause and effect.

Let us consider the 'Amperian loop' in Figure 10.22. The plane of this loop is perpendicular to the electric field E. As the tidal wave advances, the electric flux through the loop increases at a steady rate which, according to Ampères law, has the effect of generating a new magnetic field according to the following equation:

Ampère's law (plus Maxwell's extension)

$$\int \vec{B} \cdot \vec{ds} = \mu_0 I + \mu_0 \varepsilon_0 \frac{d\varphi_E}{dt}$$

Considering free space, where there are no charges and no currents, $I = 0$.

But the rate of change of the electric flux

$$\frac{d\varphi_E}{dt} = EvL$$

$$\rightarrow \quad \int \vec{B} \cdot \vec{ds} = BL = \mu_0 \varepsilon_0 LvE$$

$$\rightarrow \quad\quad v = \frac{B}{E\mu_0\varepsilon_0}$$

We conclude that $v = \dfrac{1}{v\,\mu_0\varepsilon_0} \rightarrow v^2 = \dfrac{1}{\mu_0\varepsilon_0}$

$$\rightarrow v = \left(\frac{1}{\mu_0\varepsilon_0}\right)^{1/2}$$

The speed at which the tidal wave propagates is constrained by the laws of electromagnetism to have a precise value expressed in terms of the physical constants,

Permittivity of free space $\varepsilon_0 = 8.85 \times 10^{-12}$ C²/N · m² and

Magnetic constant of permeability $\mu_0 = 4\pi \times 10^{-7} \text{NA}^{-2}$ *

$$\rightarrow \quad v^2 = \frac{1}{\varepsilon_0 \mu_0} = 8.9875 \times 10^{16}$$

$$\rightarrow \quad v = 2.99 \times 10^8 \, \text{ms}^{-1}$$

10.7 Then there was light

10.7.1 *Cause and effect — A summary*

It follows from the laws expressed in Maxwell's equations that if an electric charge is accelerated it will start an endless chain of cause and effect. From the instant the charge begins to move, and while it is accelerating, it creates a *changing* electric field which gives rise to a magnetic field. We now have a magnetic field which was not there before, but according to the fourth equation, the change in the magnetic field will produce an electric field. The fields have left their source and propagate through space perpetually maintaining themselves. As one changes it creates the other, which in turn recreates the first.

As we have seen, Maxwell was able to calculate the speed at which the fields propagate in terms of the magnetic and electric constants which had been determined independently. When he put in the numerical values, the result looked familiar — it was the same as the speed of light, which had recently been measured by Fizeau and others. In his own words: '*We can scarcely avoid the conclusion that light consists in the transverse undulations of the same medium which is the cause of electric and magnetic phenomena.*'

The medium to which Maxwell was referring was called the *ether*, an invisible and odourless medium proposed to exist merely in order to conduct light. The speed of a wave depends

---

* See Appendix 8.2 for a summary of units and dimensions.

on the elasticity of the material it passes through (Section 4.2). If light travels so very fast, then this ether must be very stiff, stronger even than steel, in order to transmit vibration at that speed. Then how come the ether seems to let ordinary objects move through it with no resistance?

This was a real puzzle and was not solved until Albert Einstein developed his theory of special relativity, which finally buried the ether idea (Chapter 15). In the meantime, it was necessary to live with the rather embarrassing concept of an improbable material, the existence of which could not be experimentally verified.

10.7.2 *Making an electromagnetic pulse*

While the circuit switch in Figure 10.23 is closed a constant current flows. Around the current carrying wire there exists a magnetic field *B* (represented by concentric lines of force), as discovered by Oersted in 1820 and illustrated by iron filings in the photograph on p. 321. The magnitude of this field decreases very quickly as we move away from the wire and becomes negligible within a matter of cm.

Maxwell's discovery of the mind came in 1864. Using additional knowledge, provided in particular by Faraday (in 1840), he deduced that something quite fascinating happens *at the moment when the current starts to flow*. The concentric magnetic circles around the wire do not appear instantaneously, but expand radially like a tidal wave in increasing circles. A segment of the

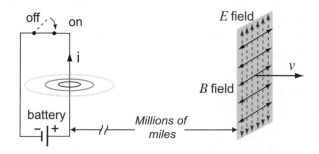

Figure 10.23 A packet of *E* and *B* fields has taken off.

leading edge of this tide is shown in Figure 10.23, consisting of crossed electric and magnetic fields which create and re-create each other in a continuous chain of cause and effect. This *electromagnetic pulse* continues to move through space even where there is no longer any trace of Oersted's steady magnetic field. The fields have taken off independently, and may now be millions of miles away propagating on their own through free space.

10.7.3 *Electromagnetic waves*

In the idealized example illustrated in Figure 10.23 the current is switched on suddenly and just once. In practice such a single sharp pulse has little significance. More interesting and relevant is the effect of a periodically alternating source of current giving rise to an *electromagnetic wave*.

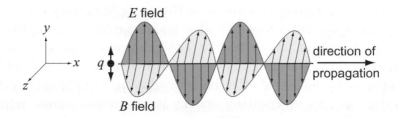

Figure 10.24 Electromagnetic wave.

The electric and magnetic fields are perpendicular to one another. They change continuously oscillating in *phase* at right angles to the direction of propagation. In the diagram above a source charge oscillates along the *y* axis. The *E* field also oscillates along the *y* direction while the *B* field vectors always point along the *z* axis.

10.7.4 *Putting theory into practice*

Maxwell's prediction of electromagnetic waves was made in 1864 and immediately there followed attempts by various physicists to produce such waves experimentally. The technology to

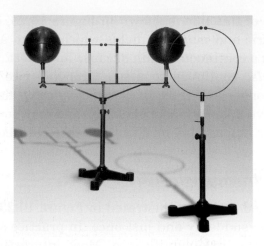

Figure 10.25 Hertz spark apparatus and detector. *Courtesy of John Jenkins, www.sparkmuseum.com.*

produce alternating currents of sufficiently high frequency was not available, and Maxwell's idea was treated with scepticism by many leading physicists. In 1887, however, more than 20 years later, the German physicist Heinnrich Hertz (1857–1894) discovered a method of effectively producing rapidly oscillating electric charges by creating sparks between two points which were at a large potential difference.

A large potential difference is created across the 'spark gap' between the two tiny spheres in the apparatus on the left of Figure 10.25. Electrons are liberated from the tips of the metallic surfaces at each side of the gap, producing an electrical discharge in the form of sparks. The sparks jump back and forth at a high frequency, creating the effect of an oscillating electric charge and producing an electromagnetic wave.

The detector consists of a wire loop with a small gap similar to the spark gap. When the electromagnetic wave reaches the circular loop it induces a current in the loop which acts as a detecting antenna, and small sparks are produced across the secondary gap. Radio communication between source and detector is established.

Hertz did not realise the practical importance of his discovery. In reply to a question as to what use might be made of the phenomenon, he is recorded to have said: 'No use whatsoever, this is just an experiment which proves Maxwell was right, we have mysterious electromagnetic waves which we cannot see with the naked eye, but they are there.'

Gugliemo Marconi (1874–1937), at the time only a teenager, read about Hertz's experiments and realised that far from being a practically useless phenomenon, Hertz's discovery could be applied to the transmission of signals. With the enthusiasm of youth and the vision of a business entrepreneur he set about developing the method and by 1895, at his father's farm in Italy, had succeeded in sending wireless signals over a

Gugliemo Marconi.
Courtesy of An Post, Irish Post Office.

distance of 2 km. By 1899 he had improved the technology sufficiently to establish wireless communication across the English Channel, and in December 1901 he sent the first signals across the Atlantic.

The US navy was particularly impressed by Marconi's inventions. Here was a method by which ships moving freely in the ocean could communicate. A problem arose as more and more messages were being transmitted simultaneously and it became difficult to distinguish individual messages. Marconi was able to overcome this by arranging that each transmitter sent waves of a particular frequency so that the receiving stations could tune in to that frequency only, thereby eliminating the background 'radio noise'. Following the tragic sinking of the *Titanic* in 1912, 24-hour wireless communication between ships at sea became mandatory.

A historical interlude: James Clerk Maxwell (1831–1879)

James Clerk Maxwell spent his childhood at the family estate Glenlair in Galloway, in the Scottish lowlands. From the age of three he showed exceptional interest in how things work, and soon his par-

ents had to answer endless questions on topics as diverse as the mechanism of a lock and the workings of the universe.

His mother was responsible for her son's early education and, under her guidance, James studied the basics of reading, writing and arithmetic, and learned and memorised passages from the Bible. By the time he was eight, she boasted that he could recite all 176 verses of the longest psalm in the Bible. This probably helped train his incredible memory; in later life he could recall information accumulated over long periods of time. This skill was apparent not only in his scientific work but also in his general knowledge on a whole range of topics. At one stage, between jobs, he was a professional memory-man in a circus.

James' mother did not live long enough to see her talented son develop. She died in 1839 and it now became more difficult to educate James at home. A tutor was employed, but this did not work out and John Maxwell had to send his son to the Edinburgh Academy. James arrived there on his first day wearing a tweed smock and square-toed shoes. With this unconventional dress and his slow, hesitant speech he was nicknamed Dafty. Used to a solitary existence, young Maxwell did not like the rough and tumble of the schoolyard and was essentially ostracised by the other pupils.

(Continued)

As the years passed Maxwell's ability and his extraordinary flair for mathematics became more and more apparent. When Maxwell was just fourteen, he completed his first scientific paper, 'On the Description of Oval Curves'. The method was original, the work probably inspired by unsuccessful attempts of a local artist to draw perfect ovals. His father showed it to John Forbes, the Professor of Natural Philosophy at the university, who had it published in the *Proceedings of the Royal Society of Edinburgh.*

After leaving school, Maxwell spent a year at Edinburgh University, where Philosophy was an integral part of the university course. The course put the emphasis on education rather than on cramming and there were no competitive examinations. His scientific training gave him the ability to approach problems with an open mind and without preconceptions, while Philosophy enabled him to peel away the superfluous parts of scientific theories, leaving only the essentials.

As a student, Maxwell had the privilege of being allowed to use Forbes' private laboratory during his undergraduate years and often assisted Forbes in experiments. He developed an interest in colour vision and was introduced to William Nicol (1768–1851) the inventor of the Nicol prism for polarisation of light. In typical fashion, young Maxwell built a crude polariser at home, did some experiments and sent watercolours of the results to Nicol, who presented him with a pair of Nicol prisms.

Maxwell moved to Cambridge in 1850. He brought a relaxed attitude towards the strict Cambridge traditions. When told that there would be compulsory church service at 6 a.m., his reaction was typical: *'Aye, I suppose I could stay up that late.'* He spent one term at Peterhouse College and then moved to Trinity College, the largest and most liberal of the colleges.

After graduating he became a Fellow in 1855 and soon afterwards obtained a Professorship at Marischal College in Aberdeen. There he was responsible for the whole of the Natural Philosophy course. The preparation of demonstrations was one of his many duties. In one demonstration he would stand a student on a platform and 'pump him

(Continued)

fu' o' electricity so his hair stood on end'. He allowed students to do their own experiments — a revolutionary innovation.

During his leisure time, Maxwell competed for and won the 1859 Adams Prize with an essay on the structure of Saturn's rings, in which he proved that they could not be gaseous but consisted of numerous small particles.

Maxwell was appointed Professor of Natural Philosophy at King's College, London, in 1860 and did much of his most important work on electromagnetism while he was there. He was very impressed by the work of Faraday, his ideas of 'lines of force' extending through space around interacting bodies. *'As I proceeded with the study of Faraday, I perceived that his method of conceiving the phenomena was also a mathematical one, although not exhibited in the conventional form or mathematical symbols'.*

In 1864, Maxwell made a presentation to the Royal Society in London, and in 1865 he published his first paper, 'A Dynamical Theory of the Electromagnetic Field', in the *Philosophical Transactions of the Royal Society.* He concludes the paper with the following observation: *'We have strong reason to believe that light itself including radiant heat and other radiations, if any, is an electromagnetic disturbance in the form of waves propagated through the electromagnetic field.'*

He left the college in 1865 and retreated to Glenlair. There he was free to do research and to rebuild the house as his father had intended.

Maxwell was persuaded out of retirement to become the first Cavendish Professor of Experimental Physics in 1871. He designed the famous Cavendish Laboratory, which opened in 1874. Maxwell's equations in the form we know them appeared in his book *Electricity and Magnetism* in 1873.

Maxwell was both a theorist and an experimentalist, an all-rounder brilliant enough to make fundamental contributions to many branches of physics. In addition to electromagnetism his name is linked with the kinetic theory of gases, colour vision, geometrical optics, and thermodynamics. Maxwell and T.H. Huxley were the scientific editors of the famous ninth edition of the *Encyclopaedia Britannica.*

(Continued)

Maxwell the philosopher had a great interest in the mechanics of the mind. In a letter to Campbell he wrote *'I believe there is a department of the mind conducted independent of consciousness, where things are fermented and decocted, so that when they are run off they become clear.'*

Maxwell died in 1879. Just after his death, David Edward Hughes discovered, almost by accident, that sparks oscillating across a loose contact in an electrical circuit created electromagnetic signals which he could detect in another room of his house in Great Portland Street, London. Soon he was even able to detect them on the street outside, and became known as 'the mad professor wandering about listening to a box by his ear'. We now realise that he was experimenting with the first mobile telephone!

David E. Hughes
(1831–1900)

On 20 February 1880, Hughes demonstrated his equipment to members of the Royal Society, who, however, were quite sceptical. Edward Stokes (1819–1903) in particular suggested that the signals were produced by induction and not by electromagnetic waves, effectively dampening Hughes' enthusiasm, with the result that his idea was never published. The first experimental production of electromagnetic waves was recognised and credited to Heinrich Hertz in 1887.

An athlete clears the bar at the Olympic Games. Electromagnetic waves which came from the sun, and happen to be reflected in the right direction, cause electrons to jiggle in a television camera. Within a fraction of a second electrons in a hundred million television sets across the globe execute a similar jiggle. The waves which entered the camera from the athlete emerge from television screens in Europe, Asia, Africa, America and Australia. We do not even need wires connecting all these sets! All

(Continued)

this is possible because the laws of Nature discovered by Maxwell have been applied in clever ways by men and women of science who followed him.

The discovery of the production and detection of electromagnetic waves has changed the history of mankind. 'Action and communicating at a distance', such as watching events on television, or directing and landing a spacecraft on a distant planet, have become commonplace. Maxwell's waves connect mobile phones, allow ships to 'see in the dark', guide aeroplanes safely to their destinations. At shorter wavelengths, X-rays penetrate matter and are used in medicine for both diagnostic and therapeutic purposes.

One can hardly express Maxwell's legacy better than in the words of Richard Feynman: *'From a very long view of the history of mankind — seen from, say, ten thousand years from now — there can be little doubt that the most significant event of the 19ᵗʰ century will be judged as Maxwell's discovery of the laws of electrodynamics. The American Civil War will fade into provincial insignificance in comparison with this important scientific event of the same decade'.*

Appendix 10.1　Energy density of a uniform electric field

Definition of electrical capacity: $C = \dfrac{Q}{V}$.

Work done in charging a capacitor C

$$dW = Vdq = \frac{q}{C}dq$$

Work done to add a charge increment dq when the charge of the capacitor is already q.

$$W = \int dW = \int_0^Q \frac{q}{C}dq = \frac{1}{C}\int_0^Q q\,dq = \frac{Q^2}{2C}$$
$$= \frac{1}{2}CV^2$$

Parallel plate capacitor

In the special case of a capacitor consisting of two parallel plates of surface area A separated by a distance d, the electric field (E) can be taken as uniform provided that the separation d is much smaller than the dimensions of the plates. It is also confined to the volume between the plates.

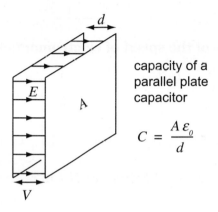

capacity of a parallel plate capacitor

$$C = \frac{A\varepsilon_0}{d}$$

Energy of the electric field $= \dfrac{1}{2}CV^2 = \dfrac{1}{2}\dfrac{A\varepsilon_0}{d}(Ed)^2$

$$= \frac{1}{2}\varepsilon_0 E^2 Ad \quad \text{(recall that } V = Ed\,)$$

Energy density $u_E = \dfrac{energy}{volume} = \dfrac{1}{2}\varepsilon_0 E^2$

Appendix 10.2 Units and Dimensions

Definition of an *ampere*, the unit of current: force per unit length between two current-carrying conductors.

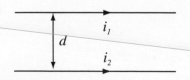

F/unit length = $\dfrac{\mu_0 i_1 i_2}{2\pi d}$

When $i_1 = i_2 = 1$ A, and $d = 1$ m, $F = 2 \times 10^{-7}$ Nm^{-1} by definition. It follows that $\mu_0 = 4\pi \times 10^{-7}$ NA^{-2} (permeability of free space).

Definition of a *coulomb*, the unit of charge: 1 A = Cs^{-1}

The dimensions of the speed of electromagnetic radiation:

Coulomb's law: $F = \dfrac{q_1 q_2}{4\pi\varepsilon_0 d^2}$

$\Rightarrow$ permittivity of free space $\varepsilon_0 = 8.854 \times 10^{-12}$ C^2 N^{-1} m^2

$\Rightarrow \mu_0 \varepsilon_0 = 8.854 \times 4\pi \times 10^{-19}$ NC^{-2} s^2 C^2 N^{-1} m^{-2}

$\Rightarrow$ speed of light $c = \left(\dfrac{1}{\mu_0 \varepsilon_0}\right)^{1/2}$ ms^{-1}

(numerically 2.998 ms^{-1})

Chapter 11

'Atoms of Light' — The Birth of Quantum Theory

The light from hot soot ...

This chapter describes a fascinating series of events in the history of science. They occurred at the beginning of the 20th century, and started quite innocuously with a disagreement between experimental evidence and theoretical prediction in the area of *'blackbody radiation'*. The distribution of colours of light from hot surfaces, such as, burning coal or molten metal, did not make sense.

Classical thermodynamics, the well-established theory of heat and heat transfer, made excellent predictions of some of the general features of the spectrum, but when it came to specific models for the process, something was decidedly wrong....

Wilhelm Wien produced a model which agreed quite well with experiment at short wavelengths, but in order to make it right at longer wavelengths, he had to 'fiddle' with the formula, or make a mathematical adjustment, for which there was no justification. Lord Rayleigh produced a more detailed model, which was right for long wavelengths, but the predicted energy of radiation shot off to infinity at the other end of the spectrum — *the ultraviolet catastrophe*.

In 1901, Max Planck made a revolutionary suggestion — an 'act of despair', as he called it. By restricting the allowed energy of light for a given wavelength, he was able to derive an exact formula, which fitted the experimental results everywhere. As a matter of fact it was just like Wien's 'fiddled' expression, but this time it was based on sound physical reasoning. Still, there was a heavy price to pay: Planck had to make the assumption that only certain levels of energy are possible for an electric oscillator, which are an integral number of quantum units *hf*.

Up to then, an inherent understanding in natural philosophy was that there are no restrictions on the possible values of physical entities. The suggestion that Nature is not continuous was so reactionary that Planck was afraid to publish his hypothesis, and it was in fact not generally accepted until the successful work of Niels Bohr in 1913. Physicists were reluctant to abandon long-established beliefs, with some notable exceptions, such as Albert Einstein, who in 1905 had developed the concept of quantisation even further in his theory of the photoelectric effect.

To follow this chapter in detail, some knowledge of basic mechanics and thermodynamics is helpful, but the reader should get a good feeling for the concepts by skimming through the more detailed parts of the discussion, particularly in Section 11.2. When you come to Sections 11.5.1–11.5.4, again skim through the mathematical discussion. If you are comfortable with the mathematics, however, you should enjoy following Planck's footsteps to one of the most fundamental discoveries in the history of physics....

11.1 Emission of energy by radiation

11.1.1 *How does matter emit electromagnetic energy?*

Matter consists of atoms and molecules containing protons and electrons, which respectively have positive and negative charge. There is always electrical activity even in an electrically neutral piece of matter, because of atomic oscillators whose motion becomes more rapid with increasing temperature. As predicted by Maxwell, and verified by Hertz in 1888, oscillating charges emit electromagnetic waves and in the process lose energy and slow down. That is one way in which a hot surface may cool down by radiating light into the space around it. Eventually, thermal equilibrium is established when the average energy emitted per second is balanced by the radiation absorbed.

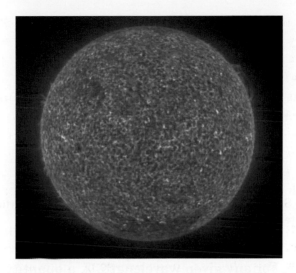

Figure 11.1 What is happening at a hot surface?

11.1.2 *Experimental results*

Experimental study of the electromagnetic radiation emitted by surfaces at high temperature relative to the surroundings, such as the sun (Figure 11.1), showed that surfaces which emit light energy most efficiently also absorb such energy completely. When all light is absorbed and none is reflected such surfaces appear black. For that reason the process is called *blackbody radiation*.

> White clothes are normally worn in hot climates because they reflect rather than absorb the heat of the sun. They also tend to retain body heat during the night. Experience tells us that it is not a good idea to wear black or dark clothes in hot sunlight.

11.1.3 *The Blackbody radiation spectrum*

The results of experimental studies of the relative amounts of energy emitted *at different wavelengths* are illustrated in

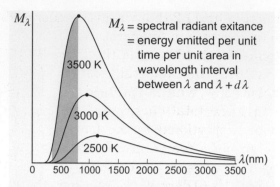

Figure 11.2 Spectral distribution of radiation from a black surface (the visible region extends from about 400–750 nm).

Figure 11.2. For any given wavelength, λ, a quantity called the *spectral radiant exitance* (M_λ) is defined which is a measure of the rate at which electromagnetic energy is radiated per unit area per unit wavelength interval at that wavelength.

The spectrum of electromagnetic energy emitted by a black surface is universal. It depends only on the temperature of the surface and is independent of the size, shape or chemical composition of the material. The figure above shows the spectral radiant exitance, M_λ, at three temperatures, as a function of the wavelength. The dimensions of M_λ are power per unit area per unit wavelength interval.

At a temperature below about 2000 K, only a small fraction of the power is emitted at wavelengths in the visible region.

11.1.4 *The Stefan–Boltzmann law*

The total power emitted, at all wavelength at a given temperature (M), is represented by the area under the corresponding curve:

$$\text{Total power emitted} = \text{radiant exitance} = M = \int_0^\infty M_\lambda \, d\lambda$$

$$= \text{area under the curve for a given}$$
$$\text{temperature } T$$

It is clear from Figure 11.2 that the area under each curve rises rapidly with an increase in temperature.

Josef Stefan (1835–1893) was the first to publish (in 1879) an empirical relation known as Stefan's law, which expressed the observation that the total energy emitted appeared to be proportional to the fourth power of the absolute temperature of the surface. About five years later Ludwig Boltzmann (1844–1906) derived the relation using thermodynamic arguments which is now known as the *Stefan–Boltzmann law*.

Josef Stefan.
*Courtesy of
Austrian Post.*

Stefan–Boltzmann law: $M = \varepsilon \sigma T^4$

M = total energy emitted per unit area per sec
ε = emissivity (= 1 for black surface)
σ = Stefan–Boltzmann constant
  = 5.67×10^{-8} Wm^{-2} K^{-4}
T = absolute temperature (K)

Ludwig Boltzmann.
*Courtesy of
Austrian Post.*

11.1.5 *Wien's displacement law; the spectral distribution*

As the surface temperature rises, the peak of the spectrum moves towards shorter wavelengths (higher frequencies). This is in line with the common experience that as a lump of coal or an iron bar or a tungsten filament gets heated, it begins to glow. Not only does it become brighter, but the predominant colour of the light changes from deep red to bright yellow to 'white-hot'.

Experiment shows that the relationship between the temperature and the wavelength at the peak of the curve follows a
simple relation known as *Wien's displacement law,* which states
that the higher the temperature the lower the wavelength at
which radiation is maximum. We can see this in Figure 11.2.
At 2500 K the maximum occurs at about 1160 nm, moving to
970 nm at 3000 K and 830 nm at 3500 K.

$$\text{Wien's displacement law: } \lambda_m T = \text{Constant}$$

λ_m = wavelength at which M_λ is maximum
T = absolute temperature

The value of Wien's constant is found by experiment to be 2.9×10^{-3} mK.

Wien's law allows us to calculate the temperature of an
object from the colour of the emitted light. For example, the
spectral radiant exitance of the sun is maximum at a wavelength of about 500 nm. Hence we can calculate the temperature
of the surface of the sun as follows:

$$\lambda_m T = 2.9 \times 10^{-3} \Rightarrow T = \left(\frac{2.9 \times 10^{-3}}{500} \right) \times 10^{-9} = 5800\,\text{K}$$

The radiation maximum falls right into the middle of the
visible spectrum. From the viewpoint of the evolution of
mankind this is not at all sur-
prising! (The temperature of the
sun's surface should not be confused with the temperature of the
core interior of the sun, which is
$\sim 1.5 \times 10^7$ K.)

There is so much
to be learned from
the colour of
hot coal!

From Wien's law we can deduce
temperatures of distant objects
without having to go there!

11.1.6 *Optical pyrometers*

A non-contact method of measuring high temperature is to match the brightness of a tungsten lamp filament with the brightness of a hot object. Instruments which use this technique, called *optical pyrometers*, are used to measure temperatures, such as those of hot furnaces, molten glass, nuclear blasts and the surface of the sun.

The advantage of tungsten is that its melting point is ≈ 3700 K, much higher than that of iron (≈ 1800 K) or copper (≈ 1350 K). The brightness of the source is reduced by a screen when one is measuring even higher temperatures, such as that of the sun's surface.

11.2 Classical theoretical models of the blackbody radiation spectrum

11.2.1 *Cavity radiation*

The classical theory of blackbody radiation used what may at first appear to be an unlikely model, consisting of an enclosed cavity, like an oven, with inside walls at high tem-

perature. The radiation emitted from the inside surfaces would bounce around the cavity, and would sometimes be reflected and sometimes absorbed by the walls. Eventually a state of equilibrium would be reached, when the energy emitted by the walls equalled the energy absorbed. The radiation inside the cavity could then be defined as the *equilibrium blackbody radiation* corresponding to the oven temperature. It does not depend on the size or shape of the cavity, nor on the material of its walls.

We can visualise the radiation criss-crossing the cavity in all directions carrying energy, as illustrated in Figure 11.3. At equilibrium there is no *net* flow of energy to anywhere, and the flux across any area element inside the cavity is the same in all directions, and balances out. Nevertheless, there is continuous traffic of energy in such an isothermal enclosure and we can express this in terms of a quantity, ψ, the *radiant energy density* at a point.

There is a small opening in the wall of the cavity. The idea is that any radiation which enters has practically no chance of finding its way out. As a result the entrance to the cavity becomes a perfect absorber. Much more important, it is also a perfect emitter, or a window through which a sample of the radiation leaves the cavity.

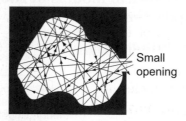

Small opening

Figure 11.3 Electromagnetic waves are bouncing around inside the cavity, reflecting off the walls. Radiant energy per unit volume in the wavelength range λ, $\lambda + d\lambda$, is $\psi_\lambda d\lambda$.

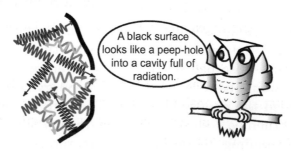

A black surface looks like a peep-hole into a cavity full of radiation.

Figure 11.4 Radiation emerging from the cavity looks light from burning coal.

11.2.2 *Thermodynamics of the cavity model*

Ludwig Boltzmann (1844–1906) created a theoretical model in which the cavity is treated as a thermodynamic engine operating in the so-called *carnot cycle*, with electromagnetic radiation as the working substance. By applying the by then well-established rules of thermodynamics, he obtained a T^4 dependence of the total energy density ψ consequently and a historical right to be perpetuated with the inclusion of his name in the Stefan–Boltzmann law.

11.2.3 *Wien's displacement law*

Wien's displacement law could also be derived from the same model. Wilhelm Carl Wien (1864–1928) noticed that the shape of the black-body radiation spectrum bore a remarkable resemblance to the distribution of speeds of the molecules in a gas. This, he thought, may not be a coincidence since if the molecules of the blackbody are thermally agitated, then their velocities and accelerations may be related to the molecular velocities of a gas which is also thermally agitated. The radiation which results from the vibrating charges might well display similar characteristics to the corresponding distribution of molecu-

Wilhelm Carl Wien

lar energies. On this basis Wien was able to show that his displacement law, $\lambda_m T$ = constant, is a special case of a more general law which states that at corresponding wavelengths the energy density of cavity radiation varies as:

$$\Psi_\lambda = T^5 f(\lambda T) \quad \text{(Wien's spectral distribution law)} \qquad (11.1)$$

where $f(\lambda T)$ is an unknown function. If we knew this function, we would know the shape of the curve. The peak of the curve depends on the temperature, and is proportional to T^5. As we saw before, the area under the curve is proportional to T^4.

Plotting M_λ/T^5 against λT, we should get a curve which is unique in the sense that each point on the curve corresponds to a given value of λT. If we choose different temperatures, i.e. different values of T, corresponding values of λ such that $\lambda_1 T_1 = \lambda_2 T_2$ will give the same point on the curve. Wien's displacement law is a special case for the point at the apex of the graph.

Each point on the x axis of the graph below corresponds to a value of λT. The maximum occurs at $\lambda T = 2.9 \times 10^{-3}$ mK.

Wien's reasoning was also based on principles established in the 19th century. It is remarkable that the powerful methods of classical thermodynamics could go so far as to predict not only the Stefan–Boltzmann law but also the existence of a function $f(\lambda T)$, as verified by the experimental results illustrated in the graph.

At that point all that is possible had been done on the basis only of pure thermodynamic reasoning. This had now to be supplemented with a more detailed model to derive a mathematical expression which gives the *shape* of the function.

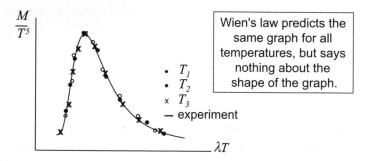

Figure 11.5 Experimental verification of Wien's displacement law. A set of measurements at one temperature determines $f(\lambda T)$. Measured points at other temperatures fall on the same curve.

11.2.4 *What is the function f(λT)?*
First try — a mathematical fit to the data

Wien proposed a formula which was essentially empirical:

$$\Psi_\lambda = \frac{C}{\lambda^5 e^{C'/\lambda T}} \quad \text{(Wien's empirical distribution law)}$$

The constants C and C' could be adjusted to agree quite well with experiment at short wavelengths, but not quite so well at longer wavelengths.

11.2.5 *Second try — a model which incorporates the wave nature of light*

Lord Rayleigh, with some mathematical input from James Jeans (1877–1946), proposed a more specific model for energy exchange at the walls of the cavity and for the modes of vibration of the electromagnetic waves. Rayleigh argued that as the radiation is reflected back and forth in the cavity, a system of standing waves will be set up for each frequency.

The shorter the wavelength λ of the radiation is, the more possible standing waves or *modes of vibration* there will be. Rayleigh derived the result that the number of modes of vibration per unit volume of wavelength in the range λ, $\lambda + d\lambda$, is

$$dn_\lambda - 8\pi \frac{d\lambda}{\lambda^4}$$

A well-established principle in thermodynamics called the *principle of equipartition of energy*, states *that in a system in thermal equilibrium at a given temperature, the average kinetic energy associated with each degree of freedom has the same value.* According to the equally well-established kinetic theory of gases, the value of this energy for each molecule is $\frac{1}{2}kT$ per degree of

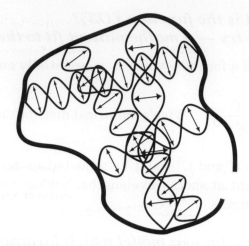

Figure 11.6 Standing waves in a cavity.

freedom, where k is a universal constant (Boltzmann's constant). Applying the same criteria to the standing waves in a cavity gives

$$\Psi_\lambda d\lambda = \frac{8\pi kT d\lambda}{\lambda^4} \quad \text{(Rayleigh– Jeans law)}$$

The increase in the number of modes as the frequency gets higher and the wavelength gets shorter is dramatic, giving the so-called 'ultraviolet catastrophe'. Assuming that each mode still has the same energy, kT, the value of Ψ_λ shoots off to infinity as the wavelength $\lambda \to$ zero. We get an infinite amount of energy radiated, which obviously is not true. Something very basic is wrong with the model.

11.2.6 *How can the theoretical models be 'half right'?*

While *Wien's displacement law* was based on convincing physical argument, and was well verified by experiment, his *spectral distribution formula* could be treated with suspicion. By judicious use of the exponential function, combined with

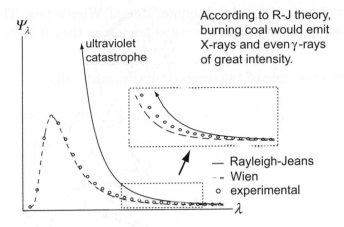

Figure 11.7 Two 'half right' theoretical models.

adjustable constants, one can find a curve to fit almost any set of experimental data. The *Rayleigh–Jeans law*, however, had to be treated much more seriously, and many 'natural philosophers' of the time were reluctant to abandon it. It was rigorously derived on the basis of what seemed to be a good model. What is more, *it agreed very well with the experimental spectrum at longer wavelengths*. There were no arbitrary coefficients; instead there appeared the fundamental physical constant k. One such natural philosopher was Max Planck (1858–1947).

11.3 Max Planck enters the scene

11.3.1 *A guess to start*

The formula, but not the explanation

Planck's first step was to try to set up an empirical formula which reduces to Wien's at short wavelengths and to Rayleigh–Jeans' at long wavelengths. Mathematically this is not too difficult. Planck achieved such a formula simply by putting an additional term into the denominator of Wien's law. The only justification for this mathematical trick was that it works!

Planck's merger of two curves mathematically

Wien's spectral distribution law   Rayleigh–Jeans law

$$\Psi_\lambda = \frac{C}{\lambda^5 e^{C'/\lambda T}}$$

$$\Psi_\lambda = \frac{8\pi k T}{\lambda^4}$$

Planck's 'lucky guess'

It works, but why???

$$\Psi_\lambda = \frac{C}{\lambda^5 (e^{C'/\lambda T} - 1)}$$

11.3.2 *Showing that Planck's 'lucky guess' formula works at all wavelengths*

Short wavelengths $\boxed{(e^{C'/\lambda T} \gg 1)}$

$$\Psi_\lambda = \frac{C}{\lambda^5 (e^{C'/\lambda T} - 1)} \quad \Rightarrow \quad \frac{C}{\lambda^5 e^{C'/\lambda T}}$$

Planck $\Rightarrow$ Wien

Long wavelengths $(C'/\lambda T$ small)

And since $e^x = 1 + x + x^2/2! + x^3/3! + \cdots \quad \rightarrow 1 + x$ for small x

$$\Psi_\lambda = \frac{C}{\lambda^5 (e^{C'/\lambda T} - 1)} \quad \Rightarrow \quad \frac{C}{\lambda^5 \left(\dfrac{C'}{\lambda T}\right)} = \frac{(\text{Const})T}{\lambda^4}$$

Planck $\Rightarrow$ Rayleigh Jeans

Planck first presented his formula as a 'comment' to the German Physical Society at its meeting on 19 October 1900. The title of his comment was simply 'On an Improvement of Wien's Radiation Law', and it was presented more as a guess than as a statement of physical significance. Nevertheless it seemed to be the correct formula. Heinrich Rubens (1865–1922) and Ferdinand Kurlbaum (1857–1927), two German physicists who had made a series of careful measurements on the blackbody spectrum, working through the night following the academy session, compared their experimental results with the formula, and reported complete agreement. As Planck himself later admitted, such experimental verification was crucial: '...without the intervention of Rubens the foundation of quantum theory would have perhaps taken place in a totally different manner, and perhaps even not at all in Germany.'

11.3.3 *Trying to crack the code*

The next step, however, was much more difficult. Fiddling with mathematical expressions is one thing; deriving the resulting formula from the fundamental principles of physics is another. What secrets were hidden at the microscopic level which could give rise to the radiation spectrum given by the *'lucky guess'* formula?

Planck tried everything he could on the basis of classical laws, but without success. These laws were tried and tested, and yet just did not lead to the perfectly matched formula for blackbody radiation!

11.3.4 *Nature's secret*

At this time many *natural philosophers* were of the opinion that all major questions in science had been answered. All that remained were a few 'loose ends' and perhaps a few more significant figures in the determination of the values of fundamental physical constants. The blackbody radiation spectrum was one such loose end.

Such an attitude, however, was not shared by Max Planck. He became convinced that something which had started innocuously as 'the colour of light from burning coal' had developed into a phenomenon with much deeper meaning. Perhaps light was a messenger telling us something about the constituents of matter from which it was coming? Were the oscillating atoms and molecules sending us signals which we could not quite interpret? Was there a deep hidden secret of nature encoded in these signals?

11.3.5 *The cause of the 'ultraviolet catastrophe'*

A central feature of the Rayleigh–Jeans model is the assumption that in free space, before we 'confine them to a cavity', all modes of electromagnetic vibration are possible. Inside the cavity, the number of modes is limited by the condition to form stationary waves, but as the wavelength gets shorter and the frequency increases, this number increases rapidly. As we have seen in Section 11.2.5, this leads to the ultraviolet catastrophe.

There can be only one conclusion:

Something prohibits higher frequencies.

11.4 Planck's 'act of despair'

Having exhausted the classical means, Planck found a hypothesis which was revolutionary at the time and contrary to the belief that the laws of Nature are 'continuous':

11.4.1 *The quantum hypothesis*

Perhaps there is a *law of nature* which forbids some high frequency modes of vibration. This law must be such that the higher the frequency, the more modes are forbidden. Then only a fraction of the high frequency oscillators will be active. This law could apply also to the vibrating charges in the walls of the cavity (and everywhere else, for that matter).

In a letter written in 1901 Planck wrote: '...the whole procedure was an act of despair because a theoretical interpretation had to be found at any price, no matter how high that might be.'

The proposed law of nature was that an oscillator can only have energy consisting of an integral number of *quantum* units. The value of such a quantum depends on the frequency of of the oscillator, and is

Planck's quantum postulate: $E = hf$,

where f = frequency of the oscillator

Planck presented his postulate, less than two months after his previous 'comment', at a German Physical Society meeting on 14 December 1900, a date now often regarded as 'the birthday of quantum theory'.

Planck obtained a value for his constant, $h = 6.55 \times 10^{-34}$ Js, which best fitted the data on the radiation spectrum then available. The modern value of this fundamental constant is

Planck's constant $h = 6.625255 \times 10^{-34}$ Js

'Much less simple was the interpretation of the second universal constant of the radiation law, which as the product of energy and time, I called the elementary quantum of action....'

11.4.2 *Quantum discrimination*

The postulate discriminates strongly against high frequencies. The unit of energy, and therefore of energy exchange, becomes greater, and 'energy transactions' become more and more difficult! Planck assumed that an electric oscillator will emit radiation of

the same frequency as that of the oscillator itself. It can only do so in discrete 'quanta' of value hf, otherwise it will be left in a forbidden energy state. Likewise, if an oscillator is the recipient of energy, it can only accept an exact number of such 'bundles'. To make a classical analogy; in a limited economy, a person who can only deal in, say, 1000-dollar bills will find it difficult to carry on business!

11.4.3 *A summary of Planck's hypotheses*

1. An oscillator can only have one of a discrete set of possible values of energy $E = nhf$.
2. The emission and absorption of radiation are associated with transitions between these levels. The energy emitted or absorbed appearing as a quantum of radiant energy of magnitude hf, which is the energy separation between the levels.

The light emitted from the hot surface comes in discrete *quanta* of energy hf, where f is the frequency of the radiation. When an oscillator emits a photon it drops from energy an level

$$E_n = nhf$$

to

$$E_{n-1} = (n-1)hf$$

The constant h is one of the fundamental constants of nature.

11.4.4 *What does the quantum do after it is emitted?*

Planck did not assume anything more specific about the quantum of energy; in particular, he said nothing about how it is distributed in space. He had no reason at the time to picture it as anything other than a wave, dissipating in space, with energy spread out over the whole wave front. As he recalled in his Nobel lecture in 1920, 'What happens to the energy after emission? Does it spread out in all directions with propagation in the sense of Huygens' wave theory, in boundless progressive attenuation? Or does it fly out in one direction in the sense of Newton's emanation theory?

We will see in the next two chapters extraordinary and sometimes contradictory evidence on whether or not such a quantum is localised!

Much later, in 1926, Gilbert Newton Lewis (1875–1946) introduced the term *'photon'* for the quantum of light. As this name is now in common use we will introduce it at this point, and use it freely for the remainder of this book, remembering that strictly speaking we are historically premature.

11.4.5 *How does the quantum hypothesis solve the 'ultraviolet catastrophe'?*

Without going into detailed mathematics we have seen that *quantum discrimination* means that Nature imposes a restriction on the energy of an atomic oscillator. Only specific energies are allowed. The higher the frequency, the greater the restriction and the further apart the allowed energy levels. The ultraviolet catastrophe is avoided by these increasing restrictions because at higher frequencies the allowed

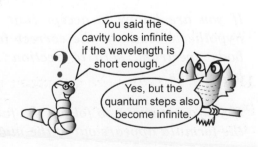

energy levels become further and further apart, while all others are forbidden, and do not exist.

Getting a feeling for the numbers

1. *What is the energy of a 'typical' photon?*

Calculating the energy of a photon of wavelength $\lambda = 500$ nm (in the middle of the visible spectrum),

$$E = hf = \frac{hc}{\lambda} = \frac{6.63 \times 10^{-34}\,\text{Js} \times 3 \times 10^{8}\,\text{ms}^{-1}}{500 \times 10^{-9}\,\text{m}} = 3.98 \times 10^{-19}\,\text{J}$$

On the macroscopic scale this is a tiny amount of energy; quantisation has a negligible effect at the macroscopic, or 'household', level. At the atomic level, however, it plays a major role.

2. *How many photons do you need to activate the human eye?*

The maximum sensitivity of the human eye is about 2×10^{-18} W. How many photons per second of wavelength 500 nm does this represent?

   The energy of each photon is 3.98×10^{-19} J. Therefore the eye is sensitive to $2 \times 10^{-18}/3.98 \times 10^{-19} \cong 5$ photons.

> *If you are happy to accept that Planck's quantisation hypothesis leads to the correct formula for the black-body spectrum, skip to Section 11.5.5, 'A reactionary hypothesis'.*
>
> *If you are curious to follow the fascinating story of how the formula appears out of the mathematics, read on.*

11.5 From an idea to a formula — the mathematical journey

11.5.1 *The classical law of energy distribution*

If we raise the temperature of a gas, for example by supplying energy to it by means of a Bunsen flame, the atoms or molecules will be excited by *thermal agitation*. This thermal energy is, however, not equally divided, and some will have more energy than others. The classical Maxwell–Boltzmann law of distribution of energies states that, in a gas in thermal equilibrium at temperature T, the number of molecules of energy E is proportional to $e^{-E/kT}$.

$$\text{The ratio} = \frac{\text{number with energy } E_1}{\text{number with energy } E_2}$$

$$= \frac{e^{-E_1/kT}}{e^{-E_2/kT}}$$

We will now apply the same distribution law to oscillators instead of molecules.

11.5.2 *The average energy*

The spectral radiant exitance is directly proportional to the average energy $\langle E \rangle$ of oscillators of frequency f. We now want to derive an expression for $\langle E \rangle$.

(a) according to classical theory, which assumes the disribution to be continuous, and

(b) using the quantum postulate of discrete energy levels.

$$\text{In each case } \langle E \rangle = \frac{\text{total energy}}{\text{total number of oscillators}}$$

Let N_T = total number of oscillators of frequency f.

11.5.3 *Classical theory*

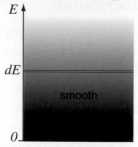

Continuous classical energy levels

The higher the energy,
the lower the population density.

Assuming that the distribution of energies is continuous, we obtain the total energy by integration. Dividing by the total number of molecules, we get the well-known classical result for the average energy kT per oscillator as used by Rayleigh–Jeans.

$$N(E, E + dE) = N_T e^{-E/kT} dE$$

$$\langle E \rangle = \frac{\int N_T e^{-E/kT} dE}{N_T}$$

$$-kT \left| e^{-E/kT} \right|_0^\infty = -kT(0 - 1)$$

Average energy $\langle E \rangle = kT$

11.5.4 *Quantum theory*

$$N_1 = N_0 e^{-hf/kT}$$

$$N_2 = N_0 e^{-2hf/kT}$$

$$N_3 = N^0 e^{-3hf/kT}$$

$$\langle E \rangle = \frac{N_0 E_0 + N_1 E_1 + N_2 E_2 + N_3 E_3 + \cdots}{N_0 + N_1 + N_2 + N_3 + \cdots}$$

$$= \frac{N_0 hf (0 + e^{-hf/kT} + 2e^{-2hf/kT} + 3e^{-3hf/kT} + \cdots)}{N_0 (1 + e^{-hf/kT} + e^{-2hf/kT} + e^{-3hf/kT} + \cdots)}$$

$$(\text{let } x = e^{-hf/kT})$$

$$\langle E \rangle = \frac{hf (0 + x + 2x^2 + 3x^3 + \cdots)}{1 + x + x^2 + x^3 + \cdots} = \frac{hf S'}{S}$$

Quantised energy levels

N_2 ——————————— $E_2 = 2hf$

$\Delta E = hf$

N_1 ——————————— $E_1 = hf$

$\Delta E = hf$

N_0 ——————————— $E_0 = 0$

N_0, N_1, N_2 = the number of oscillators at each frequency.
The higher the energy, the greater the spacing between levels.

> If only certain energy levels are possible, the integration is replaced by a series of discrete terms. The expression for average energy takes on a completely different look!

To evaluate the two infinite series, we use the binomial theorem:

$$(1+x)^n = 1 + nx + \frac{n(n-1)}{2!}x^2 + \frac{n(n-1)(n-2)}{3!}x^3$$

and note that

$$(1-x)^{-1} = 1 + x + x^2 + x^3 = S \quad \text{and}$$

$$(1-x)^{-2} = 1 + 2x + 3x^2 + 4x^3 = \frac{S'}{x}$$

$$\Rightarrow \langle E \rangle = \frac{hf\ S'}{S} = \frac{hf\ x}{(1-x)} = \frac{hf}{(x^{-1}-1)}$$

Average energy $\Rightarrow \langle E \rangle = \dfrac{hf}{(e^{hf/kT}-1)}$ Looks familiar?

One can imagine the thrill experienced by Planck when he saw the expression appear in the denominator, which bore such a close resemblance to his empirical formula. The difference of

course was that this time the expression emerged from a physical argument and not from a mathematical manipulation. The quantum hypothesis produced a fit to experimental data!

Quantum discrimination. The average energy of oscillators in thermal equilibrium is not = kT but a smaller value, depending not only on the temperature T but also on the frequency f. As the frequency increases, the number of oscillators decreases because they can only accept large quanta.

$$\langle E \rangle = \frac{hf}{e^{hf/kT} - 1} = \frac{hf}{1 + hf/kT + (hf)^2/2!(kT)^2 + \cdots - 1}$$

For *low frequencies* the average approaches the classical value.

$$\langle E \rangle \Rightarrow kT \quad \text{(if we neglect } f^2 \text{ and higher powers)}$$

Increasing $f \rightarrow$ quantum discrimination begins to bite.

$$\langle E \rangle \Rightarrow \ll kT \quad \text{(decreasing rapidly as higher powers take over)}$$

Planck's radiation law

By combining the number of modes of oscillation determined by Rayleigh–Jeans with the newly calculated mean energy, we obtain

$$\Psi_\lambda d\lambda = \frac{8\pi}{\lambda^4} \frac{hf}{e^{hf/kT} - 1} d\lambda$$

$\left(\begin{array}{l} \text{just like Planck's first} \\ \text{empirical formula, but} \\ \text{without arbitrary} \\ \text{constants} \end{array} \right)$

It can be written as

$$\Psi_\lambda d\lambda = T^5 f(\lambda T) d\lambda, \quad \text{where} \quad f(\lambda T) d\lambda = \frac{8\pi}{(\lambda T)^5} \frac{ch}{e^{ch/k(\lambda T)} - 1} d\lambda$$

Conforms with Wien's
thermodynamic equation (11.1)

11.5.5 A reactionary hypothesis

Suddenly the smug belief that all basic concepts in natural philosophy were well understood, and only some loose ends
remained to be tied up, had suffered a major blow. The idea that
natural processes come in 'jumps' and are not continuous was
completely foreign to most physicists, including Planck.

'Atomicity of matter' was one thing; it was well established
since it had been originally proposed by the Greeks. 'Atomicity
of energy' was quite another story. Planck himself was at first
reluctant to accept it and was still not quite sure whether his
quantum idea was merely a mathematical device or whether he
had discovered a fundamental law of Nature. He wrote later
that he tried for many years 'to save physics from discontinuous
energy levels', but without success. 'The quantum idea obstinately refused to fit into the framework of classical theory'. He
would surely have been even more worried had he realised the
full implications, which were later destined completely to transform the laws of physics.

A historical interlude: Max Planck (1858–1947)

Max Planck was born on 23 April 1858 in Kiel, then the Danish city
of Holstein, and died on 4 October 1947 in Göttingen. His father,

(Continued)

Julius Wilhelm, was Professor of Constitutional Law at the University of Kiel. His mother was Emma Patzig. He entered the University of Munich at the very young age of 16, to study physics. Ironically,

his professor Philipp von Jolly (1809–1884) told him at the time that Physics was a closed science, in the sense that all of the subject, with the exception of a few details, had been mastered. Accordingly, he warned, there was little prospect of future research or discovery. Little did he know how wrong he would be proved in the future by the young person he was addressing!

Planck remained in Munich until 1880, apart from one year (1877–1878) in Berlin, where he studied under Gustav Kirchhoff (1824–1887) and Herman von Helmholtz (1821–1894). He did not, it appears, enjoy his time there, describing Kirchhoff's lectures as 'dry and monotonous', while those of Helmholtz were 'badly prepared and careless, 'giving us the unmistakable impression that the class bored him as much as it did us'.

PLANCK, Max Karl Ernst Ludwig
Nobel Laureate PHYSICS 1918
© Nobelstiftelsen

On his return to Munich, Planck received his doctorate at the age of 21, with a thesis on thermodynamics, and one year later he was appointed to a teaching post at Munich, at what was again an exceptionally young age. In 1885 he obtained a professorship at the University of Kiel, and soon afterwards married his childhood sweetheart, Marie Merck. Marie died in 1909 and some time later he was to remarry her cousin Marga von Hösslin.

In 1889, Planck moved to Berlin, where he had been offered the Chair of Theoretical Physics, previously held by Kirchhoff. Planck

(Continued)

stayed in this post for a period of 38 years until 1927, during which time he produced his most brilliant work, including the discovery of his radiation formula in 1900.

In his autobiography Planck describes the lonely path he plotted, which was clouded by an anxious and uncertain mood, unlike the dramatic successes of some of his contemporaries, such as the Curies or Lord Rutherford. He called his postulate of energy quanta 'an act of desperation', because 'a theoretical explanation of black-body radiation had to be found at all cost, no matter what the price'. Even after the publication of his famous paper he tried to fit the intro-duction of his constant *h* into the framework of classical physics, but without success.

During the initial period after 1900, practically nobody seems to have realised the significance of Planck's discovery. The quantum postulate was reported by Arthur L. Day at the 1902 Washington meeting of the American Philosophical Society, but was deemed to deserve little or no reference in texts and papers published over the next few years. In short it seems to have been regarded as an expe-dient methodology without deeper physical significance.

The general skepticism may be illustrated by the statement made by Hendrik A. Lorentz during a series of lectures entitled 'New and Old Problems in Physics' which he gave in Göttingen as late as 1910: *'We cannot say that the mechanism of the phenomena has been unveiled by Planck's theory...it is difficult to see the reason for this partition of energy into finite portions, which are not even equal to each other, but vary from one resonator to another'.*

Whatever about the opinion of German physicists, quantum theory took a long time to cross the English Channel to Britain. Rutherford, in a letter to his friend William Bragg, about 1908, wrote: *'Continental people do not seem to be the least interested to form a physical idea of Planck's theory.... I think the English point of view is much more to be preferred.'* In his lecture at the presentation of the Nobel Prize in 1919, Planck gave the story of

(Continued)

the origin of quantum theory from his own unique perspective. Quoting Goethe's 'Man errs as long as he strives', he described how he encountered many pitfalls and difficulties, punctuated with steps which brought him 'conclusively nearer the truth'.

The quantum hypothesis obtained its greatest support when Niels Bohr introduced it in 1913 into his theory of the atom. Planck further recalled that *'it fell upon this theory to discover, in the quantum of action, the long sought key into the wonderland of spectroscopy...there was no other decision left to the critic, who does not intend to resist the facts, but to award to the constant h full citizenship in the system of universal physical constants'.*

The consequences of the new philosophy of science introduced by Planck continued to reach further and further as quantum mechanics was being developed. The ideas were so strange that even the originators became bewildered. Planck himself was often confused: *'If anybody says he can think about quantum problems without getting dizzy, that only shows he has not understood the first thing about them'.* In a lecture given in New York in 1949, he remarked: *'An important scientific innovation rarely makes its way by winning over and converting its opponents. What does happen is that its opponents gradually die out, and that the growing generation is familiarized with the ideas from the beginning.'*

Planck was once described by Kurt Mendelssohn as 'a spare figure in a dark suit, starched white shirt, and black tie, with the look of a typical Prussian official, but for the penetrating eyes under the huge dome of his bald head'. He had the great strength of character to blaze a new and original path despite the lack of interest and encouragement on the part of many of his peers.

Planck's life was punctuated with drama and tragedy from the time when, as a child, he watched Prussian troops take over the city of Kiel. His eldest son was killed in 1916 at the battle of Verdun. During the period of Nazi government, he was openly opposed to some of the policies, particularly the persecution of the Jews. He suffered another great personal tragedy when his other son, Erwin, was executed in 1945 for plotting to assassinate Hitler. A short time later an American army unit was sent to rescue him from the ruins of his bombed-out house in Berlin.

Appendix 11.1   Deriving the Stefan–Boltzmann law from Planck's radiation formula and calculating the value of Stefan's constant

$$\Psi_\lambda d\lambda = \int_0^\infty \frac{8\pi hc}{\lambda^5} \frac{1}{e^{hc/\lambda kt} - 1} d\lambda$$

$$= \frac{8\pi k^4 T^4}{c^3 h^3} \int_0^\infty \frac{x^3 dx}{e^x - 1}$$

Let $x = hc/\lambda kT$

$\Rightarrow \lambda = hc/xkT$

$\Rightarrow d\lambda = -(hc/x^2 kT)dx$

$$= \frac{8\pi k^4 T^4}{c^3 h^3} \left[\frac{\pi^4}{15} \right] = \frac{8\pi^5 k^4}{15 c^3 h^3} T^4$$

$k = 1.38 \times 10^{-23} \, \text{JK}^{-1}$

$c = 3.0 \times 10^8 \, \text{ms}^{-1}$

$h = 6.63 \times 10^{-34} \, \text{Js}$

$$\Rightarrow \Psi = 7.54 \times 10^{-16} \, T^4 \text{Jm}^{-3}$$

But, it can be shown that the radiant exitance $M = \dfrac{c\Psi}{4}$

$$E = 5.66 \times 10^{-8} \, T^4 \, \text{Jm}^{-2} \, \text{s}^{-1} \, \text{K}^4 \quad \text{(The Stefan–Boltzmann law)}$$

Showing good agreement with experiment.

$$\sigma = 5.70 \times 10^{-8} \, \text{Wm}^{-2} \, \text{K}^{-4}$$

Chapter 12

The Development of Quantum Mechanics

Planck's quantum hypothesis was only the beginning of a new and fascinating story. In 1913 Niels Bohr extended the idea of quantisation to the orbits of electrons in the planetary model of the atom. Light played a central role in that model in that it was postulated to be created by the energy loss of atomic electrons as they dropped from higher to lower orbits, and thus gave information on the orbits themselves. As a result, Bohr, in his acceptance speech at the Nobel Prize ceremony in Oslo in 1922, was able to say '...we not only believe the existence of atoms to be proved beyond doubt, but also we even believe that we have an intimate knowledge of the constituents of the individual atoms'.

Perhaps even more exciting than learning about the structure of the atom was to discover the laws of nature as they applied to *'the world of the very small'*. These laws depend very much on the principles of quantisation and their consequences. In 1921 Bohr founded his famous Copenhagen Institute, which was attended by practically all of the world's top physicists of the time. They helped to develop the *Copenhagen interpretation of quantum mechanics*, which completely changed our view of the most basic laws of nature — a far cry from Planck's original hypothesis.

In this chapter we follow the exciting story of those developments, a story which contains a mixture of philosophy, physics and the mathematical models which were used to interpreted the physical laws. We describe how the *mechanics* of dealing with the world where the *quantum* rules supreme was developed independently by Werner Heisenberg and Erwin Schrödinger, and how their methods, which looked at first as different as chalk and cheese, turned out later to be just two different representations of the same thing.

Paul Dirac brought it all together in 1927, combining *quantum theory* with the *theory of relativity*. He was rewarded with some unexpected predictions, including the prediction of the existence of *antimatter*, a most impressive example of a *discovery of the mind* that was later to be confirmed by experiment.

The chapter concludes with some apparent paradoxes which follow from the *Copenhagen interpretation of quantum mechanics* and the concept of a *quantum reality*.

12.1 The development of quantum mechanics

12.1.1 *From oscillators to photons to other things*

In 1901, Max Planck introduced the *quantum hypothesis* as an 'act of despair' in order to provide an explanation for the spectrum of blackbody radiation. This for him was a huge step, so new and controversial that he was reluctant to publish it for some time. Little did he realise the full consequences, which reached far beyond blackbody radiation.

In 1905, Einstein proposed that the quantum hypothesis applies not only to oscillators and to the kind of light which they emit, but to the very nature of light itself. He pointed to experiments which provided evidence that light appears to come in individual bundles of energy, or '*photons*' of energy hf. Such a quantum of highly concentrated energy could knock electrons out of the surfaces of metals, as quantified in Einstein's theory of the photoelectric effect, which we will deal with in detail in Chapter 13.

Over the next few years came the realisation that *quantisation* forms a basis for all the fundamental laws of nature. The laws of mechanics discovered by Isaac Newton work very well in the everyday 'household' world because the effect of quantisation is negligible. When we come to 'the world of the very small' of electrons, atoms and molecules, this effect can no longer be ignored; in fact it becomes dominant to the extent that it changes completely the character of natural laws. Mathematical methods had to be developed to represent these laws in the theory of *quantum mechanics*, which incorporates Newton's laws of *classical mechanics* as a special case under the conditions of the 'household' world.

12.1.2 *The planetary model of the atom*

One can say with hindsight that it was only to be expected that quantisation would play a major role in the structure of the atom. Rapid developments were taking place in the first and second decades of the century in building up an atomic model, which would help to visualise atomic structure and to explain its function. The fact that there is a structure was demonstrated in 1911 by Sir Ernest Rutherford (1871–1937). Together with his students Hans Geiger (1882–1945) and Ernest Marsden (1889–1970), he discovered that the mass of an atom is concentrated in a very small *nucleus* which is positively charged. *Electrons*, which carry negative electric charge, occupy the outer layers.

Since positive and negative charges attract one another there must be some mechanism which keeps the electrons from falling into the nucleus. Soon afterwards, Niels Bohr (1885–1962) and Arnold Sommerfeld (1868–1951) put forward the planetary model of the atom, in which the electrons orbit the nucleus in a similar fashion to planets orbiting the sun in the Copernican model of the solar system. Just as the earth does not plunge into the sun, in spite of the attractive gravitational force, electrical attraction serves to keep the electron in orbit, without crashing into it into the nucleus.

12.1.3 *The quantum enters the picture*

There is an important difference between an orbiting planet and an orbiting electron. Unlike a planet, which is electrically

neutral, the electron is charged, which should have an important effect on the stability of an orbit. Maxwell's prediction that
an accelerating electric charge emits electromagnetic radiation
must apply even in the realm of the atom. An orbiting electron
undergoes constant centripetal acceleration and should therefore radiate energy. Like a satellite, which on entering the
earth's atmosphere, begins to lose energy by friction, the electron should spiral inwards, and crash into the nucleus. Why
then does an electron not do that? Certainly, since normal atoms
are stable, there must be something which prevents the orbiting
electron from radiating energy.

Niels Bohr postulated a law of nature which ensures the
stability of atoms. This law states that only certain electron
orbits are possible. The permitted orbits are those for which the
angular momentum (L) is always an integral multiple of $h/2\pi$.

$$L = \frac{nh}{2\pi}, \quad \text{where } n = 1, 2, 3, \ldots$$

Planck's constant h appears again, this time in a different context! Here it governs the allowed angular momentum of an electron in orbit.

Since h is a universal constant, Bohr's quantisation condition
gives a numerical value for the size of electron orbits. We can use
these orbits to define the size of an atom which turns out to be
about 10^{-10} m. (See Appendix 12.1 for the atom of hydrogen.)

12.1.4 *Quantum jumps — light comes out of the atom*

The electrons in Bohr's model do not spiral inwards continuously, but can only drop from an orbit of higher energy to one of
lower energy. Rather than the continuous loss of energy expected
from classical physics, the loss is sudden, carried away by a
light photon of energy hf, exactly equal to the energy difference

between the initial and final orbits. What we see is a spectrum of discrete wavelengths corresponding to transitions between different orbits. Since the allowed orbits differ from atom to atom, we see a spectrum of frequencies characteristic of the element.

Hydrogen has a characteristic 'fingerprint'; so has helium, and so has iron when vaporised to a high temperature. By the light emitted from a sample we can recognise minute traces of elements, even if the source of light is inaccessible, such as a distant star.

Light from the atom tells us about its structure.

12.1.5 *The lowest orbit*

When $n = 1$ the atom is in its state of lowest energy, called the *ground state*. It cannot lose energy, and therefore does not emit light. There is nowhere for the electron to go and therefore it remains in the ground state indefinitely. The quantum condition ensures that we have a stable universe!

The simplest atom is the atom of hydrogen. It has just one electron in orbit around a proton. At a sufficiently high temperature the atom is in an *excited state*, with the electron in one of the higher orbits. When the electron makes a *quantum jump* from a higher to a lower orbit, a photon of light is emitted of energy

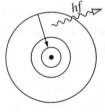

Quantum jump.

$$hf = E_i - E_f$$

where E_i = energy of initial state, and E_f = energy of final state.

12.1.6 *The demise of determinism*

The idea that an electron should drop from a higher to a lower energy state was, in itself, not unreasonable. However, applying

Newton's mechanics one would expect some reason to cause it to jump at a particular instant. There should be some inner mechanism, perhaps difficult or impossible to observe, to trigger the transition.

According to the thinking of that time, *quantum jumps* were quite mysterious. There was no evidence to suggest that there was any direct cause and effect in the process. It seemed

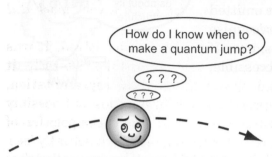

The electron is uncertain about the future!

that the electron makes the jump spontaneously, quite independently of how long it has been in a given state. Gone was the Newtonian order of determinism. A new philosophy was emerging. The laws governing the atom did not determine the occurrence of an event, but only the *probability* of the occurrence of the event. It also appeared that one knew nothing about the system during the transition from one state to another. Such a 'state of limbo' for a physical object was an equally foreign idea.

12.1.7 *A new way of thinking*

A new approach was required to the understanding of fundamental physical processes. To quote Erwin Schrödinger (1887–1961): *'the task was, not so much to see what no one yet has seen, but to think what nobody has yet thought, about that which everyone sees'.*

Old, well-established habits die hard, and many physicists found it difficult to come to terms with the new ideas. As late as 1916, Schrödinger himself was reportedly led to exclaim: *'If one has to stick to this quantum jumping, then I regret ever to have been involved in this thing'.* To which Bohr replied: *'But we others are very grateful to you that you were, since your work did so much to promote this theory.'*

There were other conceptual problems to be solved. It was clear that the model of the atom could not be taken literally. It should, rather, be treated as a mathematical representation, which could be used to represent atomic properties and possibly make predictions. Even while the atom is stable, the position of the electron is indeterminate. It was impossible to actually 'see' an orbiting electron. Perhaps a better mental picture was that of an 'electron cloud' around the nucleus representing a ring within which the electron should be!

12.1.8 *The Copenhagen interpretation*

In 1921, Niels Bohr founded the Copenhagen School which addressed the mysterious new topics which had changed the face of physics over the previous years. At one time or another practically every top physicist of that period attended Bohr's institute. They formed a 'think tank' of enormous brain power and imagination which developed the *Copenhagen interpretation* of *quantum mechanics*. It concluded that Planck's discovery, and Bohr's atomic model, were only the first steps in an exciting adventure....

12.2 Matrix mechanics

12.2.1 *Heisenberg's approach*

Werner Heisenberg (1901–1976) was one of Bohr's young and very bright students who first visited the Institute in 1924.

He had the idea that one should try to construct a theory in terms of quantities which were provided by experiment, rather than to build it up on the basis of a model which involved many quantities which could not be observed. Things like electron orbits cannot be measured directly; it is better to try to write down mathematical formulations that relate directly to observable facts.

12.2.2 *Light from the hydrogen atom*

Heisenberg started by considering the frequencies of the light emitted with the electron's transition from orbit to orbit in the hydrogen atom. These could be arranged in a matrix of numbers such that the frequency due to a transition from orbit m to orbit n would form the element occupying row m in column n. They

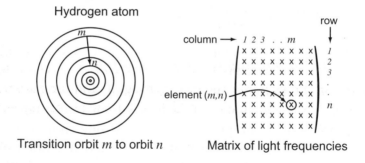

Transition orbit m to orbit n Matrix of light frequencies

are experimental numbers representing nature's way of telling us all about the hydrogen atom.

12.2.3 *A matrix for everything*

Heisenberg's idea was original, quite different from anything which had previously been suggested. He set out to construct matrices for physical observables such as energy, momentum, position, frequency and velocity. Thus there would be an 'energy matrix', a 'position matrix' and so on.

An atom would be represented not by a physical picture, but by a purely mathematical model. Instead of thinking about electron orbits one could think of the atomic model as a matrix with rows and columns of 'empty spaces'. Each space would be filled by a number as we obtained the necessary information about that particular atom.

Applying one matrix to another in a certain defined way would give a set of numbers representing the results of experimental observation. For example, applying the 'frequency matrix' to the 'hydrogen matrix' would give the observed values of the frequencies of light emitted by hydrogen. Applying the same frequency matrix to the matrix for the sodium atom would give the spectral lines emitted by sodium, and so on.

12.2.4 *Rules of the game*

The first part of the problem was to define rules for 'applying' one matrix to another; the second was to construct the relevant matrices for the systems such as 'hydrogen' and for observables such as 'frequency', which would lead to results in agreement with experiment.

Heisenberg returned to Göttingen, where he was fortunate to have as his professor Max Born (1882–1970), who was familiar with the subject of matrix algebra, which had been devised in the 19th century by the English mathematician Arthur Cayley (1821–1895). Cayley had defined a set of self-consistent rules for matrix addition, multiplication and division purely as a mathematical formalism, with no particular relevance to physics. Born drew Heisenberg's attention to the work and enlisted another gifted young post-doctoral student, Pascual Jordan (1902–1980), to help with the project.

It is interesting to note that at that time physicists either did not know what matrices were, or were reluctant to apply them to theoretical models. An illustration of the situation in 1925 is a comment by Heisenberg in a letter to Jordan: *'Now the learned Göttingen mathematicians talk so much about Hermetian matrices, but I do not even know what a matrix is!'*

12.2.5 *The laws of nature*

So far in the history of physics the language of mathematics had worked very well. The mathematics of linear algebra works well when applied to Newton's laws. The mathematics of differential calculus allowed Maxwell to predict the existence and properties of electromagnetic waves. At least in some instances nature appears to be accessible to human logic! In the words of Albert Einstein: *'How can it be that mathematics, being after all a product of human thought, is so admirably appropriate to the objects of reality?'* To which he himself replied: *'The Lord is subtle, but not malicious.'*

12.2.6 *An example of the Heisenberg method*

The matrix equation to obtain the frequencies of the radiation emitted by hydrogen (i.e. the 'spectral lines' of hydrogen) was written in the form:

$$\begin{pmatrix} \times\times\times\times\times \\ \times\times\times\times\times \\ \times\times\times\times\times \\ \times\times\times\times\times \\ \times\times\times\times\times \end{pmatrix} \times \begin{pmatrix} \times\times\times\times\times \\ \times\times\times\times\times \\ \times\times\times\times\times \\ \times\times\times\times\times \\ \times\times\times\times\times \end{pmatrix} \longrightarrow \begin{pmatrix} \times\times\times\times\times \\ \times\times\times\times\times \\ \times\times\times\times\times \\ \times\times\times\times\times \\ \times\times\times\times\times \end{pmatrix}$$

matrix representing frequency	matrix representing hydrogen	matrix of experimental numbers

The 'application' of the frequency matrix to the matrix for the hydrogen atom followed the rule of matrix multiplication according to Cayley's matrix algebra.

12.2.7 *Matrices do not commute!*

Heisenberg had not gone very far with his idea when he noticed something in his scheme which worried him. Physical phenomena had up to now been represented by an algebra in which a number A multiplied by B gave the same result as B multiplied by A. This is called the *commutative* law of multiplication, but matrices do not necessarily obey this law. If we use matrices to determine the values of two physical observables, we will in general get a different result if we change the order in which we make these determinations. Thus if matrix $[A]$ represents observable A, and matrix $[B]$ represents observable B, it is natural to represent a measurement of A followed by a measurement of B by the product $[B][A]$ ('B after A'). If $[B][A] \neq [A][B]$ we conclude that measuring B after A is not the same as measuring A after B.

Heisenberg felt that the non-commutative nature of matrix algebra gave rise to a fundamental error in his theory. However, Paul Dirac (1902–1984), who had received an early copy of the paper later published by Bohr, Heisenberg and Jordan, quickly realised that non-commutation might in fact be exactly the dominant feature governing basic natural phenomena. The theory was not wrong — quite the opposite, the matrix equations were trying to tell us something!

12.2.8 *Laws of nature must be built into the matrices*

Heisenberg's next task was to determine what kind of matrices should represent different physical observables. At this point it becomes necessary to match theory with experiment, and to put experimentally determined numbers into the theory, which so far was completely abstract.

Heisenberg's best-known and most basic discovery involved the observables *position* and *momentum*. He found that the matrices representing the position and momentum of a particle such as an electron or a proton must obey the following *commutation condition*:

$$[q][p] - [p][q] = \frac{ih}{2\pi}[I] \qquad (12.1)$$

where

> *Law of Nature* governing coordinate q and momentum p.

[p] is a matrix representing the physical observable '*momentum*',
[q] is a matrix representing the physical observable '*position*',
[I] is the *unit matrix*, and
h is Planck's constant, and $i = \sqrt{-1}$.

More specifically, if we consider a particular coordinate, say in the x direction,

$$[x][p_x] - [p_x][x] = \frac{ih}{2\pi}[I] \qquad (12.2)$$

where p_x is the component of momentum in the x direction.

12.3 Order does matter

12.3.1 *One measurement disturbs the other*

In classical physics, the properties of a physical system exist independently of the observer. For example, a particle has a certain position and a certain momentum irrespective of whether we have observed it or not. We can in principle measure its position without disturbing its momentum, and vice versa. It makes no difference if one or the other of these parameters is measured first, or if position and momentum had been measured simultaneously.

The rules of described by Equations (12.1) and (12.2) say something quite different. They say that measuring position after momentum is not the same as measuring momentum after position. The difference is quantified using the language of matrices, and is expressed in terms of Planck's constant, h. This constant of nature plays a much more mysterious and deeper role than Planck ever imagined!

Although it is possible in principle to measure the momentum or the position of a particle to an arbitrary accuracy, it is not possible to measure these two observables *simultaneously*. Position and momentum are complementary properties of a system but the theory does not permit knowledge of the two at the same time. This idea goes very much against intuition, what we 'feel' the laws of physics should be. But our intuition is based on our experience of the everyday 'household' world; the world of electrons and atoms is definitely not 'household'.

12.3.2 *A 'table top' experiment with polaroids*

We have discussed the polarisation of light in Section 8.12 from the wave theory point of view, representing the polaroids material schematically as a 'slit' which allows only the component of vibrations parallel to it to get through. The emerging light is then represented as consisting of electric field oscillations only in the direction of the slit. If the light then comes to a second polaroid with its axis inclined at an angle θ to the first, the intensity of the light emerging is reduced by a factor $\cos^2\theta$. In the case when the axes are perpendicular ('crossed polaroids'), no light gets through.

We will now consider the phenomenon in terms of the particle model. Each polaroid is represented as a mechanism for measuring a certain property of a photon called 'polarisation'. Any photon which is transmitted through a 'θ polaroid' has the property that it will pass through any other 'θ polaroid', but has only a limited probability of transmission through a polaroid oriented at any other angle. This probability is equal to $\cos^2\Delta\theta$, where $\Delta\theta$ is the difference between the inclination of the two slits. (*Malus's law*)

Let us start with just two polaroids, with axes of polarisation at 90° to one another, which we will call the 'X polaroid' and the 'Y polaroid'. A photon which has 'passed the test' of the X polaroid, and then comes to the Y polaroid, will certainly 'fail' this second test, and be absorbed. This is the standard 'crossed polaroid' effect.

Let us next introduce the third, randomly oriented polaroid — the 'R polaroid' — between the other two, and let us call its angle of orientation θ. By inserting the random polaroid we are making a new measurement which is a test to see if the 'X photon' passes through the R polaroid. If it passes this test it becomes a photon with angle of polarisation θ relative to the X direction. It 'forgets' that it had ever been a X photon and becomes a 'θ photon'.

What happens when a 'θ photon' meets the Y polaroid, i.e. the second of the two original crossed polaroids. This was the test we knew was going to fail in the original system. Now we are not sure! It may or may not pass this next test. The probability depends on the angle θ and, according to *Malus's law*, is proportional to $\cos^2 (90° - \theta)$.

By inserting another obstacle into the path of the photon, we give it a chance to pass through the system. Paradoxically the extra fence has made the passage easier and not harder (Figure 12.1).

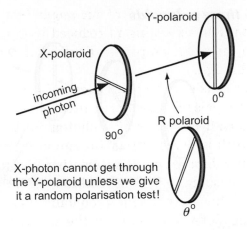

Figure 12.1 An experiment with polaroids.

What we have learned from the experiment:

1. *The order in which we make observations matters.* The result is quite different if we place the random polaroid in front of or behind the other two.
2. *The photon behaves as a typical 'quantum particle'.* An observed photon is not the same as it was before it was observed. This is something we shall discuss later.

This experiment can be performed using the simplest of equipment, and a normal light source, such as sheets of polaroid placed on an overhead projector.

12.3.3 *Experimenting with a series of polaroids*

Let us find out what happens when we send a random photon through various combinations and arrangements of polaroids.

Example 1 — three polaroids

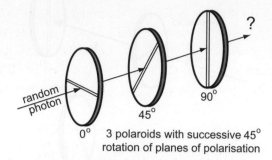

0°   3 polaroids with successive 45°
rotation of planes of polarisation

Using Malus's law,

Probability that the photon gets through all three polaroids

$$= \cos^2 (45°) \times \cos^2 (45°)$$
$$= 1/4$$

One photon in four gets through.

Example 2 — five polaroids

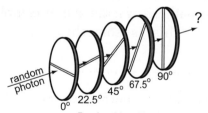

5 polaroids with successive 22.5°
rotation of planes of polarisation

Probability

$$= \cos^8 (22.5) = 0.92388^8$$
$$= 0.53$$

More than half the photons will get through, and yet the three polaroids from example 1 are included in the row of five!

Example 3 — 91 polaroids

Let us give the photon a real test. What is the probability that a random photon gets through a series of 91 polaroids as depicted in the diagram below? The axis of each successive polaroid is rotated by 1° relative to the previous one. Note again that the axes of the first and last polaroids are perpendicular to one another.

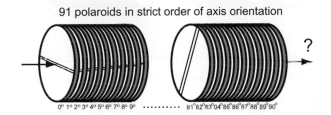

91 polaroids in strict order of axis orientation

$$\text{Probability} = \cos^{180}(1°) = 0.99985^{180} = 0.97296$$

Over 97 photons out of 100 get through this system!*

Don't shuffle the polaroids!

We must be careful that all polaroids are in their correct order. Even one out of place will change the result completely! *Order does matter.*

Example 4 — the importance of order

In this example we take the 2° polaroid in the previous example out of its rightful place and insert it near the end of the line — say, between the 88° and the 89° polaroids. The fact that there is now a gap between the second and fourth polaroids makes very little difference. (If we want to be pedantic we can replace $\cos^4(1°)$ (= 0.99939) by $\cos^2(2°)$ (= 0.9976).) On the other hand, inserting it near the other end has a dramatic effect. We are

---

* We can write a general formula for the probability of a photon passing through n polaroids relatively rotated through equal angles θ as $P = [\cos^2(\theta)]^{n-1}$.

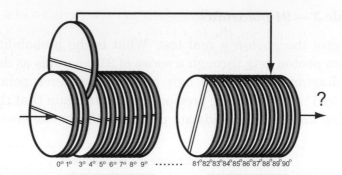

0° 1° 3° 4° 5° 6° 7° 8° 9° · · · · · · · 81°82°83°84°85°86°87°88°89°90°

effectively creating the equivalent of two crossed polaroids, making the probability of the photon getting through the system negligible.

$$P = \cos^{174}(1°) \cos^2(2°) \cos^2(86°) \cos^2(87°)$$
$$= 0.99985^{174} \times 0.99939^2 \times 0.06976^2 \times 0.05233^2 = 0.0000062$$

About six photons in a million get through!

12.3.4 *The uncertainty principle*

The intrinsic lack of knowledge which we can have of the 'quantum world' can be stated in a slightly different form which does not involve matrix operators or commutation relations. This is known as the *Heisenberg uncertainty principle*. The principle forms the basis of quantum mechanics and states that:

There is a fundamental limit with which we can simultaneously know the position coordinate q and the corresponding momentum coordinate p of a particle.

If the *uncertainty* in position is Δq, then there is an uncertainty in momentum Δp, such that the product

$$\Delta p \times \Delta q \approx \frac{h}{2\pi} \quad \text{(The Heisenberg uncertainty principle)}$$

The more exact is one's knowledge of the position of a particle such as an electron, the less is it possible to know its momentum, and vice versa. Somehow, the laws of Nature do not allow us to know these parameters precisely at the same time.

The uncertainty principle defines the limits beyond which the concepts of classical physics cannot be employed.

The commutation relation between matrices which represent position and momentum (12.1) is another way of expressing the uncertainty principle.

In Newtonian mechanics, things have an independent existence, with precise values for everything observed and not observed. That is why the theory of quantum mechanics is so strange. We are used to an unambiguous kind of physical reality. If there is uncertainty in our knowledge of any physical attribute it is due to the imperfection of our methods of measurement. Instinctively we feel that, the better our measuring instruments are, the closer we can get to the value of a physical quantity. That is why Einstein found these and other concepts in quantum mechanics hard to swallow.

Classical theory assumes that, in principle, it is possible to know everything about a physical system at a given instant. It follows that, by using such knowledge and applying the laws of physics, it should, again in principle, be possible to determine the future of such a system. For that reason Newtonian or classical mechanics is termed *deterministic*. It applies to the 'household world'.

In the world of the fundamental entities such as photons, electrons, atoms and atomic nuclei, the Heisenberg uncertainty principle takes on a dominant role. Since you cannot have exact knowledge of the present it is hardly surprising that you cannot determine the future! As a consequence, *probability* and not *determinism* is inherent in the laws which govern them.

12.4 Wave mechanics

12.4.1 *The Schrödinger approach*

Erwin Schrödinger was appointed as Professor of Theoretical Physics at Zurich in 1921. He was interested in the developments which were being reported by Bohr's Copenhagen School, but worked on his own formulation of quantum

mechanics, which at first appeared quite different from the matrix mechanics of Heisenberg.

Schrödinger's *wave mechanics* was based on an idea of *Louis de Broglie*, who, in his doctoral thesis, put forward the idea that the behaviour of particles such as the electron was governed by associated '*matter waves*'. Just as light, which behaves as a wave, also exhibits particle characteristics, should not particles equally exhibit the characteristics of waves?

Erwin Schrödinger
(1887–1961)

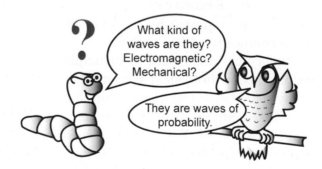

12.4.2 *De Broglie's original idea*

De Broglie had suggested that the wavelength of these matter waves, could be evaluated from the same relationship which exists between the wavelength of light and the energy of a

photon. As we know, and will further discuss in Section 13.2, a photon has energy
$E = hf$, and momentum $p = hf/c = h/\lambda$; similarly, matter waves of a particle of
momentum p should have a wavelength
$\lambda = h/p$. The propagation of these matter
waves, according to de Broglie, governs the
motion of the particle, with the consequence that under the right conditions
particles such as electrons should exhibit
'wave phenomena' such as interference and
diffraction!

Louis de Broglie
(1892–1987)

The de Broglie wavelength of even the
tiniest particles in the household world is
so short that their wave characteristics are impossible to
demonstrate. For example, let us calculate the wavelength of
the matter wave of a mosquito of mass 1/100 of a gram moving
with a speed of 10 ms^{-1}:

$$\lambda = \frac{h}{p} = \frac{6.63 \times 10^{-34}}{10^{-5} \times 10} = 6.63 \times 10^{-30}\,\text{m}$$

which is 20 orders of magnitude smaller than the diameter of an
atom.

In the case of electrons, however, the wavelength of the matter wave is similar to that of X-rays, which is of the order of the
separation of atomic planes in a crystal. A beam of electrons
should therefore undergo diffraction in the same way as a beam
of X-rays.

De Broglie presented his ideas in a doctoral thesis to the
University of Paris in November 1924. The examining committee were highly impressed by the originality of his ideas, but
did not really believe in the reality of the proposed waves. When

asked whether their existence could be experimentally verified, De Broglie replied that this should be possible in diffraction experiments of electrons by crystals. Neither he nor the examiners were aware of the fact that such evidence already existed, but had not been recognised. For example, in a paper by C.J. Davisson (1881–1958) and C.H. Kunsman on 'the scattering of electrons by nickel' published in 1921, examination of experimental figures with hindsight clearly shows a peak due to diffraction. The first official recognition of electron waves came in 1927, in a paper by Davisson and Lester H. Germer (1896–1971) entitled 'Diffraction of Electrons by a Crystal of Nickel'.

In 1928, George Paget Thompson (1892–1979), working at Cambridge, was able to secure diffraction patterns by passing narrow beams of electrons through thin sheets of matter. The electrons, when they reached a photographic plate, created a pattern of concentric rings very similar to the patterns made by X-rays. Matter waves interfere in the same way as light waves. Figure 12.2 shows the resulting patterns of X-rays (*left*) and

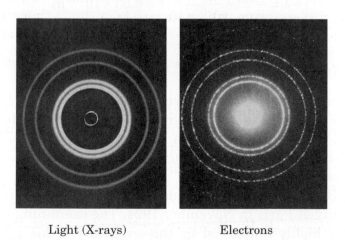

Light (X-rays)　　　　　Electrons

Figure 12.2 Diffraction patterns.

electrons (*right*) passing through the same thin sheet of aluminium foil.

12.4.3 *Adapting de Broglie waves*

De Broglie's ideas applied only to free electrons, and Schrödinger was faced with the problem of adapting the wave picture to electrons which are subject to forces, and in particular to electrons in an atom such as hydrogen. The effect of forces was represented by 'potential wells' within which the wave function had to be accommodated. An even more important and basic problem was to express the laws of physics in the form of a mathematical relation which the wave functions have to satisfy. Using clues provided by classical mechanics, and in particular the classical methods developed by William Rowan Hamilton (1805–1865), Schrödinger was able to construct an equation which described the laws of Nature in terms of the wave representation.

Over a period of just a few months Schrödinger had applied his theory to explain the spectral wavelengths of the hydrogen atom, although initially he had some difficulties in predicting the observed spectrum, which we now know has a fine structure caused by the spin of the electron. In about a year he had developed practically the whole of what is now known as *non-relativistic wave mechanics*.

It is difficult to find in the history of physics two theories which appeared to differ more radically than Heisenberg's matrix mechanics, and Schrödinger's wave mechanics, and yet both correctly interpreted the same experimental results. This could hardly be a coincidence; in fact Schrödinger soon discovered that they were just two different mathematical ways of looking at the same thing. In Spring 1926 he published a paper in *Annalen der Physic* entitled 'On the Relationship of the Quantum Mechanics of Heisenberg–Born–Jordan to mine', which proved the formal mathematical identity of the two representations.

12.4.4 *Uncertainty from another aspect*

It is interesting to see how the methods of wave mechanics lead to the uncertainty principle, albeit by a different route to that chosen by Heisenberg:

The matter wave, or *wavefunction*, of a particle of momentum p is a continuous sinusoidal wave of wavelength $\lambda = h/p$. The amplitude of the wave (or, more precisely, the square of the amplitude) is a measure of the probability of finding the particle in a given spot, but in the case of a continuous wave the amplitude does not change in space. The chance of finding the particle is the same everywhere.

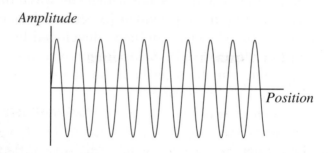

We know the momentum, but we know nothing about the position.

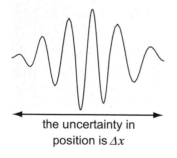

the uncertainty in
position is Δx

A localised particle is represented by a *wave packet.* This gives the most likely place to find the particle. Since the borders of the wave packet are not sharp, the width, Δx, is approximate (but can be defined more rigorously by a statistical formula).

We have seen in Section 6.8 that a square wave is made up of the superposition of sine waves of different wavelengths. Similarly, a wave packet of matter waves consists of a series of waves of different wavelengths λ, and therefore different values of momentum p. We can relate an uncertainty in position to the corresponding uncertainty in momentum using the results of a more exact mathematical calculation, known as *Fourier analysis*, which states that a combination of sine waves in the wavelength range $\Delta\lambda$ will form a wave packet of width Δx, where

$$\Delta x \,\Delta k \approx 1, \quad \text{and } k = \text{the 'wave number'} = \frac{2\pi}{\lambda}$$

$$\text{but} \quad \lambda = \frac{h}{p} \Rightarrow k = \frac{2\pi p}{h}$$

$$\Rightarrow \Delta k = \frac{2\pi\Delta p}{h}$$

$$\Rightarrow \Delta x \,\Delta p \approx \frac{h}{2\pi}$$

(the uncertainty principle)

12.5 Generalised quantum mechanics

12.5.1 *A wider view*

Paul Dirac developed a beautiful mathematical representation of quantum mechanics based on *the principle of superposition of states of a physical system*. The principle is a statement of the central feature of quantum mechanics, that an unobserved physical system can be considered as existing in a number of possible states all at once. When a measurement is made the system 'jumps' into a new state in which it has the measured value of the relevant physical quantity.

Paul Dirac (1902–1984)

Dirac's idea was to represent the state of a physical system by a vector in a space of many dimensions. The components of the vector along the many directions would then represent the various possible results of physical measurement.

Just as an 'ordinary' vector has three independent components, along the x, y, and z axes, Dirac's '*state vector*' has many, possibly even an infinite number of, components in the many directions of Dirac's mathematical space. This representation of quantum mechanics is now generally known as *Dirac's generalised transformation theory,* because it involves transforming the coordinate axes to look at the *state vector* from different frames of reference.

The representations of Heisenberg and Schrödinger fell in neatly, as special cases within Dirac's scheme, which then became, and still remains, the most complete and the most general theory of quantum mechanics.

12.5.2 *Relativity and quantum mechanics*

There was just one problematic feature in the emerging quantum mechanics: it did not encompass the laws of relativity. We will meet Einstein's theory of relativity in Chapters 15 and 16; it had been put forward at the beginning of the century, and was by now well established. If relativity and quantum mechanics are both correct representations of the laws of nature, then they must fit together, and yet there appeared to be serious inconsistencies between the two. For example, Schrödinger's equations dealt with space and time in separate ways, but according to relativity the two must be treated on the same footing.

The dilemma appeared to be resolved by Oskar Klein (1894–1977) and Walter Gordon, but not to Dirac's satisfaction, since their methods did not fit in with his generalised quantum theory. The Klein–Gordon method was based on the relativistic relation between energy and momentum, which involves the squares of these quantities. Dirac's theory dealt only with *linear* mathematical relations, and quadratic relations had no place in it. While this might be considered a technicality, it presented a serious difficulty to Dirac. Such a major flaw in his generalised quantum mechanics was simply unacceptable.

The solution to the problem came in 1927 — in Dirac's own words, 'by accident, just by playing with the mathematics'.

He found that he could express the relativistic relations in a linear way, by using matrices instead of ordinary algebraic quantities. The principles of relativity now fitted neatly into his generalised quantum mechanics, and the two basic laws of nature seemed to be represented correctly in the scheme.

12.5.3 *Triumph out of difficulty*

Just as the problem appeared to be solved, a difficulty appeared. The solutions to the equations were such that energy could have both positive and negative values. How could a particle be in a state of negative energy? Dirac felt that either the negative solutions were 'non-physical' and could be ignored, or perhaps the equations were 'trying to tell him something!'.

Having agonised over this for some time, Dirac felt that the 'difficulty' could not be ignored. His own equations were 'more clever than he was'! They were telling him about some law of nature that had not yet been discovered.

Let us quote from Dirac's own account of his conclusions in a lecture he gave at a conference in Budapest in July 1977, to mark the 50th anniversary of his discovery: '... *the difficulty was solved...by making the rather bold assumption that negative energy states exist, but normally in a vacuum they are all filled up. If such a negative energy state is not filled, there is a hole which appears as a physical particle. It would be a particle similar to the electron, and it would have positive instead of negative charge, and it would have positive energy.*'

Dirac's discovery was a 'discovery of the mind', made on the basis of logic and mathematics. It predicted the existence of a new particle which nobody had yet observed experimentally, an '*antiparticle*' to the electron (later named the '*positron*'). Within a few years came confirmation by experiment, when, in 1932, C.D. Anderson observed the track of a positively charged particle of the same mass as the electron when studying cosmic ray interactions. Antimatter really existed!

12.5.4 *Antimatter*

We now know that antiparticles exist not only for electrons, but for all fundamental particles. Indeed, one may speculate that there are distant galaxies made entirely out of antimatter. Such galaxies would emit light identical to the light emitted by galaxies such as ours we would have no way of telling that these photons came from quantum jumps of positrons rather than electrons.

It would, however, be most unwise to handle a piece of antimatter! When a particle meets its antiparticle they mutually annihilate, and literally disappear in a puff of light. The energy released in such an annihilation is carried away by photons of light.

It is not necessary to go to a distant galaxy to search for antimatter. Positrons are emitted in the radioactive decay of certain radioactive isotopes. Some such positron-emitting decays are given below:

Decay	Half-life
$_6C^{11} \rightarrow {}_5B^{11} + e^+$	$T = 20.3$ min
$_9F^{17} \rightarrow {}_8O^{17} + e^+$	$T = 1.2$ min
$_9F^{18} \rightarrow {}_8O^{18} + e^+$	$T = 110$ min
$_{11}Na^{22} \rightarrow {}_{10}Na^{22} + e^+$	$T = 2.6$ yr

12.5.5 *Positron emission tomography*

A particle of antimatter in a world of matter finds itself in a most hostile environment, and this fact has been exploited to great effect in diagnostic medicine. In *positron emission tomography* (PET), a positron-emitting isotope is chemically attached to a metabolically active molecule (such as glucose) and injected intravenously into the body. The radioisotope acts as a 'spy', building up in and identifying areas of abnormal biochemical activity within certain organs and tissue.

The positron emitted in the radioactive decay will on the average travel only a fraction of a millimetre before it meets an electron. Such a meeting of matter and antimatter will result in mutual annihilation, sending out two photons of light in opposite directions. The energy of each photon is 511 MeV, equal to the mass energy of an electron, which means that it has a wavelength of 2.43×10^{-12} m in the gamma ray region of the electromagnetic spectrum. Human or animal tissue is transparent to gamma radiation, so that this high frequency 'light' escapes the body, and serves as a perfect signal to detectors outside the patient.

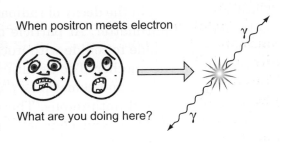

$$e^+ + e^- \Rightarrow \gamma + \gamma \quad \text{sends out accurate location signal}$$

The PET system is triggered when diametrically opposite gamma ray detectors register simultaneous 'hits'. Even a small number of such coincidences makes it possible to locate the

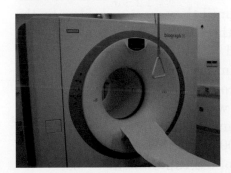

PET/CT scanner.

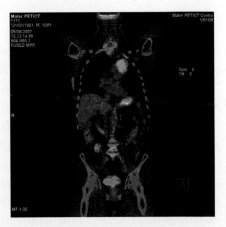

Scan of human torso.

Courtesy of Adrian Adams, Mater PET/CT centre, Dublin.

origin of the radiation with high accuracy. The result is a very accurate three-dimensional map of the area in which the isotope is located and hence the location of regions of biochemical activity. $_9F^{18}$ is the most commonly used isotope, partly because of its very suitable mean life of about two hours.

12.5.6 *Antiprotons and antihydrogen*

Positron emission in the decay of radioactive elements is not the only source of antimatter. At particle accelerators high-energy protons can be made to collide with nuclei in a target, or indeed head-on with one another. The resultant energy release creates many new particles, and sometimes particle–antiparticle pairs such as protons and antiprotons. These emerge at high speed and can be separated and channelled into beams. One such beam is the beam of antiprotons at the centre of European nuclear research (CERN) near Geneva.

Atoms of antihydrogen

In August 2002, physicists at CERN announced the first controlled production of large numbers of atoms of antihydrogen. Antiprotons were slowed down until they were very slow by the standards of accelerators, and then trapped in an electromagnetic cage. They were then mixed with about 75 million positrons from radioactive decay of sodium-22, caught in a second trap. There was now created an environment of antimatter, in which the positrons and antiprotons were 'cold', i.e. moving slowly enough to have a significant chance of joining up as atoms. The team estimated that about 50,000 antihydrogen atoms had been produced in the first few weeks of the experiment.*

The next step will be to study the light emitted from the antihydrogen, and compare its spectrum with that of normal hydrogen. If there is a difference, no matter how small, it will

---

* Some more details of the experiment are given in Appendix 17.1.

mean that nature has come up with yet another major surprise: a fundamental basic asymmetry between matter and antimatter.*

12.6 Quantum reality

12.6.1 *Critics of the Copenhagen interpretation*

Albert Einstein found it very difficult to come to terms with the new ideas. In correspondence between Einstein and Bohr, which lasted for many years, we find a lively discussion on the principles of the Copenhagen interpretation, with Einstein asking clever questions, and Bohr supplying equally clever answers. It is there that we find Einstein's much quoted saying 'I cannot believe that God will throw dice' and Bohr's riposte 'Do not tell God what to do'.

Other physicists and philosophers expressed similar scepticism. Erwin Schrödinger, who himself had played a key role in the development of quantum mechanics, could not come to terms with the indeterminate nature of *quantum reality*. How could physical reality depend on whether or not this 'reality' had been observed? He devised what is perhaps the best-known example of the apparent absurdity of the theory — the paradox of the *Schrödinger cat*. This describes an imaginary situation in which a cat is placed in a sealed box. Its fate is determined by a small amount of radioactive substance, any atom of which may decay at any moment causing the release of a deadly gas which poisons the cat. Since the probability of radioactive decay is statistical we have no knowledge of the fate of the unfortunate cat, and hence the cat has a *dual existence*, one of which is observed and the other not observed. In the not-observed existence the cat is simultaneously alive and dead!

A physically more convincing example was presented in 1935 by Einstein together with Boris Podolski (1896–1966) and

---

* By its nature, antimatter is extremely difficult to contain and observe. Some more details of the experiment are given in Appendix 17.1.

Nathan Rosen (1909–1995), which demonstrated that, according to the Copenhagen interpretation, a measurement of the angular momentum of an electron could have an immediate effect on another electron located at an arbitrary distance somewhere else in the universe — the *EPR paradox*. If quantum mechanics was a correct representation of the laws of Nature, then these laws had the amazing quality of being *non-local,* in that observation made in one place, instantaneously changed things far away. Einstein in particular was unhappy with this notion, since according to his theory of relativity no effect could ever be transmitted with a speed greater than the speed of light!

All of the above paradoxes could be resolved by postulating the existence of '*hidden variables*'. These could be described as 'hidden gears and wheels', which determine the results of a physical measurement but may not be susceptible to our techniques of detection. Hidden variables maintain the classical point of view that the basic laws are indeed deterministic, and that the apparent probabilistic behaviour of Nature is due to our lack of knowledge of these variables.

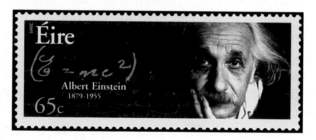

Albert Einstein never really came to terms with quantum theory.
Courtesy of An Post, Irish Post Office.

12.6.2 *Bell's theorem*

John Bell investigated the paradoxes of Bohr's quantum mechanics and originally set out to establish Einstein's assertion that the *Copenhagen interpretation of quantum mechanics was incomplete.* He was able to pinpoint a method for solving the question of whether or not hidden variables exist by invoking the so-called

Bell's inequality, a mathematical statement which applies to systems with 'realis-
tic' properties. Theories with hidden
variables satisfy Bell's inequality; quantum
theory does not. Experiments to test Bell's
inequality could resolve the question once
and for all.

During the period from 1980 onwards
many experiments have been performed to
test Bell's inequality. The first and perhaps most notable experiments were those
with photons of light, by Alain Aspect and
his group at Orsay. The overwhelming evi-

John Stewart Bell
(1922–1990).
*Courtesy of Queen's
University, Belfast.*

dence is that the laws of nature do not obey Bell's inequality.
Quantum mechanics and the Copenhagen interpretation agree
with experiment. The photon does not have hidden variables
which determine how it will behave. We will see further examples of the strange behaviour of light when we discuss experiments on individual photons one at a time in the next chapter.

12.6.3 *A precursor of quantum reality*

George Berkeley (1685–1753) was a
philosopher and an Anglican bishop.
His book *A Treatise Concerning the
Principles of Human Knowledge*,
published in 1710, proposed a philosophy that the existence of physical
objects consists in being perceived.
The physical world depends only on
the mind, and physical substance
does not have an independent existence. Not surprisingly, his philosophy did not gain general acceptance
at the time, but can now be viewed
in the light of the philosophical
implications of quantum mechanics!

*Courtesy of An Post Irish
Post Office.*

He emigrated from Ireland to America, where the city of Berkeley, California, and its university are named after him.

A historical interlude: Niels Bohr (1885–1963)

Niels Bohr was born on 7 October 1885 in Copenhagen, Denmark. His father, Christian, was then *Privatdozent*, and later Professor of

BOHR, Niels Henrik David
Nobel Laureate PHYSICS 1922
© Nobelstiftelsen

Physiology at the University of Copenhagen. His mother was Ellen Adler.

Bohr did well in school without being brilliant. He excelled in physical education and was a good soccer player, although not quite as good as his brother Harald, who won a silver medal playing for Denmark in the 1908 London Olympics.

Bohr entered the University of Copenhagen in 1903 to study Physics. In 1905, while only a second year student, he decided to try for the prize essay of the Royal Danish Academy, on the subject of 'the vibration of liquid jets'. He constructed his own apparatus, and impressed the judges to such an extent that he was awarded a Gold Medal in January 1907.

Bohr received his master's degree in 1909, and his doctorate in 1911. A grant from the Carlsberg Foundation enabled him to go to Cambridge in the same year, to study under Sir J.J. Thomson. He brought with him a copy of his thesis. Thomson was preoccupied with his own work and gave little of his time to students. In one of his letters to Harald, Niels complained that his manuscript lay unread under a pile of papers on Thomson's desk. Within less than a year

(Continued)

Bohr left Cambridge to join Ernest Rutherford's research team in Manchester.

Rutherford's laboratory was a hive of activity. Rutherford himself had just published what has since been recognised as one of the classic papers in the history of physics, showing that the mass of an atom was concentrated in a tiny nucleus. In another letter to Harald, Niels wrote: *'You can imagine it is fine to be here, where there are so many people to talk with...and this with those who know most about these things....'*

Bohr had married Margarethe Norlund in 1912, who accompanied him to Manchester and was expecting their first child when the Bohrs returned to Denmark in 1916. They had six sons, one of whom, Aage, followed his father into physics, and eventually into the ranks of Nobel Prize winners (in 1975).

Soon after his return from Manchester, Bohr was appointed to the Chair of Theoretical Physics at the University of Copenhagen. In 1917 he was elected to the Danish Academy of Sciences, and obtained the necessary backing to establish the Institute of Theoretical Physics in Copenhagen. He became its director from the opening in 1921, a post which he held for the rest of his life.

The institute soon attracted physicists from all over the world. The list reads like a who's who in physics at the time: Lise Meitner (Austria), Georg Hevesy (Hungary), Peter Kapitza, Hendrick Casimir, George Gamow and Lev Landau (Russia), Paul Dirac (England), Werner Heisenberg (Germany), Robert Oppenheimer (USA), Kazuhiko Nishijima (Japan), Albert Einstein.... The atmosphere was one of great excitement. The structure of the atom was being unravelled and the fascinating laws which govern the world of the very small was being explored.

In his acceptance speech at the Nobel Prize ceremony in 1922, Bohr was able to say: *'We not only believe in the existence of atoms, but we believe that we have an intimate knowledge of the constituents of the individual atom. All physical and chemical properties of substances are now clear.'*

(Continued)

Discussion of physics went on most nights until it was very late, but there still was time for games and banter. Bohr, who liked Westerns, produced a theory as to why the hero shoots faster even-through the villain draws first, and they tried out the theory with toy pistols. Bohr's theory was that the premeditated action of the villain was not as fast as the hero's automatic reaction. Bohr, who always played the hero, was usually faster, but it was not clear if this was due to Bohr's natural athleticism or the more profound reasons given by the theory.

One night, when returning from a party, Gamow, Casimir and Bohr tried to climb the wall of the Bank of Copenhagen building. The police, on arriving, took no action — *'Oh, it is only Professor Bohr!'*

There was a horseshoe above the door of the institute. When asked if he believed in luck, Bohr replied: *'Certainly not, but they say it does bring luck even to those who don't believe in it'*.

Considerable financial support for the Danish Royal Academy came from the Carlsberg brewery. There was no problem with accommodation, as Carl Jacobsen, whose father had founded the brewery, had specified in his will that his mansion in Copenhagen was to be used as a residence by the 'most distinguished Danish scientist, at that time, Niels Bohr.

Bohr was extremely exact and meticulous in every word he wrote or spoke. In fact his colleagues and students invented the term *'Bohrspeak'*, defined as a closely knit language in which every word counted and which required very careful listening. Bohr's sentences were carefully constructed to express his thinking. As he said himself: *'You can never express yourself more clearly than you can think.'*

On 11 February 1939, two Austrian physicists, Otto Frisch (1904–1979) and Lise Meitner (1878–1968), published a paper in *Nature*, in which they interpreted the results of earlier experiments by Meitner and Otto Hahn (1879–1968) as a new process which they called *nuclear fission.* When uranium is bombarded by neutrons, it breaks into 'two nuclei of roughly equal size'. The reaction involves the uranium nucleus, the tightly knit core of the atom, hitherto considered impenetrable. Much energy is released in the process

(Continued)

together with more neutrons, which can in turn trigger the reaction in neighbouring uranium nuclei. Bohr had been informed of the results in advance, and announced their conclusions at a meeting of the American Physical Society, in Washington DC, on 26 January 1939.

Following the outbreak of World War II and the occupation of Denmark by Germany, Bohr received a visit from his former German student Werner Heisenberg. There is no historical record of what transpired during that visit. It is possible that Heisenberg told Bohr that Germany was on the point of using fission in uranium to make an atomic bomb. Perhaps he was trying to pick Bohr's brains on the critical mass of uranium 235 required to sustain a chain reaction. A play by Michael Frayn, entitled '*Copenhagen*', which opened in London's West End in 2001, puts forward the above two alternatives, as well as a third conjecture, that Heisenberg came to warn Bohr of the danger of remaining in Denmark, and to urge him to escape.

It was later revealed at the Nüremberg trials that it was the original intention to arrest Bohr immediately after the day of the occupation of Denmark, 29 August 1943. It was decided to postpone the order for about a month, when such an arrest 'would attract less attention.'

On 29 September 1943, as it later transpired in the nick of time, Bohr and his family were smuggled by fishing boat to Sweden. On the very next day Bohr, who had already arranged the reception of many refugees to Sweden, met the Secretary of State and King Gustav V of Sweden to try to arrange the acceptance of more ships of Jews 'for internment in Sweden'. Unfortunately, at this stage, the Swedes were powerless to comply.

From Sweden, Bohr and his son Aage were flown to England in the bomb bay of an RAF Mosquito, nearly dying from the cold and oxygen starvation in the process. Apparently the flying helmet was too small to fit Bohr's rather large head, resulting in a faulty connection in the oxygen supply tube, so that he was actually unconscious when the plane landed. On his arrival in England he heard of the development of fission research in the US.

Bohr had believed that the rare U-235 isotope of uranium, which made the chain reaction possible, was too difficult to

(Continued)

separate in quantities large enough to make an atomic bomb. But now news had come that Peirls and Frisch in Birmingham had shown that the critical mass of U-235 was about 1 kg, and not 1 ton, as previously believed. The US had imported 100 tons of uranium ore from the Belgian Congo. The implications were clear — the Americans believed that the manufacture of a bomb was possible.

Bohr immediately wrote to Winston Churchill: *'This project will bring either disaster or benefit to mankind on a scale hitherto unimaginable.'* He tried to persuade Churchill to take the development of nuclear power seriously, to ensure that it would serve mankind as a whole, and renewed his request that the development of an atomic bomb be dropped. Churchill was not impressed, and it was clear to Bohr that he was making little impact.

Shortly afterwards the Bohrs sailed secretly to the Los Alamos Atomic Laboratories in the US, where Bohr became known as 'uncle

Nick' as his security name was *Nickolas Baker.* He saw the enormous plant for the separation-of-uranium part of the Manhatten project under his former student Robert Oppenheimer. The great energy derived from the atomic nucleus was clearly to be used as a weapon of war and nothing could be done to stop the project. Bohr admitted: *'They don't need me to make an atomic bomb.'*

When World War II ended in Europe on 8 May 1945, Bohr's primary concern was to prevent a nuclear arms race. Nuclear power could lead to the greatest boom for mankind, but also to

Jumbo atomic device for the 'Trinity test'.

man's greatest disaster. Bohr travelled to London to meet Churchill on 16 May 1944 to try to impress on him that control of nuclear weapons

(*Continued*)

was vital to world security, but met with a frosty reception. Churchill clearly distrusted him. A meeting with Roosevelt on 26 August 1944 was considerably more amiable, but still he could not stop military plans to use atomic bombs against Japan. Two bombs were exploded on Hiroshima and Nagasaki on 6 and 9 August 1945.

Nagasaki explosion, August 1945.

At the end of August 1945 Bohr returned to Denmark and continued where he had left off. Unpretentious as ever, he rode every day to the institute on his bicycle. The year 1952 saw the establishment of the centre of nuclear research (CERN) in Geneva, with Niels Bohr as its first chairman. Bohr devoted the rest of his life to promoting peaceful uses of atomic energy. He died on 18 November 1962 in Copenhagen.

Appendix 12.1   Calculating the radius of atomic orbits for hydrogen

Let us assume that the electron moves in a circular orbit around a proton which is at rest.

Classical angular momentum $mvr = \dfrac{nh}{2\pi}$, $n = 1, 2, 3....$

The quantum condition

Centripetal force $\dfrac{mv^2}{r} = \dfrac{e^2}{4\pi\varepsilon_0\, r^2}$

Coulomb attraction

$$\Rightarrow v^2 = \frac{e^2}{4\pi\varepsilon_0\, mr} = \left(\frac{nh/2\pi}{mr}\right)^2$$

$$\Rightarrow r = \frac{4\pi\varepsilon_0 n^2 h^2}{4\pi^2 m e^2} = \frac{\varepsilon_0 n^2 h^2}{\pi m e^2}$$

$n = 1$ for the stable hydrogen atom $\Rightarrow r = 0.529 \times 10^{-10}$ m.

Planck's constant, in combination with previously known physical constants, enabled Bohr to calculate the size of electron orbits, which can be considered as defining the size of an atom. We have detailed information about something that is small almost beyond imagination.

Putting the size of an atom into perspective

Consider a cube of size 1 cm³. We could fit 10^8 atoms along each side. This would mean 10^{24} atoms squashed into the tube.

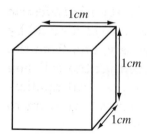

Compare that with the Pacific Ocean: 10,000 km × 10,000 km × 1 km = volume 10^8 km³ = 10^{23} cm³.

Thus there are 10 times as many atoms of hydrogen in a 1 cc box as there are cubic centimetres of water in the Pacific Ocean!

Chapter 13

Atoms of Light Acting as Particles

The energy of light comes in quantum units of value hf. In this chapter we go a step further, and look at evidence that it comes as a bundle, concentrated at a point, and behaves just like a particle. We will describe experiments that show that the *photon* has not only energy but also momentum.

First, we will show that the photon can behave like a bullet causing considerable disruption to surfaces of materials, particularly metals, whose atoms have 'loose' outer orbit electrons which are vulnerable to such bombardment. In the *photoelectric effect* an electron is knocked out from the surface of a metal by a photon. Because this happens instantaneously, and from the energy of the ejected electron, we can deduce that the collision is a 'one-on-one' encounter. There are many practical applications; we can use the photoelectric effect, whenever we want to translate a light signal into an electric current.

Next, we will look at an experiment which demonstrates that the photon has another characteristic of a particle, namely *momentum*. In the *Compton effect* the photon bounces off an electron, and the two then go off at different angles. We can apply the laws of conservation of energy and momentum to calculate angles and energies just as we do for a collision on a snooker table. In this experiment we look at the photon after it has been scattered in the collision.

There is one interesting difference between a scattered photon and a scattered snooker ball. The photon cannot slow down;

and always travels at the speed of light. It loses energy and momentum by decreasing its frequency and correspondingly increasing its wavelength, without changing its speed. The exact way in which this should happen is calculated in Appendix 13.1, while experimental evidence that it does happen is presented below.

13.1 The photoelectric effect

13.1.1 *Evidence for the particle nature of light*

As the new ideas presented by Planck were being discussed, Albert Einstein turned his attention to the specific nature of the quantum of light. As he pointed out, Maxwell's classical theory had been truly tried and tested with respect to the *propagation* of light; it had nothing to say, however, about how radiation interacts with matter, how it is emitted or absorbed. Is the energy in a quantum concentrated at a point, or is it spread over the whole wave front? If it were concentrated, it would have a much greater effect locally. If, for example, it were to interact with an electron, could it deliver to it a 'short sharp shock'?

13.1.2 *Short sharp shocks*

In the everyday 'classical' world, energy delivered in a series of short sharp shocks has a more dramatic effect than the same amount of energy delivered continuously. You may wash a car with a water hose, which gives a relatively smooth bombardment of the surface. Sand-blasting, however, is not recommended! What may be the same amount of total energy concentrated in individual grains of sand, gives a series of localised sharp shocks, and may well penetrate and damage the paintwork. There would be even more local damage

Grains of light can give you short sharp shocks.

if all the energy were concentrated in a single steel pellet, but the rest of the surface would remain unaffected.

How about light shining on a surface? Everyday 'classical' instinct tells us that the distribution of energy will be smooth. This conclusion seems justified in that, for example, it can be tolerated by as delicate a surface as human skin, at least in moderation. One would not expect any effect on a metallic surface except perhaps an increase in temperature if the light were intense enough. It is therefore quite surprising that when surfaces of certain metals such as zinc, rubidium, potassium or sodium are illuminated by light of sufficiently low wavelength, we get what is called the *photoelectric effect*, an *immediate release of electrons* which can then be drawn away to produce an electric current.

13.1.3 *An accidental discovery*

The *photoelectric effect* was first discovered, quite by accident, in 1887, by Heinrich Hertz (1857–1894), who was working on the production of radio waves by means of an electrical spark discharge between two metal spheres. He noticed that when he illuminated the spheres with ultraviolet light, the sparks which came across were bigger and brighter. At the time he did not realise that this was due to electrons liberated from the metal by the light. A year later Wilhelm Hallwachs (1859–1922) confirmed the phenomenon for various other metallic surfaces.

There are 'grains' in the light beam

The fact that the photoelectric effect works at all is already strong evidence for the particle-like nature of light. Bits of matter thrown out look more like the result of sandblasting! Particularly convincing is the fact that there is no measurable delay between exposure to light and the appearance of photo-electrons. Light of intensity as low as a few microwatts per square metre produces an immediate, detectable photoelectric current. The effect is practically instantaneous (the delay is less than 10^{-9} s).

13.1.4 *How long would we expect to wait? An order-of-magnitude calculation*

The minimum energy required to pull an electron out of the metal is of the order of 3 eV. (The electron volt, eV, is the unit of energy commonly used in atomic physics, equal to the energy of an electron which has been accelerated through a potential difference of 1 volt, and is equivalent to 1.60×10^{-19} J.) Let us assume that there is one free electron per atom and that the photoelectron comes from one of the ten layers of sodium atoms closest to the surface. Assuming a light beam is spread out evenly and smoothly with an intensity of 5×10^{-6} Wm^{-2}, we can estimate the time required for an electron to receive the necessary energy as follows:

Estimating the time taken if the light energy is evenly divided

A monolayer of sodium atoms contains $\approx 10^{19}$ atoms m^{-2}

$$\Rightarrow \approx 10^{20} \text{ atoms in 10 layers}$$

5×10^{-6} Wm^{-2} divided among 10^{20} atoms $= 5 \times 10^{-26}$ W per atom

$$= 5 \times 10^{-26} \text{ Js}^{-1} \text{per atom}$$

Required to liberate electron $3 \text{ eV} = 3 \times 1.6 \times 10^{-19}$ J

$$\Rightarrow \text{average wait} = \frac{3 \times 1.6 \times 10^{-19}}{5 \times 10^{-26}} \text{s}$$

$$\approx 10^7 \text{ s} \approx 4 \text{ months}$$

Allowing for the very approximate nature of the above estimate, it is clear that the observed emission of photoelectrons is too fast by at least 16 orders of magnitude. The situation can be compared to the power of the sea which gradually erodes the shore. The energy of the waves is distributed over a large area, and over a long period of time. If this power were concentrated in quanta of analogous propor-

tions, we might find the Rock of Gibraltar ejected suddenly and without warning.

13.1.5 *The 'lucky' electron*

A model consistent with experimental evidence of the ejection of electrons from a metallic surface by photons is as follows: the atoms of the metal have one or two outer electrons which are relatively 'free' and can be considered as constituting an *electron gas* within the solid material. These are candidates for ejection not only from the atom but from the surface of the metal. We will assume that the photon collides with an individual electron and gives it enough energy to overcome the potential barrier and escape. Only a tiny fraction of the available electrons is released in the whole process, and which electrons are hit by the photon is entirely a matter of chance. The 'lucky' (or, perhaps, unfortunate) electron receives *a whole quantum of energy* while its neighbours remain undisturbed.

Some of the liberated electrons are attracted back into the metal, while more escape. In this way a steady state of

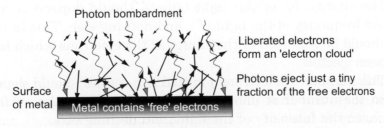

Figure 13.1 'Electron gas'.

equilibrium is established with an 'electron cloud' near the surface. The electrons move randomly within the cloud (Figure 13.1).

13.1.6 *Einstein's photoelectric equation*

In 1905 Albert Einstein (1879–1955) published a paper modestly entitled 'On a Heuristic Point of View Concerning the Generation and Transformation of Light'. The word 'heuristic' implied that the matter was not necessarily in its final form and that alternative explanations were possible. He proposed that if the photon was localised in space it could transfer its entire energy to an electron. This energy could then be used by the electron to escape from the metal. Whatever energy was left over would appear as kinetic energy of the 'liberated electron' according to the equation

$$hf = W + (KE)_{max} \quad \text{Einstein's equation}$$

where hf is the energy of the incoming photon, W is the *work function* of the metal, which is the minimum work needed to overcome the restraining forces which bind the electron to the metal, and KE_{max} is the energy of the most energetic emitted electron.

Einstein was a great thinker, but he did not do experiments. It remained for others to submit the equation to the rigorous test of experiment. If the equation was correct, one would have to observe the following:

1. The energy, hf, of the 'light bullets' should depend only on the frequency of the light (f), and nothing else. This in turn should be reflected in the energy of the electrons which have been ejected.
2. The number of electrons emitted per second should depend on the number of photons per second striking the surface, i.e. on the intensity of the light, and nothing else.
3. Emission would be instantaneous.

The American physicist Robert A. Millikan (1868–1953) is perhaps best known for his work which showed that electric charge comes in discrete indivisible units. An electron carries one such fundamental unit of charge, which he measured in his 'oil drop experiment' published in 1912. It was natural that he should also turn his attention to a quantitative study of the characteristics of electrons emitted in the photoelectric effect. His first step was to measure the energies of electrons emitted by light of various frequencies shining on different metals. As early as 1908, at the Boston meeting of the American Physical Society, he had announced that he intended to measure the kinetic energy of these electrons by a determination of the *stopping potential*. This is the adverse electrical potential required to prevent them from reaching a second metallic plate at the end of an evacuated tube. Essentially he would give the electrons an 'electrical hill' to climb, and investigate how high this hill must be to completely stop the photoelectric current, for different values of the frequency of the incoming light.

At first Millikan was quite sceptical of Einstein's theory, and his original intention was as much to disprove as to confirm the validity of the equation. 'I scarcely expected that the answer, when it came, would be positive, but the question was very vital, and an answer of some sort had to be found.'

The principles of Millikan's experiment were quite straight-forward, but in practice the work was long and tedious, and it took some years to complete. When his results confirmed Einstein's theory Millikan he was generous in giving due credit to Einstein and admitting that his original scepticism had been unfounded.

13.1.7 *Millikan's Experiment*

Millikan's method was to set up a 'workshop *in vacuo*' by enclosing the photosensitive surface in a highly evacuated quartz tube. Light shining on the surface liberates electrons. By applying an electric field in the right direction the electrons can be continually swept away, forming an electric current. They are now 'out in the open' and their behaviour can easily be studied.

Figure 13.2 illustrates an apparatus such as was used by Millikan. The photoelectrons, which have been ejected from the metal, are now in a vacuum, and can be subjected to electric forces for analysis. Since they are negatively charged, the photosensitive plate, or *emitter*, is made the *negative* electrode. At the opposite end, the positive electrode, is the *collector* of the electrons.

The electric field accelerates the electrons away from the emitter creating a current through the tube and around the circuit.

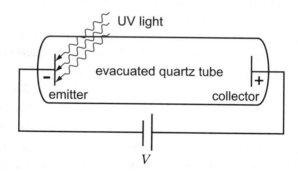

Figure 13.2 Experimental set-up to draw the electron cloud across an evacuated tube.

	Adverse field:		Accelerating field:	
	Emitter	Collector	Emitter	Collector
	+	−	−	+

We are now ready to study the properties of the electrons emitted when the photosensitive surface is illuminated by light of different frequencies.* For each frequency the resulting

---

* Even when light is behaving like a particle, we can describe the photon in terms of wave characteristics such as frequency and wavelength.

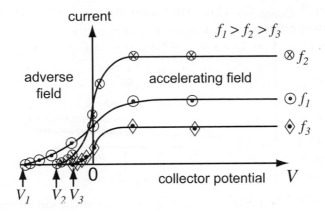

Figure 13.3 Current for different frequencies of the incoming light plotted as a function of the voltage across the tube.

current is plotted as a function of the voltage across the tube and the results are presented in Figure 13.3.

As the accelerating voltage is increased the current increases untill it reaches the *saturation current* where all the electrons are swept across the tube as they are liberated. In a separate experiment, it can be shown that the magnitude of the saturation current depends on the intensity of the incoming light. (In this example the light of frequency f_2 happens to be the most intense.)

13.1.8 *Current flowing uphill*

When the external voltage is reduced to zero, there is still a current; some electrons get across to the collector without help. Even when the polarity of the source is reversed some current remains. The energy of some of the liberated electrons is high enough to overcome the retarding potential and reach the collector. This is apparent from the left side of the graph, which shows the presence of remnants of current despite an adverse electric field. We have the equivalent of spring water flowing uphill against the force of gravity.

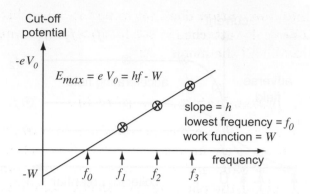

Figure 13.4 Experimental data giving the value of the cut-off potential as a function of the light frequency. The points lie on a straight line, exactly as predicted by Einstein's equation.

Eventually, when the adverse voltage reaches a certain value called the *cut-off* or *stopping potential*, even the most energetic electron is repelled, and the current stops. In Figure 13.3 the stopping potentials for photocurrents liberated by light of frequencies f_1, f_2 and f_3 are V_1, V_2 and V_3 respectively. The higher the frequency of the incoming light, the greater the cut-off potential; the exact relationship is plotted in Figure 13.4.

The most energetic photoelectron has an energy eV, just enough to overcome an adverse potential V. Figure 13.4 shows this energy as a function of the frequency of the light shining on the photosensitive surface. According to Einstein's equation it is equal to the energy of the photon minus the *work function W*, the energy used up by the electron in getting out of the metal.

All the experimental data on the photoelectric effect agree with the predictions of one simple equation! Abandoning his former scepticism Millikan conceded *'this is a bullet-like, not a wave-like effect'*. As a bonus, the slope of the graph in Figure 13.4 provides an independent measurement of Planck's constant, h.

Note that *momentum* does not come into play, because the electron is basically attached to the lattice, which is much heavier, and absorbs all the momentum.

13.1.9 *The photoelectric work function*

The work function W is a measure of the minimum amount of work required to draw an electron out of the metal. It varies from metal to metal, and determines the minimum photon energy required for the photoelectric effect to take place at that metallic surface.

The most energetic photons of visible light are at the violet end of the spectrum. It can be seen from the table that such photons have enough energy to liberate electrons from caesium, potassium and sodium, but not from aluminium, copper, mercury or lead.

Table 11.1 Photoelectric work function of some elements.

Element	Al	Cs	Cu	Hg	K	Na	Pb
W (eV)	4.28	2.14	4.65	4.49	2.30	2.75	4.25

$1 \text{ eV} = 1.60 \times 10^{-19}$ coulombs $\times 1$ volt $= 1.60 \times 10^{-19}$ J

Wavelength (nm)	400	550	700
Photon energy, hf (eV)	3.11	2.26	1.61

An example with numbers

When light of wavelength 480 nm strikes a photosensitive surface, the emitted electrons have a stopping potential of 0.45 V.

What is the photoelectric work function of the metal? (1 eV = 1.60×10^{-19} J)

$$\lambda = 480\,\text{nm} \Rightarrow f = \frac{c}{\lambda} = \frac{3 \times 10^8}{480 \times 10^{-9}} = 6.25 \times 10^{14}\,\text{s}^{-1}$$

$$\Rightarrow hf = 41.44 \times 10^{-20}\,\text{J} = 2.59\,\text{eV}$$
$$\Rightarrow W = hf - \text{eV} = 2.59 - 0.45 = \boxed{2.14\,\text{eV}}$$

Looks like caesium!

13.1.10 *Practical applications*

Quite apart from its theoretical importance, the photoelectric effect has numerous practical applications. It makes it possible to convert a light signal into an electric current. Television cameras, burglar alarms, barcode readers and light sensors of every description are based on the photoelectric effect.

Perhaps the most remarkable practical example is the *photomultiplier*, which amplifies the electrical signal to such an extent that it is possible to detect the arrival of a single photon. The use of this instrument in some fascinating fundamental experiments with individual photons will be discussed in Chapter 14.

In 1907, a Russian scientist, Boris Rosing, realised that light from a cathode ray tube could be thrown onto a screen and made into a picture. One of his students, Vladimir Kosma Zworykin (1889–1982), was fascinated by this notion and applied for a patent for what was in effect the idea for the first television camera. The patent application was lodged in 1923, but Zworykin had no working model at the time.

Zworykin went to work for Westinghouse Electric and Manufacturing Company in America and began to develop a practical version of the camera tube called the *iconoscope*. Its face consisted of a thin sheet of mica on which thousands of grains of photosensitive silver caesium compound had been deposited, creating a mosaic of tiny photoelectric cells. When

light illuminated one of these
cells, it liberated electrons which
left the cell positively charged. A
scanner passing across the cells
collected the charge and turned it
into an electric signal. At first
Zworykin's attempts were unsuccessful, to the extent that his
employers at Westinghouse told
him to abandon the project and
turn his attention to 'something
more useful'.

Vladimir Zworykin.
Courtesy of Restelli Collection,
http: // history.net.

Zworykin left Westinghouse and in 1929 went to work for
the Radio Corporation of America, established by another
Russian émigré, David Sarnoff. He persuaded Sarnoff that
the idea would work and that a commercially viable television
system could be established. Sarnoff joked afterwards about
Zworykin's skills as a salesman. *'He told me it would cost*
$100,000, but RCA spent 50 million dollars before getting a
cent profit.' Zworykin managed to transmit successful television images in 1933. In 1938, fifteen years after it was first
lodged, his patent for the iconoscope was eventually granted.

The picture shows the first
demonstration of the transmission of a TV drama. This was
written by Eddie Albert and
called *The Love Nest*. The broadcast was made by the National
Broadcasting Company (NBC)
from the top of the Empire State
Building on 6 November 1936
and was received on the 62nd
floor of the RCA building.

Courtesy of Restelli Collection,
http: // HistoryTV.net.

Zworykin never claimed to be
the sole inventor of television. He always insisted that '... *hun-*
dreds contributed, each added a rung to the ladder which

enabled others to climb a little higher and see the next problem better'.

One such person was Philo Taylor Farnsworth (1906–1971), a 14-year-old prodigy who independently conceived the idea of an electronic camera. It is said that the concept of a line scanner first came to him as he was going up and down ploughing a field at his father's farm in Utah. He showed a diagram of his plans to his teacher Justin Tolman, who was greatly impressed and encouraged and supported him for the rest of his life. Farnsworth spent two years at Brigham Young University, and then set up his own laboratory in a loft in San Francisco. In 1927, at the age of 21, he succeeded in transmitting a picture from one room to another, impressing some sponsors, who helped him to found Farnsworth Television Inc in 1929.

There was to follow litigation, which lasted for some years, between RCA and Farnsworth Inc as to who owned the rights to the manufacture and sale of television sets. In 1934 the US patent office awarded priority of invention to Farnsworth, as a result of which RCA had to pay them royalties until the beginning of World War II, when the sale of all such sets was suspended.

13.2 The Compton effect — more evidence for the particle nature of light

13.2.1 *Real bullets have momentum*

Einstein's equation for the photoelectric effect, and indeed Planck's original postulate, involve conservation of energy only and make no reference to momentum. The next important question to be answered was whether or not the particle representation of light can be carried further to show that a photon possesses *momentum as well as energy.*

Snooker with a photon as the cue ball.

13.2.2 *The Compton effect*

The challenge of showing that the photon has momentum was answered in 1923, when Arthur H. Compton (1892–1962) made a discovery which was to earn him the Nobel Prize in 1927. He set out to demonstrate that a photon behaves exactly like a particle when it makes a collision with an electron, and both energy and momentum are conserved in the process.

In order to penetrate a carbon target, and interact with atoms and their orbital electrons, Compton needed more energetic photons than those of visible light. Accordingly he bombarded graphite with X-rays of sharply defined wavelength λ. Some X-rays could be expected to scatter from electrons in orbit around the carbon atoms, and others from the much heavier carbon nucleus.

13.2.3 *Collision dynamics revisited*

Energy transfer

When a particle collides elastically with another particle the laws of conservation of momentum and energy are strictly observed. (Since the collision is elastic, 'energy' means kinetic energy.) However, both energy and momentun are in general *transferred* from one particle to the other. How much of each is transferred depends on the relative masses of the colliding particles and on the angle of scatter.

In snooker, when the cue ball strikes a stationary ball of equal mass *head-on*, all the energy and all the momentum are transferred from the moving ball to the target ball. The cue ball stops dead, and the target ball continues with the cue ball's original velocity. If the moving ball only makes a glancing collision, it keeps most of its momentum and energy. The greater the change of direction of the moving ball, the more energy is transferred to the target ball.

Bouncing off a big target

When the target is much bigger than the projectile, it may receive a lot of momentum but very little kinetic energy. For example, if we bounce a tennis ball off a bus, the ball will bounce back with the same speed and practically no loss of KE. The momentum of the ball is reversed. To compensate, the bus recoils with twice the original momentum of the tennis ball. The momentum transfer may be larger than the original momentum of the moving particle, yet practically no KE is transferred.

13.2.4 *Collision of X-ray photons*

In the scattering of X-rays from an atomic nucleus, we may treat the impact as the collision of a photon with a very large object. Practically no energy is lost by the photon although it will generally change direction, and lateral momentum transfer will take place. No matter how large the scattering angle, the energy hf remains almost exactly the same, i.e. the frequency and therefore the wavelength of the emergent photon are the same as before.

In the case of a collision with a free electron, the mass of the target is much smaller. The dynamics of the collision can be compared with the classical collision of two snooker balls:

13.2.5 *The photon loses energy but does not slow down*

From the laws of conservation of momentum and energy it is possible to calculate the momentum of the scattered particle in terms of the scattering angle, the masses of the projectile and target particles, and the momentum of the incoming particle.

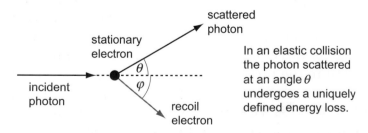

Figure 13.8 A photon scatters off an electron.

The situation here is somewhat different from that in classical mechanics, in that the photon has *zero mass and always travels with the speed c*. The scattered photon will lose energy, but its speed remains the same. The energy loss, appears as a change in *frequency* of the photon ($hf \rightarrow hf'$), and is transferred to the target particle as kinetic energy.

Compton assumed that a photon has:

$$\text{energy} = hf = \frac{hc}{\lambda}$$

$$\text{momentum} = \frac{hf}{c} = \frac{h}{\lambda}$$

Assuming that there is an elastic collision of a photon ('cue ball') of energy hc/λ, and momentum h/λ with a stationary electron of mass m_e, we can calculate the change in wavelength of the photon for a given scattering angle from the relation:

$$\Delta\lambda = \frac{h}{m_e c}(1 - \cos\theta)$$
$$= \lambda_c (1 - \cos\theta)$$

> Compton effect: change in wavelength of X-rays due to scattering by an electron

The derivation of this formula is given in Appendix 13.1.

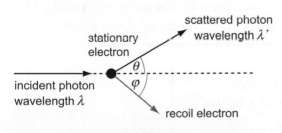

incident photon
wavelength λ

stationary electron

scattered photon
wavelength λ'

recoil electron

Note that the change in wavelength is independent of the wavelength of the incident radiation. It depends inversely on the mass of the target particle, confirming that energy transfer is negligible for X-rays scattered from an atomic nucleus. The quantity $h/m_e c$ is a universal constant called the *Compton wavelength* of the electron and has a value $\lambda_c = 2.44 \times 10^{-12}$ m $= 2.44 \times 10^{-3}$ nm.

The maximum wavelength change occurs for $\theta = 180°$, i.e. a (unlikely) head-on collision in which the photon bounces straight back. Even then $\Delta\lambda = 0.004885$ nm, which is difficult to detect unless the original wavelength is itself small and the wavelength change forms a significant fraction of the original wavelength (another reason for using X-rays).

13.2.6 *Experimental verification*

Classical analogy

Bouncing a highly elastic ball off a stone wall. No energy is transferred to the wall. The ball will come back with the same speed, at an angle determined by the angle of incidence.

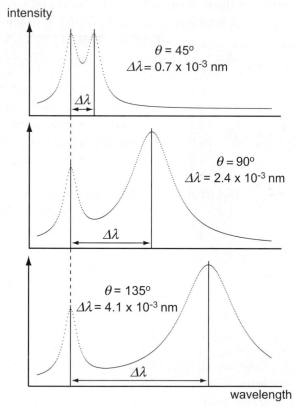

Figure 13.9 Wavelength displacement of X-rays from a molybdenum K_α line scattered from carbon.

Colliding in mid-flight with another ball. The original ball will lose energy. The greater the angle of scatter, the greater the energy transferred to the other ball.

The photon has both energy and momentum, just like any other particle. After the experimental confirmation of the theoretical model of Compton scattering, the existence of light quanta could hardly be questioned.

A historical interlude: Robert A. Millikan (1868–1953)

Robert Andrews Millikan was born on 22 March 1868 in Morrison, Illinois, USA. His grandfather had been one of the American pioneers of European descent who settled in Western Ohio and adopted their own self-sufficient lifestyle. He ran his own farm, grew his own food and tanned the hides of cattle to make shoes for his family, while Robert's grandmother spun yarn and made the clothes.

MILLIKAN, Robert Andrews
Nobel Laureate PHYSICS 1923
© Nobelstiftelsen

Robert's father was the Reverend Silas Franklin Millikan, and his mother was Mary Jane Andrews. In 1875 Millikan's family moved to Maquoketa, Iowa, a town of 3000 inhabitants. The name Maquoketa means 'big timber', and in fact it was situated at the edge of the primitive woods of the Midwest.

The family owned about an acre of grounds, on which they grew potatoes, corn and melons, and on which the three boys and three younger sisters helped during vacations from school. Robert milked the cows, tended to a neighbour's horse and mowed the lawns. His greatest thrill was to be

(Continued)

allowed to ride bareback on wild colts to break them in for riding. It is little wonder that Robert grew to be strong, fit and athletic. Later he earned most of his way through college by acting as a student gymnasium director. He remained interested in fitness and sport for the rest of his life, and was a good tennis player and enthusiastic golfer.

Millikan graduated from Maquoketa High School in 1885 and then entered Oberlin College (Ohio), where he studied Greek and Mathematics and also took a 12-week course in Physics. On the basis of this he was later appointed a tutor in Physics at Oberlin for a period of four years. In 1893, on the recommendation of his Oberlin professors, he obtained a Fellowship in Physics at Columbia University.

After Millikan had graduated from Columbia University in 1895 his mentor, Dr M.I. Pupin, advised him to go to Europe for a year or two, believing that American science had fallen far behind Europe, from which great new ideas were coming at that time. Dr Pupin even lent him $300, a not-inconsiderable amount of money, to help him on his way. Millikan found his year abroad a great experience. He met, and attended lectures by, some of the greatest physicists of that time — Poincaré in Paris, Planck in Berlin, Nernst in Göttingen. On the downside, he found the continuous drilling and parading of troops in Germany quite disturbing, and also the prevailing attitude to class distinctions of the 'strutting, overdressed and overaccoutred officers.' Another feature, which he considered a blemish specifically on the university life, was the 'bellicose attitudes among the students' which sometimes could even result in duelling, which he witnessed a number of times in Göttingen.

In the summer of 1896, when Michelson offered him an assistantship in his department in Chicago, he immediately set out to return to America. Being rather short of cash, he hocked his luggage with the captain of a transatlantic transport ship, telling him that he would pay the fare before removing his belongings on arrival in New York.

The job in Chicago had a salary of $800 per annum, compared to an offer of $1600 he had just received from Oberlin; nevertheless,

(Continued)

the opportunity to work with Michelson and to be part of the stimulating research atmosphere easily made up for the loss of earnings. One of the first seminars that Millikan prepared was a review of J.J. Thomson's paper on 'Cathode Rays'. This is now generally considered the first paper on electron theory. Thomson showed that cathode rays were not waves in the ether but consisted of tiny negatively charged particles which he called *corpuscles* of mass more than a thousand times smaller than the atom of hydrogen. This marked the beginning of Millikan's interest in the electron.

On 10 April 1902, Millikan married Greta Blanchard, a recent graduate of the University of Chicago. The ceremony took place in the evening — in his autobiography he confesses that he spent time until late that day reading the proofs of his book on mechanics in the office of the publishers and almost missed his own wedding! All ended well, however, and the young couple spent the next seven months touring Europe on their honeymoon. They were to have three sons, Clark, Glen and Max.

Millikan's first experiments to determine the charge of the electron were made in 1906, when he built a 10,000 V small cell storage battery designed to create an electric field strong enough to balance the force of gravity and hold charged water droplets in a vapour cloud suspended in the air. He soon improved the method by using oil drops and watching individual drops through a low power microscope.

If the number of electrons carried by the droplet happened to change as an electric charge fell off, or an extra charge was added, the oil drop would suddenly begin to move. By putting each drop through a series of up and down trips by varying the strength of the electric field and timing each trip accurately, Millikan discovered that the speeds of a given drop always changed by a series of steps which were a multiple of a certain unit. This unit change took place, he concluded, when the drop lost or gained a single electron as a passenger.

To get the data on one droplet usually took hours, and Mrs Millikan often had to change social arrangements at the last minute, because her husband had been detained 'watching a drop'.

(Continued)

Millikan was so convinced by the experiment that he wrote: *'One can count the number of free electrons in a given small electrical charge as one can attain in counting the number of one's fingers or toes.'* He presented his final results on the charge of the electron at the 1912 meeting of German physicists in Berlin, quoting a value of $4.806 \times 10^{10} \pm 0.005$ electrostatic units ($=1.603$ Coulombs). This was a remarkably accurate result when compared to the best modern value of 1.60217 Coulombs.

Millikan's work on the photoelectric effect began in 1912 and occupied practically all his research time for the next three years. While he accepted Planck's theory that the energy of light was quantised, he did not believe that it was also localised, moving through space in tight bundles just like a point particle. In an earlier paper on Planck's constant he had emphatically distanced himself from 'Einstein's attempt to couple photo effects with a form of quantum theory'. He wrote: *'The photon is a bold, not to say the reckless, hypothesis — reckless because it is contrary to such classical concepts as light being a wave propagation phenomenon.... A corpuscular theory of light is quite unthinkable, unreconcilable with the phenomena of diffraction and interference.'*

Later Millikan was the first to admit that he had been wrong. The evidence of his own experiments showed clearly that the hypothesis of localised photons was correct. In a paper published in the *Physical Review* in 1914 he wrote: *'After years of testing, changing and learning, and sometimes blundering... this work resulted, contrary to my own expectation, in the first experimental proof of the Einstein equation.'*

Millikan received the Nobel Prize for Physics in 1923 for 'his work on the elementary charge of electricity and on the photoelectric effect'. His acceptance speech contained an eloquent exposition of his belief in the complementarity of theory and experiment: *'Science walks forward on two feet, theory and experiment. Sometimes one foot is brought forward first, sometimes the other, but continuous progress is only made by the use of both, theorizing and testing.... I have been attempting to assist in bringing the experimental foot up to parallelism at least with the theoretical field of ether waves....'*

(Continued)

Throughout his life Millikan's researches were by no means confined to the physics of atoms, electrons and photons. In 1910 a former student of his by the name of Jewett, who had become an employee of the Bell Telephone Company, approached him with a problem in the amplification of telephone signals. At that time it was possible to talk between New York and Chicago, a distance of about 800 miles, but the technology was not ready to transmit undistorted speech signals over a much greater distance. The heads of the corporation wanted to have a line from New York to San Francisco to be ready in time for the 1914 San Francisco Fair. Using their experience with electron beams, Millikan and his research students developed an 'electron tube amplifier, capable of practically distortionless amplification of human speech'. In his own words: *'The electron, up to that time largely a plaything of the scientist, had entered as a potent agent in supplying man's commercial and industrial needs.'*

At the outbreak of World War I, Admiral Griffin of the US Navy requested Millikan's help to develop and construct listening devices for the detection of German U-boats, which were causing great damage to Allied ships. Millikan responded immediately by hand-picking a group of about ten university physicists to address the problem. The most promising suggestion, made by Max Mason of Wisconsin, was to mount a whole row of 'receiver-ears' on the opposite sides of the hull of a ship. The lengths of tubes running to the listener from each pair of such ears were adjusted, so that all impulses coming from a distant source in front of the ship would come together in phase. Essentially, the principle of this system was analogous to that of an optical transmission grating. Such an experimental 'trombone receiver' was quickly constructed and found to work, but the Navy was not able to build such ships quickly enough. They enlisted Millikan's help in interesting Henry Ford to adapt his vast car-manufacturing resources to supplement the effort by constructing small fast boats fitted with Mason detectors.

Soon the armistice came, and Millikan returned to civilian life. A couple of years later he was persuaded to change his allegiance from the University of Chicago and accept a full time appointment at Pasadena, where he had already spent several months a year as

(Continued)

Director of Physical Research. Arthur Fleming had instituted a multi-million-dollar trust for the purpose of creating 'the strongest possible department of physics'. Millikan was to become the director of an institution which was to be named the California Institute of Technology (Caltech). Caltech grew from strength to strength. The cream of European physicists came to visit and work there, including Einstein, Lorentz, Bohr, Schrödinger and Heisenberg. In the 1930s the range of research directed by Millikan expanded far beyond theoretical physics, and included genetics, biochemistry, aeronautics and cancer therapy.

In Millikan's view, Caltech existed to provide America's scientific leadership. In more recent years, Nobel Prize winners such as Richard Feynman and Murray Gell-Mann, among others, have continued the tradition in physics.

Millikan had deep religious convictions and published a number of books on philosophy, and on the reconciliation of science and religion. To quote from his autobiography: *'Religion and science, in my analysis, are the two great sister forces which have pulled, and are still pulling mankind onward and upward...you cannot possibly synthesize nature and leave out of it its most outstanding attributes, consciousness and personality...those which you know that you yourself possess... in other words. Materialism as commonly understood is an altogether absurd and utterly irrational philosophy, and is indeed so regarded, I believe, by most thoughtful men.'*

Robert Millikan died on 19 December 1953, in San Marino, California.

Appendix 13.1 Mathematics of the Compton effect

Collision of a photon with a stationary electron; deriving the Compton formula relating change in wavelength to scattering angle of the photon:

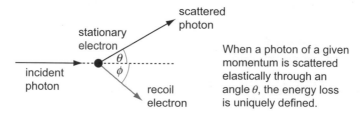

When a photon of a given momentum is scattered elastically through an angle θ, the energy loss is uniquely defined.

Assume the incoming photon has energy $E = hf$ and momentum $p = hf/c$, the scattered photon has energy $E' = hf'$ and momentum $p' = hf'/c$, and p_e is the momentum of the recoil electron.

Momentum vector triangle:

From the cosine rule and multiplying across by c^2,

$$p_e^2 c^2 = (hf)^2 + (hf')^2 - 2h^2 ff' \cos\theta \tag{i}$$

Classical derivation

If T is the kinetic energy of the recoil electron, we can write p_e^2 in terms of T using the classical formulae

$$p_e = m_e v \quad \text{and} \quad T = \frac{1}{2} m_e v^2$$

$$\Rightarrow p_e^2 = 2m_e T$$

$$\Rightarrow 2m_e Tc^2 = (hf)^2 \left[1 + \left(\frac{f'}{f}\right)^2 - 2\left(\frac{f'}{f}\right) \cos\theta \right]$$

$$= (hf)^2 \left[1 + \left(\frac{f - \Delta f}{f}\right)^2 - 2\left(\frac{f - \Delta f}{f}\right) \cos\theta \right]$$

$$= (hf)^2 \left[1 + \left(\frac{1 - \Delta f}{f}\right)^2 - 2\left(\frac{1 - \Delta f}{f}\right) \cos\theta \right]$$

$$= (hf)^2 \left[1 + 1 - \frac{2\Delta f}{f} + \left(\frac{\Delta f}{f}\right)^2 - 2\left(1 - \frac{\Delta f}{f}\right) \cos\theta \right]$$

$$\cong (hf)^2 \left[2\left(1 - \frac{\Delta f}{f}\right) \right] (1 - \cos\theta)$$

(making the approximation that the term $(\Delta f/f)^2$ can be ignored)

$$\Rightarrow T = \frac{(hf)^2}{m_e c^2}\left[1 - \frac{\Delta f}{f}\right](1 - \cos\theta) \quad \text{but} \quad 1 - \frac{\Delta f}{f} = 1 - \frac{f - f'}{f} = \frac{f'}{f}$$

$$\Rightarrow \boxed{T = \frac{h^2 ff'}{m_e c^2}\,(1 - \cos\theta)}$$

Energy lost by the photon = kinetic energy gained by the electron

$$hf - hf' = T$$

$$hf - hf' = \frac{h^2 ff'}{m_e c^2}(1 - \cos\theta)$$

$$\Rightarrow \quad m_e c^2(hf - hf') = h^2 ff'(1 - \cos\theta)$$

$$\Rightarrow \quad \left(\frac{m_e c}{h}\right)\left(\frac{f}{c} - \frac{f'}{c}\right) = \frac{ff'}{c^2}(1 - \cos\theta)$$

$$\Rightarrow \quad \left(\frac{m_e}{h}\right)\left(\frac{1}{\lambda} - \frac{1}{\lambda'}\right) = \frac{1 - \cos\theta}{\lambda\lambda'}$$

$$\Rightarrow \boxed{\lambda' - \lambda = \left(\frac{h}{m_e c}\right)(1 - \cos\theta)}$$

Compton scattering formula

Relativistic derivation (gives the same formula)

The Compton scattering formula has been derived above using the *classical* expressions for *kinetic energy and momentum* of the electron. If the electron velocity v is significant compared to the speed of light, the corresponding *relativistic* expressions must be used.

We will meet the methods of relativity later in this book (Appendix 16.2), but as happens when applied to Compton scattering, the relativistic way is simpler, requires no approximation, and gives the same final result. It is given here for completeness:

$$E^2 = p^2c^2 + m_0{}^2c^4$$

$$\Rightarrow \quad p_e^2c^2 = (T + m_0c^2)^2 - m_0{}^2c^4$$
$$= T^2 + 2m_0c^2T$$

but $\quad T = hf - hf'$ as before

> Electron:
> m_0 = rest mass
> E = total energy
> p = momentum

$$\Rightarrow \quad p_e^2c^2 = (hf)^2 + (hf')^2 - 2h^2ff' + 2m_0c^2\,(hf - hf')$$

but $p_e^2c^2 = (hf)^2 + (hf')^2 - 2h^2ff'\cos\theta$ as before

$$\Rightarrow \quad 2h^2ff' + 2m_0c^2\,(hf - hf') = -2h^2ff'\cos\theta$$

$$\Rightarrow \quad 2m_0c^2\,(hf - hf') = 2h^2ff'\,(1 - \cos\theta)$$

$$\Rightarrow \left(\frac{m_0c}{h}\right)\left(\frac{f}{c} - \frac{f'}{c}\right) = \frac{ff'}{c^2}(1 - \cos\theta)$$

$$\Rightarrow \left(\frac{m_0c}{h}\right)\left(\frac{1}{\lambda} - \frac{1}{\lambda'}\right) = \frac{1 - \cos\theta}{\lambda\lambda'}$$

$$\Rightarrow \quad \boxed{\lambda' - \lambda = \left(\frac{h}{m_0c}\right)(1 - \cos\theta)}$$

Compton formula
(via relativistic
derivation)

Similar the classical formula, with m_0 = rest mass instead of m_e = classical mass

It is interesting to note that the Compton formula calculated relativistically is identical to the formula calculated using classical mechanics. The classical method uses the approximation that the square of $\Delta\lambda/\lambda$ be ignored while the relativistic method is exact. The higher the photon energy, the less valid the approximation as $\Delta\lambda$ becomes a greater fraction of λ. The distinction between m_e and m_0 becomes meaningful only at very high energy, when the speed of the electron approaches the speed of light.

Chapter 14

Atoms of Light Behaving as Waves

We have established that light has the properties of a wave and discussed the evidence in detail. It spreads out like a wave, can bend around corners like a wave, and exhibits the properties of *diffraction and interference*. It must be a wave.

Later we described particle properties of light. The light quantum behaves like a bullet. It can knock electrons out of the surface of a metal in the *photoelectric effect*, and bounce off electrons in the *Compton effect;* it has energy and momentum. It must be a particle.

Now that we have got used to the fact that light sometimes looks like a wave and sometimes like a particle, we meet something even more mysterious. We consider what happens when we use a very dim light source so that the *photons* may be separated from one another by as much as one kilometre. We are definitely dealing with individual photons, and yet somehow they have not forgotten their pedigree of a wave! When we send them through apparatus such as Young's slits, we still see interference! No photons land where there had been a dark fringe, when we had studied interference with a brighter source, even though this time two photons are never together so that they might possibly interfere. Somehow the solitary photon behaves as if it had gone through both slits at the same time! *The photon is definitely an isolated particle, yet it behaves as a wave!*

As if this were not strange enough, we can arrange things in such a way that the alternative paths of the photon are not close together, as they are in Young's slits but separated by an unlimited distance. Now we discover something which defies logic. Let us say that the photon has two possible paths, A and B. When we close the alternative route B, miles away, it somehow has an effect on the photon which is travelling along route A. This is *action at a distance,* even more mysterious than that due to electric or gravitational forces!

We finish the chapter with a glimpse of *quantum electrodynamics (QED)*, a theory which deals with the most fundamental behaviour of light and matter.

14.1 Photons one at a time

14.1.1 *The human eye*

The sensitivity of the human eye is quite remarkable. When fully adapted to darkness it can detect a signal consisting of as little as 5 or 6 photons. To put this into perspective, a small bulb on a Christmas tree emits about 10^{20} photons per second. Perhaps it is just as well that we have not evolved so far as to see a single photon, as then, even in the darkest night, stray photons would appear as a series of intermittent tiny flashes with which our brain might find it difficult to cope.

14.1.2 *Detecting a single photon*

We can detect a single photon using a *photomultiplier*. This instrument is based on the photoelectric effect and on the fact that an electron, when it strikes a metal plate, can liberate two or more electrons, each of which in turn can repeat the process at successive plates.

An avalanche set off by a single photon.

A photomultiplier tube is an evacuated glass vessel enclosing a series of electrodes called *dynodes*. A voltage difference of between 1000 V and 2000 V is maintained

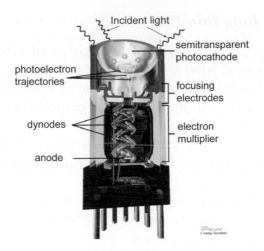

Figure 14.1 Photomultiplier tube converts 'light into electricity'. *Courtesy of Dr Sanjiv Sam Gambhir, Stanford University.*

between the photocathode (emitter) and the anode. The first dynode is a couple of hundred volts less negative than the cathode, the second dynode is a couple of hundred volts less negative than the first dynode and so on.

When a photon strikes the photocathode, it releases an electron which is accelerated to the surface of the first dynode where it liberates low energy 'secondary' electrons. Each of these may be accelerated to the next dynode where further secondary electrons are liberated, and so on.

The process continues 10 or 12 times, until billions of electrons strike the last plate, the anode. The photon has triggered a chain reaction leading to an electrical pulse which can be amplified by a regular amplifier. This can be translated into an audible 'click' by connecting to a speaker.

Albert Einstein received the Nobel Prize for the discovery of the law for the photoelectric effect. *Courtesy of An Post, Irish Post Office.*

14.1.3 *The long thin line*

By selecting a very dim source of light, and attenuating the beam by means of filters, we may obtain a beam of separated photons and study them essentially 'one at a time'. For example, consider a 1 mW helium–neon laser emitting a thin beam of light of wavelength 632 nm.

Let us calculate how many photons are emitted per second, and the average distance by which they are separated in space.

For λ = 632 nm, photon energy hf = 3.14 × 10^{-20} Joules
1 mW = 10^{-3} Joules per sec = 3.2 × 10^{16} photons per sec

If the beam passes through a filter which attenuates it by a factor 10^{12} this leaves 32,000 photons per second.

Thirty thousand photons per second may sound like a lot of photons, but they are spread over a distance which light travels in one second, i.e. 30 million metres. The line of photons emitted in one second would wrap around the earth more than seven times. This means that in the attenuated beam the photons will be an average distance of 10,000 metres apart.

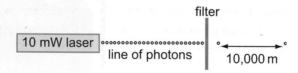

Figure 14.2 The filtered beam photons are about 10,000 m apart.

14.1.4 *Single slit diffraction*

We have seen lots of evidence that light behaves as a wave. Other experiments show just as convincingly that light behaves as a particle. It is interesting to find out what happens when we try to combine the two aspects. The challenge is to devise experiments which show light as a beam of particles and then to determine if these particles also show wavelike behaviour.

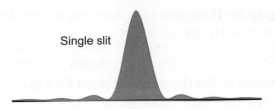

Single slit

Figure 14.3 Intensity distribution of 'photon hits' on the photographic film. Photons make the same pattern as waves when they come out of a thin slit! (cf. Figure 8.8.)

Working in a completely dark room, we reduce the intensity of a beam of light until the photons arrive 'one at a time'. In our first simple experiment we shine this beam on a narrow slit with a photographic plate behind it, in place of a screen. With such a low light intensity we will have to wait for quite a long time. Eventually, when we develop the film, we find exactly the same pattern as we got when using a beam of normal intensity.

This result is somewhat surprising, but we can probably construct some explanation. Perhaps, as the photons come through one by one, they are attracted by the edges of the slit, and not all continue in a straight line. That might explain the width of the centre of the diffraction pattern. The secondary maxima are more difficult to explain — but if we use a lot of imagination we might come up with something!

14.1.5 *Double slit diffraction and interference*

Our theory of 'edge attraction' fails completely, however, when we go one step further, and send our 'beam of separated photons' through two thin slits close together. Again we get the same result as we got with light of full intensity. As in the 'standard' Young's experiment, the region of overlap of the two beams consists of bright and dark fringes. The dark fringes are regions of destructive interference of light from the two slits. But this seems impossible. This time we can hardly talk about two beams. Photons arriving one at a time, separated by 10 km, can hardly interfere with one another!

Single photons make the same pattern as waves when they come through two thin slits!

⸕14.1.6 *Measuring 'clicks' as photons arrive one by one*

We now try an even more sophisticated experiment, in which we replace the screen by an array of photomultipliers. Each time a photon arrives we note the time and position. The hairline slits have reduced the beam intensity even further. In general only one photomultiplier goes off at a time, and we hear a series of discrete clicks. The light definitely behaves as a stream of particles.

Each photon passes through the apparatus 'on its own', the next photon is at a distance of about 10 km. It is difficult to construct a theory of interference between photons. There is no other photon or wave with which it could interfere!

The photons appear to be arriving in a random manner, rather like a shower of hailstones.

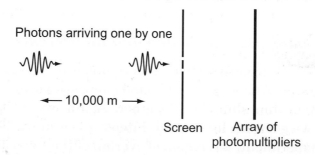

Figure 14.4 Young's experiment using very dim light and photomultiplier detectors. (As before, the diagram is not drawn to scale. The slits are much narrower and closer together.)

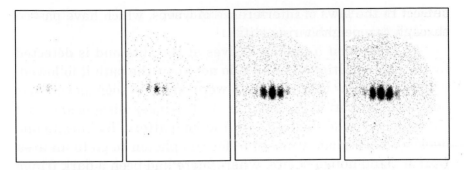

Figure 14.5 'The pattern grows'. Results of an experiment reported by Delft University. (The 'dark' fringes are actually the areas where photons landed.)

A research project on discrete photon effects which was carried out by two third year students at Delft University of Technology, Niels Vegter and Thijs Wendrich, working under the supervision of Dr S.F. Pereira, obtained the interference patterns shown in Figure 14.5. (The experimental arrangement involved an image intensifier coupled to a highly sensitive photographic film, using gradually increasing exposure times — a slight variation on the method described above.)

At first just a few photons land on the screen. They seem to be distributed in a random manner. Gradually the pattern of 'hits' builds up, and we note a surprising result. There are areas where photons never land. We have 'dark fringes', as predicted by the wave theory of light. Eventually we finish up with bright and dark fringes, exactly as in the 'regular' Young's slit experiment.

The particles somehow follow the interference pattern of waves. There is a law of nature which prohibits the photons from landing in certain areas. Even though each photon arrives on its own, it appears to be

subject to the laws of interference of waves, which have passed through two neighbouring slits!

We know that a photon arrives at the slits and is detected by the photomultiplier, but we do not know the path it followed. All we can tell is that both slits were open. As soon as we close one slit, the pattern of regular dark fringes disappears, and we get back to the single slit diffraction pattern. By '*closing* one door' we have made it possible for the photon to go to an area were it could not go before, where there had been a dark fringe when both doors were open.

According to wave theory, the waves emerging from the two slits interfere and give rise to the fringes. But how does the photon manage to be in two places at the one time and to interfere with itself?

Suppose that we try to resolve the matter by making a small adjustment to our apparatus. We insert a small detector beside each slit to tell us which way each photon goes. When a signal from one of the two detectors coincides with a hit on the screen we will have tracked the photon. We will know which way the photon went. Apart from inserting the detectors, we have not interfered with the experiment in any other way. The result is remarkable. The interference pattern on the screen changes. Instead of interference fringes typical of Young's double slit experiment, we get back to characteristic single slit diffraction.

The act of making the measurement has interfered with the result of the experiment. Many experiments of this type have

been done with electrons and every single one has confirmed this result. The answer is that we cannot determine where any particular electron or photon goes, because the very act of making a measurement changes the situation.

14.1.7 *Separating the possible paths*

It may be argued that interference effects can occur only if the possible paths of the photon are very close together. It is possible, however, to make the alternative paths as far apart as we like by an interferometer type of arrangement as illustrated in Figure 14.6.

Let us first consider what happens from the point of view of the wave representation. The light beam arrives at mirror 1 and splits into two components; approximately half is reflected (beam I), the other half is transmitted (beam II). The two rays emerge at right angles and then travel along different paths. When each beam meets the half-silvered mirror 4, a similiar thing happens as at mirror 1. Each beam splits into reflected and transmitted components. The result is that the beams recombine. The recombined beam has two components which emerge from mirror 4 at right angles, each consisting of a 50/50

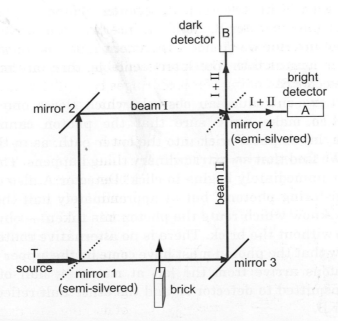

Figure 14.6 Typical interferometer experiment.

mixture of I and II. One goes towards detector A, and the other towards detector B.

Let us assume that the interferometer is adjusted so that the two light beams leaving mirror 4, are out of phase going in the direction of detector B (the 'dark' detector), and in phase in the direction of the 'bright' detector A. The fact that no light arrives at detector B is explained by the wave theory of light on the basis of destructive interference of the two combined beams.

Next, let us try the particle representation. As we did before, we can reduce the intensity of the laser beam, until the photons are so far apart that a photon usually has long left the apparatus before the next photon enters. We continue the experiment, and discover that the result is exactly the same as in the wave representation. Even though we know that this time the photons go through the apparatus one at a time, the 'dark detector' registers no clicks. No photons manage to reach it. All the photons go to detector A.

Somehow, with two routes open, the photon is prevented from reaching detector B. Note that the two alternative routes are far apart, but we do not know which route the photon has taken. So long as both routes are available, there are no clicks from detector B!

Next, we introduce an obstacle which closes one of the routes. To make quite sure that the photon cannot get through, let us put a brick into the lower path, as in the diagram. We find that an extraordinary thing happens. The dark detector immediately begins to click! Detector A also continues registering photons, but at approximately half the rate. We now know which route the photon has taken — obviously the one without the brick. There is no alternative route open; we know that the photon must have come by the 'upper' route. All photons arrive from the left at mirror 4; half of them are transmitted to detector A, and the other half reflected to detector B.

14.1.8 *'Delayed choice'*

A more sophisticated experiment was proposed by John A. Wheeler (1911–) of the University of Princeton. He suggested that you can determine what kind of condition you want to set up at a time when the photon is already 'on its way' through the apparatus. If you decide to offer it one path only, it will behave as a particle, showing no interference. If you decide that it could have gone 'along both paths at once', it will behave as a wave, and will obey the rules of interference.

In the *delayed choice* extension of the experiment, the route is randomly closed or opened when the photon has almost completed its journey towards mirror 4. The photon 'does not know' whether the 'lower route' is closed or open when it comes to mirror 1.

Obviously, this must be done faster than by moving a brick, but it can be done electronically. The experiment was carried out successfully by Carroll Alley and co-workers at the University of Maryland in 1986.

Instead of a 'brick', Alley used an optical switch which could open or close a route within 5×10^{-9} s (5 ns). The length of time for a photon to go through the apparatus was about 30 ns. The result was as predicted by Wheeler. It does not matter when the route is closed. If the photon could have taken either route, interference occurs, and detector B remains dark, whereas if one route had been closed, even in the middle of its journey, there is no interference and detector B registers 'clicks'.

14.2 Feynman's 'strange theory of the photon'

14.2.1 *Partial reflection*

We saw in Chapter 3 that Isaac Newton was deeply puzzled by the fact that when light strikes a boundary between two media, part of it is reflected and part of it enters the second medium. Looking at this from the point of view of a photon, its fate is not pre-determined. It has a certain probability of 'bouncing back',

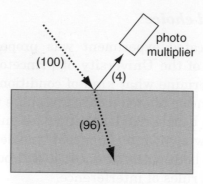

Figure 14.7 Partial reflection of single photons.

and a certain probability of penetrating into the second medium. This is a typical quantum-mechanical situation.

Let us follow Richard Feynman's account of his formulation of quantum electrodynamics by considering the partial reflection of single photons. Such photons, well separated in space, arrive at a boundary between air and glass, as illustrated in Figure 14.7.

Suppose we find that out of 100 photons which strike the glass surface, on the average, 96 are transmitted and 4 are reflected.

Let us consider in detail what seems to be happening. The photons arrive at the surface one by one. They are all identical. What causes a particular photon to be reflected and another to be transmitted? Perhaps there are some local imperfections in the glass surface, grains that reflect the photon and holes that let it through? Let us stick with that theory for the moment.

 A possible theory: If a photon hits a grain of microscopic or atomic dimensions, it is reflected; otherwise it continues on into the glass.

Let us now replace the block of glass by a thin sheet of glass. We find that when the photons come to the second surface, a similar thing happens. Some photons are transmitted and get back out into the air while others are reflected back into the

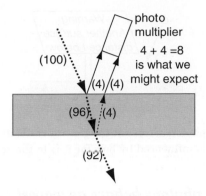

Figure 14.8 What we expect is not what we get.

glass. The relative percentage reflected and transmitted is the same as at the first surface.

The photomultiplier counts photons reflected from both surfaces. This figure is similar to Figure 8.13 in Chapter 8, when we were dealing with the reflection of light waves from thin films. This time, however, we are dealing with individual photons; we hear a 'click' as each photon enters the photomultiplier.

If 4 photons out of 100 are reflected at the first surface, we might expect 8 photons or slightly less to reach the photomultiplier after 2 reflections. (Slightly less because the original beam of 100 photons has decreased to 96 photons reaching the second surface, and then gets attenuated again when it meets the first surface again on the way back.)

The result of this experiment negates any simple theory of grains or imperfections at each surface; it is not at all what we expected. Sometimes the photomultiplier registers no clicks whatsoever, and sometimes there are as many as 16 photons! *It all depends on the thickness of the glass.*

The new arithmetic:
4 + 4 = 0 to 16
What we expect is not what we get!

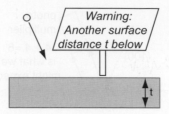

Figure 14.9 When a photon arrives at the glass surface, its decision on whether to reflect or continue is influenced by how far it is to the second surface.

Even individual photons behave as waves!

In Chapter 8 we were able to explain the phenomenon of reflection of light by thin films on the basis of interference between the two waves reflected at the first and the second surface. We now find the same result when we are dealing with individual photons! They register as clicks of the photomultiplier, i.e. behave as particles, yet at the same time they follow the rules of waves. The photon has two faces — we can see one or the other, depending on how we look at it!

Remember the
analogy?

14.2.2 *The strange theory of the photon*

We have seen that not only can light behave as a particle, but that this particle of light, the *photon*, has properties like no other particle. It seems to go from place to place by more than one route at the same time; it seems to 'know' if another route is closed or open, even when it apparently has not travelled by that route. Its rules for reflection follow a strange kind of arithmetic.

We can extend Young's experiment with two slits to three or four or more slits, in fact to any number of slits, until we have a *diffraction grating* with tens of thousands of slits. (Chapter 8). A single photon now has all these possible routes available

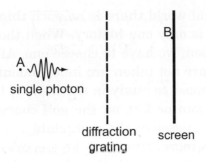

single photon

diffraction
grating

screen

Figure 14.10 A single photon faced by a diffraction grating. There are many routes available to go from A to B.

through the diffraction grating, and somehow they all contribute to the interference pattern on the screen!

14.2.3 *A 'sum over histories'*

To explain how all the routes available determine the probability for a photon to go from A to B, Richard Feynman proposed in 1948 a new formalism now known as the *sum-over-histories approach,* which can be summarised by saying:

> *'Anything which might have happened influences that which does happen.'*

14.2.4 *The rotating amplitude vector*

In Feynman's formalism, every path that the photon can take is represented by an 'arrow', technically known as the *amplitude vector*. In the case of the diffraction grating there is an amplitude to get from A to B via each of the slits, including the ones which are not directly in line. The probability that that photon will get from A to B *is obtained by taking the sum of the amplitudes over all possible paths.*

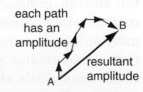

each path has an amplitude

resultant amplitude

Amplitudes are added like ordinary vectors to get the probability amplitude for a photon going from A to B

In the classical world there is no such thing as a 'sum over histories'. There is only one history. When there are a number of possibilities open, we have to choose one. All the other possible paths which are not taken are independent of the one actually taken. If I choose to study in the library, I will never know if I would have won or lost on the golf course. Neither will I know what I had missed at the nightclub!

There is one more aspect of Feynman's theory which is important. As we travel with the photon along any path, the

amplitude vector *rotates rapidly*. As time goes on, it keeps pointing in a different direction. We can visualise the vector as the hand of an imaginary stopwatch, which measures the time taken for the journey. We stop the watch at the instant the photon arrives at point B. The direction of the vector depends on the time taken for the journey. It is the direction *at the end of the journey*, which is all-important when we calculate the probability that the light took a particular path.

In the *household* world we can follow only one of many possible histories.

However, there is one difference between the amplitude vector and an ordinary stopwatch. The hand of the stopwatch rotates relatively slowly. At the end of the race it will point in more or less the same direction for every competitor. The amplitude vector rotates so rapidly that it will have made billions of revolutions while the photon was moving. When the photon arrives at B the vector could be pointing in any direction!

If we take the slits away altogether, there are still an unlimited number of paths for the photon to go from A to B and *there is an amplitude for each one*!

Figure 14.11 Rotating amplitude vector.

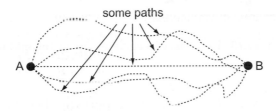

Figure 14.12 Each path from A to B has a finite amplitude.

The worm is asking a very relevant question which deserves more than the owl's snappy answer. We know from experiment that in free space light goes from A to B in a straight line, so how can we have amplitudes for all the other paths?

Why light appears to travel in straight lines

We can do no better than to let Feynman answer as he does in his book *QED: The Strange Theory of Light and Matter**: 'When all possible paths are considered, each crooked path has a nearby path of considerably less distance, and therefore much less time (and substantially different direction of the arrow). Only the paths near the straight line path have arrows pointing in nearly the same direction, because their timings are nearly

---

* Princeton University Press (1985), ISBN 0-691-08388-6.

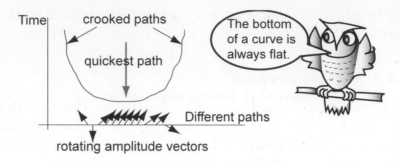

Figure 14.13 The slope of a curve is zero at the minimum.

the same. Only such arrows are important, because it is from them that we accumulate the large final arrow.'

In summary, paths near the minimum take almost identical times. For these paths the 'stop-watch' stops in nearly the same place, therefore amplitudes reinforce. Amplitudes for much longer paths far out point in random directions, as times taken differ a lot.

14.2.5 *How can we believe all this?*

The model of a sum over histories, and rotating vectors, is quite mysterious and unbelievable. Feynman explains: '*Physicists have learned to realize that whether they like a theory or not is not essential to the question. Rather it is whether or not the theory gives predictions that agree with experiment.... The theory of quantum electrodynamics describes Nature as absurd from the point of view of common sense. And it agrees fully with*

*experiment. So I hope you can accept Nature as She is —
absurd.'*

14.2.6 *It all comes together*

Nobody could be blamed for becoming confused about all these
aspects of light. We based geometrical optics on Fermat's principle of least time. In wave optics we considered the wave nature
of light. Now we are talking about vector amplitudes!

In fact, these methods are equivalent mathematically.
Vector amplitudes in Feynman's formalism add and subtract in
an identical manner to the superposition of waves. The vector
amplitudes representing paths of the same single photon are
more mysterious, but the mathematical result is the same.

Instead of talking about 'histories' and 'rotating arrows' let
us visualise two trains of waves, which follow different paths
and come together again. Paths which are near to the quickest
path are almost equal in length. The waves arriving at B are
almost exactly in phase, and the resultant amplitude at B is
large. Waves which take longer, more 'roundabout' routes can
arrive at B with any phase value. In general a large number of
such waves coming together will cancel out.

What then is new?

The new and very powerful theory developed by Feynman extends
the formalism to include the most basic particle of matter, the
electron. His theory not only considers how (i) a photon can go
from place to place, but also (ii) *how an electron can go from
place to place*, and (iii) *how an electron can emit or absorb a*

Figure 14.14 When the path lengths differ significantly, there is a random
phase difference between waves arriving at B.

photon. These, according to Feynman, are the three basic actions from which all the phenomena of light and electrons arise. They form the basis of *quantum electrodynamics.*

14.2.7 *Quantum electrodynamics*

Quantum electrodynamics (QED) was not a new theory, it was initially developed in 1929 to describe the interaction of light and matter. In particular, light photons interact not with matter as such, but with electrons, which are fundamental carriers of electric charge.

In 1929, Paul Dirac incorporated the theory of special relativity into quantum mechanics, and developed a relativistic theory of the electron. The theory says that the electron has a *magnetic moment*, which means that it behaves like a little magnet, and interacts with magnetic fields. The strength of that magnetic moment in especially chosen units was exactly 1. Experiments showed that this prediction was very nearly correct. In 1948 very accurate measurements gave a very slight but significant deviation from the predicted value, giving a value 1.00118 with a tiny experimental error of 3 in the last place of decimals. The deviation, it was concluded, was due to the interaction of electrons with *light.*

Feynman's formalism not only predicted this deviation, but predicted its value with fantastic accuracy. The calculated magnetic moment was 1.00115965246, compared to the very accurate experimental value of 1.00115965221. The accuracy of these numbers can be compared to the thickness of a human hair in a distance from London to New York!

In the theory of quantum electrodynamics the force between electric charges is communicated through the exchange of *virtual photons.* It combines the following three basic-actions: '*an electron can emit a photon*', '*an electron can absorb a photon*' and '*a photon can go from place to place*'. The process is represented by a diagram invented by *Feynman* and is illustrated in Figure 14.15.

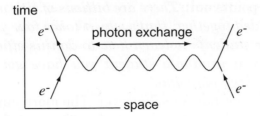

Figure 14.15 Feynman diagram illustrating photon exchange between two electrons which gives rise to the electromagnetic force between them.

Feynman diagrams illustrate what is an extremely complex process in a deceptively simple way. Let us take as an example these three Feynman diagrams:

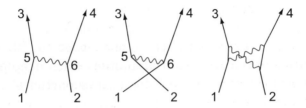

We are considering the possible ways in which we can go from the state with two electrons, one at point 1 and one at point 2 in space–time, and finish with one electron at point 3 and one at point 4. There are many ways in which this can happen. First of all, they may 'ignore' one another and go directly from 1 to 3 and from 2 to 4; alternatively, there is the 'cross-over way', from 1 to 4 and from 2 to 3. We have not bothered to draw diagrams for these.

The first two Feynman diagrams refer to ways in which the electrons go to some intermediate points 5 and 6, exchange a photon, and then continue to either 3 or 4. The third diagram represents a process in which two photons are exchanged.

The complexity of the problem arises in that the intermediate points can be *anywhere* in space–time. In each diagram the amplitudes for all these routes have to be calculated and added.

As Feynman points out: *'There are billions of tiny arrows which have to be added together, that's why it takes four years of graduate work for students to learn how to do this efficiently... but when you are a graduate student you have got to get your degree so you keep on going.'*

The task is not hopeless, however. The more junctions in the diagram, the smaller the amplitude. The probability rapidly decreases to one part in a million and even much smaller, and soon the more complicated ways can be ignored. Computers can be programmed to help (many of the calculations are repetitive, ideal for computers).

Once the method has been mastered, the rewards are great. Quantum electrodynamics has proved to be a most beautiful theory, giving predictions which agree with experiment and results with accuracy which is unprecedented.

Richard Feynman, together with Julian Schwinger and Sin-Itiro Tomonaga, won the Nobel Prize in 1965 for 'their fundamental work in quantum electrodynamics, with deep-ploughing consequences for the physics of elementary particles'.

A historical interlude: Richard Feynman (1918–1988)

Richard Feynman was born in a small town near the outskirts of New York called Far Rockaway. He never lost his unpretentious manner and his strong Brooklyn accent, and disliked pomp and ceremony of any kind. Even when his contribution to *quantum electro-dynamics* earned him the Nobel Prize in 1965, he accepted it graciously but somewhat reluctantly. He was not interested in prizes: *'I already had the prize — the discovery of such a wonderful law of Nature.'*

Feynman lived in Rockaway until he was 17 years old; he then went to MIT for four years and afterwards as a postgraduate to Princeton until 1939. In 1941 he married Arlene Greenbaum, who taught him the motto 'What do you care what other people think?'.

(Continued)

Arlene died of tuberculosis in 1946, but he kept the motto for the rest of his life.

FEYNMAN, Richard P.
Nobel Laureate PHYSICS 1965
© Nobelstiftelsen

In 1943, Feynman went to Los Alamos to work on the Manhattan Project, the highly secret American atomic research programme, and he stayed until 1946. After that he obtained a teaching post at Cornell University and worked there until 1951. He then spent a short time in Brazil, from where he went to the California Institute of Technology, where he remained as Professor of Theoretical Physics. He married Gweneth Howerth in 1955 and they had two children, Carl and Michelle.

Feynman's great contribution to physics has of course been QED, but that has not by any means been his only contribution. There is practically no area in physics to which he has not brought his unique insight. His ability to transmit knowledge was extraordinary, and generations of students and colleagues were inspired by him. Feynman's lectures were like shows, entertainment from beginning to end. His famous undergraduate course at Caltech was recorded by Robert Leighton and Matthew Sands, and the resulting three-volume textbook is arguably one of the most original and inspiring publications of its kind.

From the time he was still at school, Feynman devised ingenious little experiments on a great variety of topics. One such was to study 'the mind of the ant', how ants knew where to go, and whether they could communicate. He ferried ants, which happened to walk on strips of paper, to and from a source of sugar, high up on an isolated platform. He then brought some of them back to where they came from, and soon there were myriads of ants scurrying about at the

(Continued)

'ferry terminal' — obviously the news had got around that this was where one went to get sugar. Next, he tried an extension of the experiment, by picking up ants at the terminal, and bringing them up to the sugar, but bringing them back to a different place. The question was, if an ant wanted to get to the sugar a second time, would it go back to the original place from which it had been transported, or to the place where it landed on the way back? The result of the experiment apparently was that the memory of an ant is rather short — the ants remembered only their last port of call. Most went back to the final 'Arrivals' landing place rather than the original 'Departures' terminal.

Feynman was interested in literally everything. When in Brazil he learned to play samba music, and marched in the *Carnaval* in Rio de Janeiro. He became good friends with many artists, among them Jirayr Zorthian, who taught him how to draw. He signed all his artwork 'Ofey'; some of it was good enough to be sold and these paintings are now collectors' items. When already quite ill in 1982, he played the part of Chief of Bali Hai in a production of *South Pacific*.

Mischief was never far from Feynman's mind. At Los Alamos he was working on the highly secret Manhattan Project to separate isotopes of uranium for an atomic bomb. During his spare time he amused himself trying to crack the codes of safe combinations. Soon he was able to open safes and filing cabinets containing highly classified information. He caused consternation when, instead of going through official channels to get certain information he wanted, he opened a filing cabinet and left a piece of paper on which was written: 'Borrowed document no. LA4312 – Feynman the safecracker.'

Rejected by the army

During a psychiatric examination for the army to serve in the post-war occupational forces in Germany, Feynman decided to answer questions accurately and honestly. The result was remarkable, if not

(Continued)

totally unexpected: he was classified D, for 'deficient', as his replies apparently showed symptoms of insanity. Examples:

'Do you think people talk about you?' — *'Sure, when I go home my mother often tells me how she was telling her friends about me.'*

'Do you ever hear voices in your head?' — *'Sometimes when I have listened to a person with a foreign accent; for instance I can hear Professor Vallarta say Dee-a, dee-a eelectric field-a.'*

'How much do you value life?' — *'Sixty-four !! 'How are you supposed to measure the value of life?'* — 'Yes, but why did you say *sixty-four*, and not *seventy-three* for instance?' — *'If I had said seventy-three, you would have asked me the same question.'*

Feynman's attitute to organisation and authority is illustrated by a $10 bet he made with Victor Weisskopf, the then-director of CERN, that he, Feynman, would never accept a *'responsible position'*. The bet was duly made in writing, with the definition of a 'responsible position' as one 'which by reason of its nature, compels the holder to issue instructions to other persons to carry out certain acts, notwithstanding the fact that the holder has no understanding whatsoever of that which he is instructing the aforesaid persons to accomplish'. Feynman collected his $10 ten years later, in 1976. Evidence could not be found that he had given instructions that he could not understand himself.

The space shuttle Challenger

On 28 January 1986, the launch of the space shuttle *Challenger* ended in disaster, as seconds after take-off the shuttle exploded, killing all seven astronauts on board. William Graham, the head of NASA (National Aeronautics and Space Administration), who was a former student of Feynman, asked him to join a commission to investigate the accident. Reluctantly Feynman accepted, and then

(Continued)

Courtesy of NASA.

went about the task in his usual fashion. The commission, which was chaired by William Rogers, the former Secretary of State, held numerous official meetings (both public and private) and formal discussions, with lots of bureaucracy. Feynman went off on his own, talked to technicians and engineers, and learned details about the construction of the rocket from workers on the shop floor. He found that there was a lack of communication between technicians and management, and that their worries never seemed to reach the ears of those at the top. One particular worry was that the seals around airtight joints in the rocket engines might leak under violent vibration.

At a public meeting, shown on national television, Feynman removed an O-ring from a model of the rocket engine and dipped it into a glass of iced water. He showed that rubber used in the insulating rings had no resilience when squeezed at the temperature of iced water. The temperature on the morning of the launch had been two degrees below zero, whereas the coldest previous launch had taken place at +12°C. At the critical moment, the seals had failed and allowed rocket fuel to leak and go on fire. The partial cause of the accident had been demonstrated with a clamp

(*Continued*)

and pliers which Feynman had bought at a hardware store on his way to the television studio!

When the commission published its final report, Feynman added an appendix in which he gave reasons for his personal view that the deterioration in cooperation between scientists and engineers on the one hand and management on the other contributed to the failures which caused the *Challenger* disaster.

Despite serious illness during the last years of his life, Feynman never lost his spirit of adventure. An unusual ambition, at the back of his mind despite his many other activities, was to visit Tannu Tuva. Since childhood he remembered seeing trian-  gular and diamond-shaped postage stamps from Tuva, a small Soviet republic situated north-west of Mongolia and surrounded by mountains. A peculiar reason attracted his attention — the capital, Kyzyl, was the only five-letter word he had come across that was made up only of consonants! Laboriously, using a combination of Tuvan–Russian and Russian–English dictionaries, he and a friend, Ralph Leighton, composed a letter asking for permission to make a cultural visit to that country.

The climate of Soviet–American relations was not sufficiently amenable for such a visit at that time, and no permission was forthcoming over a period of some ten years. His students at Caltech queued in hundreds to donate blood during his final major operation, but Feynman died on  15 February 1988, a few days before a formal invitation arrived from Moscow, addressed to 'Feynman and his party', to visit Tannu Tuva.

Chapter 15

Relativity

Part 1: How It Began

Empty space is the same everywhere and for everyone ...

The *special* theory of relativity deals with space devoid of all matter and with the propagation of light in such empty space. The subsequent *general* theory of relativity introduces matter as a distortion of space leading to a geometrical representation of the universal law of gravitation. We will confine ourselves to the special theory.

Many of the principles which led to the theory of relativity were recognised long before Einstein. In the 17th century Galileo had argued that any experiment carried out in the cabin of a moving ship would give exactly the same result as the same experiment in a stationary laboratory, and therefore the motion of the earth could not be detected by experiments with falling bodies. The Dutch physicist Hendrik Lorentz and the French mathematician Henri Poincaré, who were contemporaries of Einstein, independently contributed to the mathematical formulation of special relativity. There can be no doubt, however, that the final distillation of these principles and the train of logic which led to observable predictions must be credited to Albert Einstein.

Just as quantum theory changed our understanding of the world around us, the theory of relativity 'wiped the slate clean of preconceived prejudices' and proposed new basic philosophical principles. The seemingly obvious concepts of space and time, energy, and matter are to be viewed from a different perspective. Light plays a central role in the new theory.

While working at the patent office in Zurich, Einstein spent his spare hours thinking about *space* and *time*. He was not trying to explain the results of particular experiments; this would come later as a by-product of his work. He tried to imagine how to construct the world in the most perfect way, as *he would have done if he had been God!* His basic criteria were symmetry and simplicity and elegance embodied in principles which apply equally to every aspect of the universe.

Einstein started with two basic postulates, which seem logical, if not obvious: empty space is absolutely uniform and symmetrical, and the speed of light is the same for everyone. Building on these postulates with one logical step after another, he came to conclusions which are not at all obvious.

We follow these steps slowly, one at a time. Surprisingly, they are not impossibly difficult. The relative nature of time is probably the most challenging, and strangest, concept to digest.

15.1 Space and time

It is not easy to talk about space itself without reference to material things in it or around it. One can certainly visualise a vacuum inside an empty box, or stars and planets in empty space — but take away the box, the stars and planets, then what is left?

15.1.1 *Space and the ancient philosophers*

As far as we can judge, the ancient philosophers regarded the concept of space as self-evident. Their perception seems to have been the intuitive one of a domain within which there exist concrete things. The earth is *'a sphere poised in space'*. The earth, which is a concrete thing, must exist somewhere. *Space* is where it 'lives'.

Pythagoras (with manuscript) and Aristotle. Detail from a fresco by Raphael, *The School of Athens.*

There could be other spaces, such as heaven — which, according to the Babylonians, is '*a hemispherical dome supported by the distant mountains*'. It seemed reasonable to imply that the space of heaven is different from the space occupied by the earth. The early Greeks merged the two spaces, and pictured a round earth surrounded by concentric transparent spheres, each carrying a heavenly body. They seemed not to worry unduly as to what was in the space between the spheres.

Pythagoras (≈ 550 BC) was not comfortable with the concept of empty space, and believed that it was composed of a 'chain of integral finite numbers'. Of course, numbers whether integral, finite or not, are abstract entities, and in this sense Pythagoras' explanation was not particularly helpful or verifiable. He wanted to inject substance into space and, being a mathematician numbers were the most obvious things to use.

Aristotle (384–322 BC) believed that one can identify a 'place' in space. '*Every sensible body is by its nature somewhere*'. He asserted that there is such a thing as 'absolute motion' from one place in space to another. Furthermore it follows that one can also define *absolute rest* as 'staying in the same place'. Aristotle's law of inertia can be stated as: '*A body, not acted upon by any force, remains at rest.*'

15.1.2 *Space — the intuitive view*

Perhaps surprisingly, our intuitive view in more recent times is basically the same as that of the ancients. It appears almost self-evident that the earth, sun, stars, and galaxies, even down to atoms, electrons and quantum wave packets, occupy and move in some part of a universal space. The universe is an enormous stage. It provides the *space* within which all material things exist.

The answer to the question 'What is left if we take away all material things?' is not at all obvious. We are left with an empty stage. Nothing is there — no atoms, no electrons. Pythagoras' idea of an infinite sea of finite numbers may be elegant but it is abstract, and hardly sufficient to satisfy an enquiring mind!*

15.1.3 *Space and time — according to Isaac Newton*

Is it meaningful to talk about *empty space*? Isaac Newton not only addressed the question of the fundamental nature of *space*, but also the nature of *time*, which up to then had been either ignored, or considered as a completely separate issue. His ideas are of historical rather than scientific interest, in that they do not give rise to verifiable predictions.

In his *Principia* Newton begins by stating *'I do not define time, space, place, and motion, these being well known to all'*, and then proceeds to contradict himself by offering definitions as follows:

Absolute time — 'of itself and by its own nature flows without relation to anything external'.

Absolute space — 'By its own nature and without relation to anything external remains always similar and immovable'.

---

* We will not discuss here our present view of continuous activity such as electromagnetic oscillations, or modern concepts of creation and annihilation of virtual particles in space which is devoid of physical matter.

In contrast to 'absolute space', Newton defined *'relative space'* as 'a movable dimension of absolute space, for instance space in a cavern which will at one time be part of absolute space, at another time another part of the same'. He defined a *'place'* as 'that part of space taken up by a body which can be absolute or relative depending on what kind of space a body takes up'.

Newton's concept of the *'motion'* of a body was 'the translation from one place to another, and could be absolute or relative depending on whether these places were absolute or relative'.

It is not surprising to find oneself confused by these definitions. Newton himself admits to being confused, by saying: *'It is however a matter of great difficulty to discover, and effectually to distinguish the true motions of a particular body from the apparent'.*

There seems little point in attempting to interpret Newton's definitions. *Relative space* is ill-defined and confusing, in that the use of the word 'relative' begs the question — 'Relative to what?'. It makes better sense to forget about Newton's *absolute space and time*, and to take Einstein's advice to 'wipe the slate clean', and follow a chain of reasoning in a series of logical steps.

15.2 'Dogmatic rigidity'

15.2.1 *Starting with a clean slate*

Einstein did not mince his words when referring to the conservative attitudes prevalent among natural philosophers of his time — '... in spite of great successes in particulars, dogmatic rigidity prevailed in matters of principles.' It took a great step to break free from preconceived ideas about space and time, and to develop a chain of logic built on completely new foundations. The principles of *symmetry* emerged as a basic prerequisite of the laws of Nature. When asked 'How do you know, Professor Einstein, that your theories are right?', Einstein replied: *'If they are not right, then the Lord has missed a wonderful opportunity.'*

One such preconceived idea refers to measurement in general, and measurement of velocity in particular. Take the example of a passenger walking through a carriage of a moving train. His or her speed depends on the point of view. It will be quite slow as it appears to another passenger seated inside the carriage, faster as viewed by someone on the platform of a station as the train rushes past, and possibly faster still as viewed from outside the earth. We have an in-built prejudice that somewhere out in space there is an absolute point at rest from which one could determine the correct absolute motion, and also an absolute clock which keeps universal time.

15.2.2 *Frames of reference — defining a point of view*

Whether implicitly or explicitly, physical measurements involving space and time are made from a given perspective, mathematically referred to as a *frame of reference*. We picture ourselves as 'sitting' at the origin of the frame of reference and all displacements, velocities and momenta are measured with respect to that frame. Of course, normally nobody talks about coordinate systems or reference frames; nevertheless, there is an implication which may not be apparent, but is taken for granted. The results are critically dependent on the reference frame.

Speed in different frames of reference

Let us go back to the moving train. In the example below we might choose any one of many logical frames of reference when

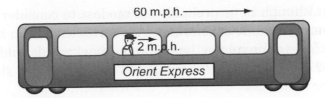

Figure 15.1 What is the ticket collector's 'true' motion?

describing the motion of the ticket collector as he walks through a carriage of the *Orient Express*. His speed will depend entirely on our point of view.

The stationmaster's point of view

By saying that *'the train is travelling at a speed of 60 mph'* we imply a reference frame outside the train, with its origin fixed perhaps at a railway station. The observer might be the stationmaster standing on the platform. From the perspective of the stationmaster, the conductor, who is walking forwards on the train, is moving that bit faster than the train; his velocity is 62 mph.

The passenger's point of view

'The conductor is walking towards the front of the train at 2 mph' implies a reference frame situated in the carriage. The observer could be a passenger sitting somewhere in the carriage. Of course, in the frame of reference of the passenger, the passenger himself is stationary, while the stationmaster, and indeed the whole station, are moving at 60 mph in the opposite direction.

The ticket collector's point of view

The ticket collector equally is entitled to view things from his own perspective, in which he is stationary and everyone else is moving (although he is probably too modest to consider himself as defining the frame of reference). Nevertheless, from his point of view the stationmaster has whizzed backwards and passed him at 62 mph while the passenger is receding more slowly, at 2 mph.

A universal point of view

Finally, the whole scenario could be expressed from the point of view of an astronaut whose frame of reference is outside the earth, or indeed outside our galaxy. Since the earth is rotating and also orbiting the sun, and indeed our whole galaxy is in motion relative to other galaxies, the motions of the ticket collector, passenger and stationmaster will be the resultant of all these motions taken from the astronaut's point of view.

We can conclude that every kind of motion can only be specified in relation to some frame of reference. One may feel instinctively that some frames of reference could be more important than others, and in particular that there exists a unique fundamental frame of reference with respect to which we can define 'absolute motion'. Until one can prove otherwise there is no evidence for the existence of such an 'absolute frame'. Einstein proposed that in order to start with a clean slate, the first step must be to examine current abstract concepts and to determine which of them are, in fact, 'preconceived prejudices'. He wanted to analyse these concepts in a concrete way and make logical deductions leading to a definite set of predictions, *which could be tested*.

15.2.3 *Specifying the prejudices*

Absolute space

One prejudice, popular among natural philosophers at the time, was that one could define somewhere an absolute and immovable set of axes, a frame of reference which determines absolute space. This 'absolute' frame of reference provided a fundamental, stationary reference point and gave a meaning to 'absolute motion'. Physical constants, such as the speed of light, would have their 'true value' measured with respect to that absolute frame. To make the idea more concrete, it was assumed that an imaginary substance with no mass, called the '*ether*', pervaded the universe, defining a universal stationary medium and providing such an absolute reference frame. It would also give substance to what would otherwise be completely empty space, with nothing in it. One notable argument for the ether was based on the fact that light waves obviously can travel across the universe. It was asserted that these waves, like sound waves, needed some kind of medium through which to propagate.

Absolute time

A second prejudice concerned the notion of *time*. According to Newton, time is *absolute*. To quote him: '*it flows by its nature,*

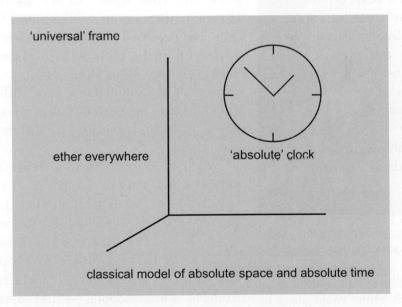

Figure 15.2 The 'obvious' model of the universe.

without reference to anything external'. One could imagine somewhere a 'master clock' which shows 'universal time', the same for all.

Pythagoras thought that space was filled with integral finite numbers. Natural philosophers thought that it was filled with 'ether'. They all believed that *something* must be there.

15.3 Looking for the ether

15.3.1 *The Michelson–Morley experiment*

During the 1880s two Americans, Albert Michelson (1852–1931) and Edward Morley (1838–1923), carried out a series of experiments to measure the velocity of the earth through space. If

Albert Michelson. With Edward Morley he carried out a series of delicate experiments to look for evidence of light being 'swept along by an ether wind'.

the universe was filled with a stationary *'ether'*, we would expect to experience an *'ether wind'* as we hurtled through space in our orbit around the sun. Light would appear to go faster with a following ether wind and more slowly when struggling against it. This should be reflected as a difference in the measured value of the speed of light travelling in different directions.

Michelson and Morley built an optical apparatus to determine if the speed of light did depend on direction. The effect they were looking for was so small and the experimental set-up so delicate that, in order to reduce vibration, all trams in the city of Los Angeles were stopped whilst measurements were being taken.

15.3.2 *Timing the ferry*

The principle of the Michelson–Morley experiment can be understood by taking the analogy of two identical ferries which bring tourists on river trips. One ferry makes return trips across a wide river and the other makes return trips along the same river (Figure 15.3). Both ferries start from and return to the same jetty. There is a current flowing in the river with velocity v from left to right, as illustrated below. Each ferry is capable of the same speed, V, relative to still water and travels a total distance of $2D$, but will the two trips take the same time?

Intuitively, the answer is not obvious. Is it more difficult to fight the cross-current both going and coming back, or to have the advantage of a following current on the outward journey, but then pay the penalty of having to struggle head-on against it on the way back?

In order to travel directly across the river, ferry 1 will have to aim at an angle upstream to allow for the current, so that the combination of the velocity v of the current and the velocity V of the ferry gives a resultant velocity directed straight across the river.

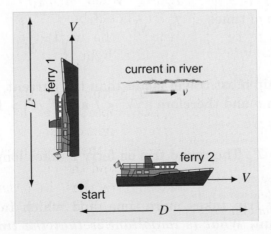

Figure 15.3 The two ferries.

Ferry 1

The vector triangle gives the magnitude R of the resultant velocity of v and V.

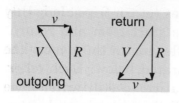

$$R = \sqrt{V^2 - v^2}$$

Total distance $= 2D$.

Total time taken $T_1 = \dfrac{2D}{\sqrt{V^2 - v^2}}$

Ferry 2

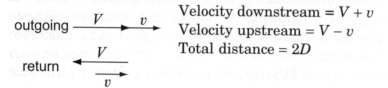

Velocity downstream $= V + v$

Velocity upstream $= V - v$

Total distance $= 2D$

Total time taken $T_2 = \dfrac{D}{V + v} + \dfrac{D}{V - v} = \dfrac{2DV}{V^2 - v^2}$

Ratio of times $\quad \dfrac{T_2}{T_1} = \dfrac{V}{\sqrt{V^2 - v^2}} = \dfrac{1}{\sqrt{1 - \dfrac{v^2}{V^2}}}$ \hfill (15.1)

Since the ship must travel faster than the current, V is always greater than v and therefore $v^2/V^2 < 1$ and $T_2/T_1 > 1$.

$T_2 > T_1$. The round trip on ferry 2 takes longer.

Which ferry trip takes more time, and which takes less, is not important. *What is important is that the times are not equal.*

15.3.3 *Details of the experiment*

Michelson and Morley had the clever idea of applying the 'timing the ferry' principle to light travelling through the imaginary ether. Their apparatus (now known as the Michelson interferometer) has already been described in Chapter 8, but is briefly described here for convenience.

A light beam is 'split' into two by a semi-silvered mirror. The two beams subsequently travel at right angles to each other. Each beam is reflected at a second mirror and returns to the semi-silvered mirror, where the beams recombine.

Suppose that the ether wind is blowing from left to right so that the light which travels via mirror 2 makes the round trip of ferry 2, downwind and against the wind, while the route via mirror 1 taken by the other half of the beam is a 'crosswind' route in both directions (Figure 15.4). If there is a time difference between the two journeys, there will be a phase difference between the two components of the recombined beam, and an interference pattern will be seen at the detector.

Michelson and Morley then rotated their apparatus slowly. They expected the phase difference to change as the crosswind route gradually became the downwind route. As the apparatus rotated, the fringes would then move across the null mark. They tried the experiment at different times during the day and night

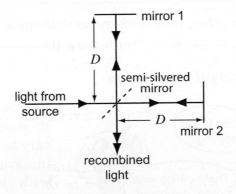

Figure 15.4 Michelson interferometer.

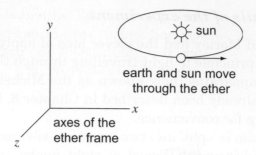

Figure 15.5 The earth moves through the ether, which is stationary and provides an absolute frame of reference.

and repeated on many occasions throughout the year. *No fringe shift was ever observed, and thus the existence of the ether was not proven.*

Note that any rotation should move the fringe pattern, so it is not necessary to know the direction of the ether wind to do the experiment.

The orbital velocity of the earth about the sun is 30 kms⁻¹.

The sensitivity claimed in the original Michelson–Morley experiment ≈ 7–10 kms⁻¹.

The sensitivity of modern experiments (with X-rays) ≈ 3 ms⁻¹ (jogging speed).

If there is an ether, why is there no detectable ether wind?

15.3.4 *A powerful conclusion*

Einstein asked why it should be necessary to fill space with integral numbers, with an imaginary ether, or with anything else, for that matter. Can we not speak of space itself without anything in it? The results of Michelson and Morley were in fact not 'negative'; the experiment was a great success. There was nothing there to measure. There is no ether and no ether wind.

15.4 Symmetry

15.4.1 *Space is uniform*

Einstein developed his ideas about the basic nature of space and time whilst working in the patent office in Zurich. It became obvious to him that if the ether does not exist, then we have '*empty space*'. In empty space nothing can distinguish one place from another and we have to abandon the possibility of 'marking' a point in space to distinguish it from any other point. Einstein's next logical step was to apply the principle of *symmetry*, stating that any place in empty space devoid of all matter is indistinguishable from any other. There are no milestones in empty space! There is no, meaning in going from 'here' to 'there', because there is no difference between here and 'there'! A dictionary definition of *motion* is 'change of position or place'. There are no *places* in empty space, and therefore no such thing as absolute motion or absolute velocity.

We can of course talk about distances between material objects, like the earth and the moon. Speed of travel relative to such objects makes sense, but it does not make sense to talk about an 'absolute speed' relative to empty space. One cannot define an absolute stationary set of coordinates, or an absolute velocity relative to such coordinates. Einstein 'wiped the slate clean' and, starting with the assumption of complete symmetry, set out to build a chain of logic leading to conclusions which could be tested.

15.4.2 *The new model*

Figure 15.2 is not a good representation of the universe. An infinite empty page would be better. There is *no centre, no edge, no preferred directions, and no absolute reference point.* The clock and the absolute coordinate frame are figments of our imagination; they are examples of *'preconceived prejudices'*.

We can of course put matter, or indeed ourselves, into the model. We can put in stars and planets and describe their motion relative to one another. We can also talk about motion from one planet to another. But absolute motion or motion relative to space itself has no meaning.

15.4.3 *Postulates of special relativity*

Einstein put forward two fundamental postulates which summarise his reasoning on the uniformity of space:

The special theory of relativity

1. All inertial (unaccelerated) frames of reference are equivalent.
2. The speed of light is the same for all observers.

15.5 The first postulate

15.5.1 *Nature does not discriminate*

We usually make physical measurements with respect to the most convenient *reference frame*, placing ourselves at the origin. An observer sitting in a laboratory, in an aeroplane or in a train will consider himself or herself to be at rest, while the rest of the world passes by. We often refer to these reference frames by symbols such as S, S′ or S″. In Figure 15.6, the laboratory is

Who is right, and who is wrong?

All are equally right and wrong.

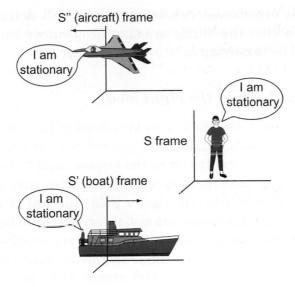

Figure 15.6 All inertial frames are equivalent.

called the S frame. There is no mystery here — just a matter of notation!

The first postulate states that the *laws of physics apply equally to all unaccelerated frames of reference*. It is quite legitimate to construct your own 'frame of reference' placing yourself at the origin at all times. No frame of reference can be defined as being absolutely at rest and therefore somehow superior to other frames.

It is certainly in line with common experience to eat your dinner in an aeroplane, walk up and down in a train, or perhaps even play snooker on an executive jet, without noticing that you are travelling. Provided that there are no bumps or stops and starts (accelerations), everything is the same as in the airport lounge.

Another experience most of us have had, at one time or another, is to look out of the window of a train in a station and think that, at last, our train is leaving. A few seconds later we suddenly realise, as the last carriage of a neighbouring train passes by, that it is the other train that is leaving, and we are still in the station! According to the first postulate, if we 'close the blinds', and have no communication with the world outside, no physical experiment can be done that will detect uniform motion. Whether the blinds are shut or open, we can only say that one of us is moving *relative* to the other.

15.5.2 *Galileo had the right idea!*

Galileo Galilei (1564–1642) was far ahead of his time when he formulated his own 'principle of equivalence'. He wrote: 'Shut yourself up with some friend in the largest room below decks of some large ship, and there procure gnats, flies, and such other small winged creatures. Also get a great tub full of water and within it put certain fishes; let also a certain bottle be hung up, which drop by drop lets forth its water into another narrow-necked bottle placed directly under it. Having observed all these particulars as long as the vessel stands still, how the small winged animals fly with like velocity towards all parts of the room, how the fishes swim indifferently towards all sides, and how the distilling drops all fall into the bottle placed under-neath...now make the ship move with what velocity you please, so as the motion is uniform and not fluctuating this way and that....You will not be able to discern the least alteration in all the above-mentioned effects, or gather by any of them whether the ship moves or stands still.' He stopped short of questioning what exactly was meant by 'standing still'!

Galileo's Gedanken experiment

Galileo used a thought experiment to illustrate his claim that the earth is moving around the sun, and that this fact would not affect experiments involving falling objects. He claimed that the water droplets will continue to fall directly into the bottle below, no matter whether the ship is standing still or moving with a steady velocity in any direction.

In current terminology we say that the experiment will work identically in all *inertial (Galilean) frames of reference*.

15.5.3 *The Galilean transformation*

The mathematical method of comparing how things appear to observers in different frames of reference is called a *transformation* of coordinates from a set of axes centred in one frame of reference to another.

The *Galilean transformation* is the 'normal', classical transformation between two inertial frames, such as a ship which is stationary at the dockside, or a ship which is moving with a speed v in the x direction. The Galilean transformation merely involves the addition of a term to the x coordinate which is changing with time at a constant rate.

$x' = x - vt$
$y' = y$
$z' = z$
$t' = t$

The Galilean transformation gives the relationship between space coordinates in frames S and S'; it also states the 'obvious' fact that the time is the same in both.

When we compare phenomena *within* two different frames of reference, this transformation gives results which are in line with *Newton's laws of motion* in *both* frames. Forces are proportional to changes in velocities (accelerations) and the addition of a constant component to all velocities will leave the behaviour of physical systems unaffected. Insects and fishes dart about in random fashion, and water drops fall from one bottle into the other, just the same as before. However, looking from the *outside*, the magnitudes of velocities in another frame will appear to be different. For example, fish in a tank being transported in a cargo jet will all have the systematic velocity vector of the plane added to their random dart velocities.

> If two observers view the same object, each using a different frame of reference, and these frames of reference are in relative motion, they will not agree on the speed of the object. It will move at different speeds relative to the two observers.

15.5.4 *The speed of a bullet*

Imagine a marksman on earth armed with a super-rifle. He fires a shot at an astronaut who is travelling away from him in a spaceship. Luckily the shot misses its target. The astronaut, a cool customer, measures the speed of the bullet as it passes the spaceship.

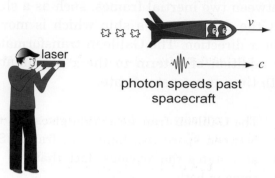

photon speeds past spacecraft

Earthman and astronaut measure the speed of the same photon.

The earthman and the astronaut disagree about the speed of the bullet. As far as the astronaut is concerned, the bullet is trying to overtake him and it will appear to him to move more slowly.

In terms of reference frames:

(See Figure 15.7 on next page)
S: Earth frame of reference
S′: Astronaut's frame of reference
v = velocity of S′ relative to S (the velocity of the spacecraft as measured by the marksman)

Let v' = velocity of the bullet as seen by the astronaut.

Let us compare the speed of the bullet as measured by the astronaut in his 'laboratory', with the speed of the same bullet as measured on earth. In the earth frame of reference S we see the astronaut moving with a velocity v in the x direction. The bullet, which is moving with a velocity V, also in the x direction, will overtake the astronaut at a relative speed of $V - v$. As far

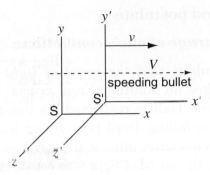

Figure 15.7 The astronaut and the earthman measure the speed of the *same* bullet.

as the astronaut is concerned, the speed of the bullet is $V - v$ in his frame of reference S′.

Speed of bullet in the earth laboratory = V
Speed of bullet in the astronaut laboratory = $v' = V - v$

We can obtain the same result more formally by applying a *Galilean transformation*.

Galilean transformation of velocity

v = speed of S′ frame relative to S frame

$$\frac{dx}{dt} = V \qquad \text{(speed in S frame)}$$

$$\frac{dx'}{dt} = v' \qquad \text{(speed in S′ frame)}$$

but $\qquad \dfrac{dx'}{dt} = \dfrac{dx}{dt} - v$

The speed of the bullet as it appears to the astronaut is $v' = V - v$, as before.

15.6 The second postulate

15.6.1 *The courage of one's convictions*

The second postulate, that *the speed of light is the same for all observers*, takes an additional and crucial step beyond the postulate of Galileo. Galileo restricted his thoughts to mechanical laws governing falling droplets or flying insects. The propagation of light did not enter into his thoughts — not surprisingly, since at that time the speed of light was considered to be infinite.

Einstein went beyond darting fishes and flying bullets. His second postulate is a logical extension of the first. It may look like a small step but could be likened to Neil Armstrong's *'giant leap for mankind'*. It led to a new understanding of the symmetry and universality of the laws of physics.

If we are convinced that all frames are equivalent, it must follow that we will get exactly the same result for the speed of light, whether we measure it in a laboratory on the earth, in another laboratory in a jet plane, or indeed in a spaceship. Otherwise, of course, a measurement of the speed of light would distinguish one frame of reference from another. For light in free space there is no *upstream* or *downstream*. All directions are equivalent. That is why the experiment of Michelson and Morley gave a negative result, or at least gave a result which was not in line with current thinking at the time. Galileo was not aware that he had an even stronger argument, using *light* rather than drops of water in his Gedanken experiment!

15.6.2 *An imaginary experiment with light*

Let us go back to the last example but instead imagine that we have light rather than a bullet chasing the astronaut. It is quite irrelevant whether we talk about a wave or a photon; let us just say that instead of a bullet, we have a light signal. Again, both the observer on earth and the astronaut measure the speed in their respective reference frames S and S'. c = speed of light relative to earth and v = speed of astronaut relative to earth.

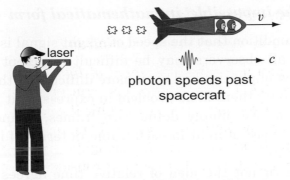

photon speeds past
spacecraft

Earthman and astronaut measure the speed of the same photon.

Both the observer on earth and the observer in the rocket measure the speed of a light signal, by timing it, as it crosses a marked distance in their respective laboratories.

This time, however, they *both get the same result*!

15.6.3 *A paradox?*

How *can* they possibly get the same result?

'*Common sense' tells us* that as the light signal is overtaking the astronaut, it will pass him, the same as the 'chasing bullet', at a lower relative speed than its speed in the earthframe. Einstein's second postulate, 'The speed of light is the same for all observers', tells us something different. We either accept it or we do not. If we accept it, we must also accept the consequences, strange as they might seem!

Paradox (Chambers dictionary definition): *A statement that is apparently absurd but is, or may be, really true.*

Light hits you at the same speed, whether head-on or from behind. That's a paradox.

We have to accept that the contradiction here is not between the laws of Nature themselves, but between the laws of Nature *as they are*, and the laws of Nature *as we think they should be.*

15.6.4 *'The impossible' in mathematical form*

While the condition that the speed of a light signal is the same relative to all observers may be difficult to accept from the point of view of physics and even more difficult on the basis of 'common sense', there is no problem in expressing it in mathematical form. We simply define two frames, S and S′, and express the speed of light in each frame in terms of its respective coordinates.

Whether or not the idea of relative time agrees with our intuition, each frame of reference has not only its own space coordinates but also *its own time coordinates* as illustrated in Figure 15.8.

Speed of light in frame S $(=c)$ Speed of light in frame S′ (also $=c$)

$$c^2 = \frac{x^2 + y^2 + z^2}{t^2}$$

$$\Rightarrow \quad x^2 + y^2 + z^2 - c^2 t^2 = 0$$

$$c^2 = \frac{x'^2 + y'^2 + z'^2}{t'^2}$$

$$\Rightarrow \quad x'^2 + y'^2 + z'^2 - c^2 t'^2 = 0$$

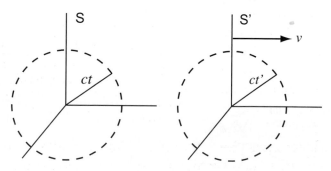

Figure 15.8 The same light signal in two reference frames.

If the speed of light is the same in the two frames of reference, then

$$x^2 + y^2 + z^2 - c^2t^2 = x'^2 + y'^2 + z'^2 - c^2t'^2 \qquad (15.3)$$

Note that space and time are treated on the same footing. However, since its square is negative, t is imaginary and also has different units. (See Chapter 15.7.4.)

The observer on earth uses the frame of reference S, while the astronaut uses S'. Assume that the origins of the two frames in Figure 15.8 coincide at $t = t' = 0$. A light source at the common origin sends out a light signal at that instant, which spreads out in all directions (like an enormous, rapidly growing, spherical balloon) from the origin of each frame of reference.

This is where Einstein's second postulate comes in, and things become counter-intuitive. Each observer has his own opinion and his own point of view. Even when the frames become separated, each says that the light is spreading out from the origin of *his* frame. Each says that the radius of the 'sphere of light' equals the speed c multiplied by the time elapsed *in his frame of reference*. The counter-intuitive part is that we cannot use a 'universal clock', but must assume a different rate of time flow in the two frames. Another part, not easy to accept, is that the two are equally right.

The astronaut, using the clock in his spacecraft (frame of reference S'), obtains the same result for the speed of light as the observer on earth using his own clock in frame S. The  earth observer says that the photon has travelled a distance $\sqrt{x^2 + y^2 + z^2}$ in a time t while the astronaut measures the distance as $\sqrt{x'^2 + y'^2 + z'^2}$ and the time as t'. They agree that the speed of the photon is equal to c, but they do not agree on the values of the distances or of the times. To the earth observer, the marked distance over which the astronaut is timing the light appears to have 'shrunk', and the astronaut's clock appears to be slower than his own. However, the ratio of the distance travelled by the light to the time taken remains the same and equals c.

Time dilation is the key to the resolution of the paradox. To 'solve the impossible' we must wipe out any preconceived ideas about time, and introduce a dramatic new concept. We must assume that *time is not the same when measured in the two frames of reference*. Newton's concept of *absolute time*, *'which, by its nature, flows without reference to anything external'*, has to be abandoned.

Time is relative: *The logical consequence of the relativity postulate*

The timing apparatus used by the earthman and by the astronaut may be the same, and both may be working correctly, but each is located in a different 'time frame'. Time flows at a different rate in their respective frames of reference. There is nothing to say that one is 'right' and the other is 'wrong'. There is no central arbiter, and no universal clock. The astronaut's biological clock, his pulse, and his sense of time, will be related to the passage of time in his own frame of reference.

15.6.5 *The Lorentz transformation*

We have already stated that the Galilean transformation from frame S to frame S′ does not give the same value for the speed of a bullet, and neither will it give the same value for the speed of light. The transformation which preserves the value of the speed of light in both frames of reference is named in recognition of the Dutch physicist Hendrik Anton Lorentz (1853–1928).

It is simplest to reverse the logical order of the argument by stating the equations of the Lorentz transformation, and later showing that they satisfy the requirement that the speed of light remain constant for all unaccelerated observers.

The Lorentz transformation

$$x' = \gamma(x - vt)$$
$$y' = y$$
$$z' = z$$
$$t' = \gamma\left(t - \frac{xv}{c^2}\right) \qquad (15.4)$$

where $\gamma = \dfrac{1}{\sqrt{1 - \dfrac{v^2}{c^2}}}$ is the relativistic gamma factor.

The basic difference between the Galilean transformation and the Lorentz transformation is that the measure of time is not the same in the two frames ($t' \neq t$).

Verifying that the Lorentz transformation satisfies Equation (15.3)

Using x' and t' from above, we substitute into Equation (15.3)

$$x'^2 + y'^2 + z'^2 - c^2 t'^2$$

$$= \frac{(x-vt)^2}{1 - \dfrac{v^2}{c^2}} + y^2 + z^2 - \frac{c^2\left(t - \dfrac{xv}{c^2}\right)^2}{1 - \dfrac{v^2}{c^2}}$$

$$= \frac{1}{1 - \dfrac{v^2}{c^2}}\left(x^2 - 2xvt + v^2 t^2 - c^2 t^2 + 2xvt - \frac{x^2 v^2}{c^2}\right) + y^2 + z^2$$

$$= \frac{1}{1 - \dfrac{v^2}{c^2}}\left[x^2\left(1 - \frac{v^2}{c^2}\right) + c^2 t^2\left(1 - \frac{v^2}{c^2}\right)\right] + y^2 + z^2$$

$$= x^2 + y^2 + z^2 - c^2 t^2 \qquad \text{Q.E.D.}$$

15.6.6 *The gamma factor*

For relative velocities which are small compared to the speed of light, the relativistic gamma factor $\gamma \to 1$. The *Lorentz transformation* becomes the *Galilean transformation* and relativistic effects become negligible.

$$\gamma = \frac{1}{\sqrt{1 - \dfrac{v^2}{c^2}}}$$

In practice, astronauts travelling in satellites in orbit around the earth or space vehicles to the moon never reach speeds even remotely close to relativistic speeds. Concorde was a supersonic aircraft, by far the fastest means of travel for fare-paying passengers, with the capability to travel at more than twice the speed of sound. The gamma factor, however, played no role in the lives of those on board, since the speed of sound in air ($\approx 300 \, \text{ms}^{-1}$) is about a million times slower than the speed of light, and the term $v^2/c^2 = 10^{-12}$.

On the other hand, fundamental particles in cosmic radiation and in accelerators travel at speeds approaching the speed of light (to within a fraction of 1%), and relativistic effects such as time dilation become critically important (see table below).

Table 15.1 The relativistic gamma factor.

v	Speed of sound in air = 300 ms⁻¹ = $10^{-6}\,c$	Orbital velocity of earth about the sun = 30 kms⁻¹ = $10^{-4}\,c$	$0.5\,c$	$0.9\,c$	$0.99\,c$	Speed of a proton at an accelerator = $0.999\,c$
γ	1.0000000000005	1.000000005	1.15	2.3	7.1	22.3

15.6.7 *Addition of velocities — a classical example*

A fighter plane travelling at a speed v fires bullets forwards with a muzzle velocity V. What is the resultant velocity of the bullets relative to the ground?

Classically, there is no problem. The velocity relative to the ground is the vector sum of the two velocities. Since they are in the same direction, we simply add the two velocities, giving the resultant $V_R = v + V$.

Again classically, we may use the slightly more sophisticated classical method of applying the *Galilean transformation* to transform from the frame of reference of the pilot to that of the earth observer, to compare how the speed of the bullet appears to each.

V = velocity of bullet in frame S'. What is the velocity of the same bullet in frame S?

Taking all velocities to be in the x direction,

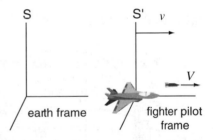

earth frame   fighter pilot frame

$$x' = \gamma(x - vt) \qquad \frac{dx'}{dt} = \frac{dx}{dt} - v$$

$$\boxed{\begin{aligned} V &= \frac{dx'}{dt} \\ V_R &= \frac{dx}{dt} \end{aligned}} \quad \Rightarrow \quad V = V_R - vt \quad \text{or} \quad V_R = V + vt, \quad \text{as before}$$

Classically there is no limit to the sum of the velocities. For example, it may happen that the resultant velocity of the bullet is greater than the speed of sound.

15.6.8 *Addition of velocities when the speeds are relativistic*

Let us now go a million times faster. Let us imagine a hopefully fictional situation of faster and faster planes firing faster and faster rockets. What happens if the velocities when combined give a result greater than the *speed of light*?

The Galilean transformation gives a result which is impossible. We must use the transformation which is in accordance with the relativity principles — the Lorentz transformation.

The Lorentz transformation of velocities

The basic difference between the classical and the Lorentz transformation is that the time scales in the two frames are different. Rates of change in S are calculated with respect to t, whereas in S' they are calculated with respect to t'.

$$x' = \gamma(x - vt)$$
$$y' = y$$
$$z' = z$$
$$t' = \gamma\left(t - \frac{xv}{c^2}\right)$$

$$dx' = \gamma(dx - vdt) \qquad dt' = \gamma\left(dt - \frac{vdx}{c^2}\right)$$

$$\frac{dx'}{dt'} = \frac{dx - vdt}{dt - \dfrac{vdx}{c^2}} = \frac{\dfrac{dx}{dt} - v}{1 - \dfrac{v^2}{c^2}\left(\dfrac{dx}{dt}\right)}$$

$$V = \frac{V_R - v}{1 - \dfrac{vV_R}{c^2}}$$

$$V_R - v = V\left(1 - \frac{vV_R}{c^2}\right)$$

Relativistic addition of velocities: $V_R = \dfrac{v + V}{1 + \dfrac{vV}{c^2}}$ (15.5)

(V = muzzle velocity, v = plane velocity, V_R = resultant velocity)

15.6.9 *Playing with the formula*

The following examples, which get progressively more unreal, test the relativistic addition of velocities according to the formula we have just derived. We never get an answer greater than c!

Example 1. *Two velocities which should 'add up'*
to more than the speed of light

Speed of (imaginary) plane = 0.9 c
Speed of (imaginary) bullet = 0.9 c

The absolute speed limit is c.

$$v = 0.9c, \quad V = 0.9c$$

$$V_R = \frac{0.9c + 0.9c}{1 + 0.9^2} = 0.9945c$$

Example 2. *Trying to add to the speed of light.*
An imaginary object travelling at the
speed of light fires a bullet forwards
with a speed V.

$$v = c \quad V = V \quad V_R = \frac{c + V}{1 + \frac{cV}{c^2}} \quad V_R = \frac{c + V}{1 + \frac{V}{c}} = c$$

$$\boxed{c + \text{anything} = c}$$

Example 3. $c + c = ?$ *An imaginary photon fires another*
photon forwards.

$$v = c \quad V = c \quad V_R = \frac{c + c}{1 + \frac{c^2}{c^2}} = c$$

$$\boxed{c + c = c}$$

Example 4. ***$c - c = ?$ An imaginary object, travelling with a velocity c, fires a bullet backwards with muzzle velocity c.***

$$v = c \quad V = -c \quad V_{\mathrm{R}} = \frac{c - c}{1 + \dfrac{-c^2}{c^2}} = \frac{0}{0}, \text{ which is indeterminate.}$$

Such a calculation requires care. Let us assume a more physically plausible situation, in which our imaginary object *approaches the speed c* in the positive x direction and fires a photon backwards, i.e. in the negative x direction.

$$v = c - \Delta v \quad V = -c \quad V_{\mathrm{R}} = \frac{c - \Delta v - c}{1 + \dfrac{-(c - \Delta v)c}{c^2}} = \frac{-c^2 \Delta v}{c^2 - c^2 + c \Delta v} = -c$$

In the other frame the photon also travels with speed c in the negative x direction.

15.7 The fourth dimension

We can express the mathematics of relativity neatly by employing a new geometry, which embodies time as a fourth dimension.

Pythagoras' theorem. *Courtesy of Greek Post Office.*

Mathematicians have been familiar with multi-dimensional geometries for a long time, but now we use it to represent physical reality.

The theorem of Pythagoras is carried forwards into this geometry. The concept that the length of a straight line can be expressed in terms of the sum of the squares of perpendicular components is expanded to four dimensions. The idea that such a length is 'invariant', i.e. independent of the coordinate axes, plays an important role in practical calculations.

15.7.1 *Definition of an 'event'*

An *event* in space–time is designated by four coordinates, x, y, z and t. *Event* 1, for example, might be the whistle to start a football match in Dublin, Ireland. *Event* 2 might be the striking of a match by an astronaut on his way to a distant galaxy. The two events may be specified by coordinates in the earth reference frame S, or in the frame of the spaceship S', or indeed in *any other* frame of reference S'' (for example, that of another spaceship) moving relative to both the S and S' frames.

The observer on earth and the two astronauts all get different values for the space and time components of the interval between event 1 and event 2, but is there an absolute value of the interval, independent of the observer?

Earth frame (S)	Spaceship frame (S')	Frame of another spaceship (S'')
$\Delta x = x_1 - x_2$	$\Delta x' = x_1' - x_2'$	$\Delta x'' = x_1'' - x_2''$
$\Delta y = y_1 - y_2$	$\Delta y' = y_1' - y_2'$	$\Delta y'' = y_1'' - y_2''$
$\Delta z = z_1 - z_2$	$\Delta z' = z_1' - z_2'$	$\Delta z'' = z_1'' - z_2''$
$\Delta t = t_1 - t_2$	$\Delta t' = t_1' - t_2'$	$\Delta t'' = t_1'' - t_2''$

15.7.2 *The invariant interval*

Is there anything on which all observers agree?

Going back to Equation 15.3, we recall that $x^2 + y^2 + z^2 - c^2t^2 = 0$ represented the *special case* of an interval between two events, which happened to be connected by the passage of a light signal.

Event 1 was the emission of a light signal from the origin of both frames as the two frames coincided. *Event 2* was the same signal arriving at a certain point in space and time. The coordinates of that second event were x, y, z, t in frame S and x', y', z', t' in frame S'. The values of *both space and time coordinates* were different in the two frames, but the value of the expression remained always the same.

The invariant interval in space–time

It is not difficult to show that, by transforming the coordinates as before using the Lorentz transformation, we can write a more general expression for the interval between *any two events* not necessarily connected by the passage of a light signal. In general its value is not zero, but whatever the value, it is *the same* in all frames of reference. *All observers agree* on the value of the expression $\Delta x^2 + \Delta y^2 + \Delta z^2 - c^2t^2$, which is the interval in space and time (space–time) between any two events — which in more technical language is described as *being invariant under a Lorentz transformation*.

15.7.3 *Pythagoras revisited*

Measuring distance on a map

On the map in Figure 15.9(a), the x and y axes of the S coordinate frame are drawn in the east–west and south–north directions. (The sides of the grid squares are approximately 100 km and we are assuming that the map is flat, i.e. ignoring the earth's curvature.) In Figure 15.9(b) a coordinate frame has

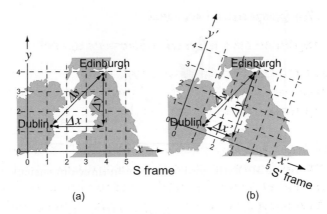

Figure 15.9   Invariant interval in space–time.

been chosen (S′ frame) which has been rotated and displaced by an arbitrary amount.

The distance between Dublin and Edinburgh is of course independent of the grid, but the *components* of that interval in the directions of the axes will depend on the orientation of the grid used.

Applying the theorem of Pythagoras,

$$\Delta S^2 = \Delta x^2 + \Delta y^2 = \Delta x'^2 + \Delta y'^2$$

The actual components in the diagram are (approximately)

S frame $\Delta x = 2.7$, $\Delta y - 2.7$ S′ frame $\Delta x' = 1.6$, $\Delta y' = 3.5$
$\rightarrow \Delta S = 3.82$ $\rightarrow \Delta S = 3.85$

in good agreement with the measurement error.

Extending the theorem of Pythagoras to three dimensions

The formula applies equally well to three dimensions, the distance between any two points being expressed in terms of the sum of the squares of three perpendicular components.

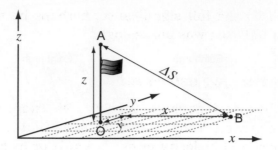

Figure 15.10 Pythagoras' theorem in three dimensions.

$$AB^2 = OA^2 + OB^2 \quad \text{and} \quad OB^2 = \Delta x^2 + \Delta y^2$$
$$\Rightarrow \Delta S^2 = \Delta z^2 + \Delta x^2 + \Delta y^2$$

ΔS is an invariant interval in three-dimensional *Euclidean*[†] space.

15.7.4 *Time as a fourth dimension*

We now extend the theorem of Pythagoras to the four dimensions of *space–time*:

$$\Delta x^2 + \Delta y^2 + \Delta z^2 - c^2 \Delta t^2$$

This expression, already given in Equation (15.3), is similar to the 'sum of the squares' Pythagoras formula, except that the fourth term, namely the square of the time component, is negative. In other words, the time component Δt is represented by an imaginary number $i\Delta t$, where $i = \sqrt{-1}$.

The interval can have both space and time components.

Everyone agrees on the value of the interval, but observers using different frames of reference will not agree on the relative magnitudes of the space and time components.

It is said that Pythagoras offered 100 oxen in sacrifice to the Muses in thanksgiving for letting him make his great discovery.

[†] Euclidean space can be defined as a space in which Euclid's definitions and axioms apply.

Had he realised the full significance, perhaps he might have decided that 100 oxen was not enough!

15.7.5 *The smoking astronaut*

To see how the invariant space–time interval expression can be applied to relativistic calculations, suppose that a space pilot travels at very high velocity either towards or away from the earth. He smokes a cigar, which to him appears identical to the same brand of cigars he smoked on the earth. In particular, the time taken for the 'smoke' is the same. A NASA observer on the earth watches the pilot with a super telescope and times the smoke. He is checking that the NASA pilot does not exceed the permitted recreation time. (In making this timing he makes due allowance for the difference in time taken for light to reach his telescope at the begining and the end of the smoke).

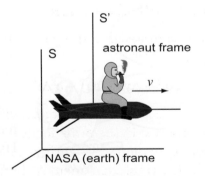

Event 1: astronaut lights cigar   *Event 2*: astronaut finishes cigar

How long does it take him to smoke the cigar?

Invariant interval:

Earth frame: $\Delta S^2 = \Delta x^2 + 0 + 0 - c^2 \Delta t^2$ where $\Delta x = v \Delta t$.

Astronaut frame: $\Delta S^2 = 0 + 0 + 0 - c^2 \Delta t'^2$ where $\Delta t' = \Delta t_0$.

Time in the rest frame of cigar and astronaut is called the 'proper time'.

To the astronaut there is no space interval between the two events. Both lighting and finishing the cigar are happening 'here'.

To the earth observer the astronaut has travelled a distance Δx in the x direction during the 'smoke'.

$$v^2 \Delta t^2 - c^2 \Delta t^2 = 0 - c^2 \Delta t_0^2$$

$$(v^2 - c^2)\Delta t^2 = -c^2 \Delta t_0^2$$

$$\Delta t^2 = \frac{-c^2 \Delta t_0^2}{v^2 - c^2} = \frac{\Delta t_0^2}{1 - \dfrac{v^2}{c^2}}$$

$$\Delta t^2 = \gamma^2 \Delta t_0^2 \;\Rightarrow\; \boxed{\Delta t = \gamma\, \Delta t_0} \qquad\qquad (15.5)$$

Time in the other frame = γ(proper time) $\gamma > 1$ always.

They move about in slow motion inside that spaceship!

And they say the same about us!

To *us* the astronaut's cigar (*and his life!*) appears to last longer, and his actions appear in 'slow motion'. The astronaut himself feels no difference. His clock shows his own *proper time*.

> Things move more slowly in 'the other frame'.

15.8 A philosophical interlude

What is the meaning of 'simultaneous'?

The Oxford dictionary defines 'simultaneous' as 'occurring at the same time'. Translated into the language of relativity, this should mean that Δt, the time component of the interval between two events, is zero. But we know that time is only one component of the interval; it is not the same for all observers. What is simultaneous for some will not be simultaneous for others. The word 'simultaneous' has no meaning when applied to events separated in space–time.

There is no absolute 'now' for all places in the universe.

Let us go back to our 'practical' examples. The start of a football match in Dublin, and the striking of a match by an astronaut travelling somewhere a million miles away, are two events separated in space and time. Observing them from some other frame of reference, one might find that the interval between them has only space components, and the time component is zero. In this frame the events are 'simultaneous'. Other observers, in other reference frames, find that the interval has both space and time components. For them the events occur at different times. None can be said to be 'absolutely' right or wrong.

The concept of simultaneity is relative.

A historical interlude: Hendrik A. Lorentz (1852–1928)

Hendrik Antoon Lorentz was born on 18 July 1853 in Arnhem, Holland. His mother, Geertruida, died when he was four years of age. He was very late in talking, which shows that early language skills may bear little relation to mental ability in later life. At the age of six he entered Master Swater's school, which was highly regarded in the city of Arnhem, and soon became first in his class.

(Continued)

LORENTZ, Hendrik Antoon
Nobel Laureate PHYSICS 1902
© Nobelstiftelsen

Being shy by nature he found this a source of embarrassment, so much so that he considered making mistakes in Mathematics on purpose so that he could become inconspicuous. However, when he considered what possible mistakes he could make, they all seemed so ridiculous that he soon gave up the idea. Instead, he bought a book of logarithms with his own money, and taught himself how to use them at the age of nine.

Hendrik's father, Gerrit Fredrik Lorentz, remarried when Hendrik was ten years old. His stepmother was Luberta Hubkes, with whom he seems to have got on well.

In 1870, Lorentz entered the University of Leyden to study Mathematical Physics, obtaining his doctorate the early age of 22. Three years later he was appointed to the Professorship of Theoretical Physics — a meteoric rise by any standards! He remained at Leyden for the rest of his life.

From the beginning of his studies, Lorentz was fascinated by James Clerk Maxwell's theory of electromagnetism, and in his doctoral thesis he adapted the theory to deal with the phenomena of reflection and refraction. In 1878, he extended this work in a paper which discussed the reason why light slows down when passing through a material medium and derived a relationship between the speed of light in a medium and the composition of that medium. Almost simultaneously the same formula was proposed by a Danish physicist, Ludwig Lorenz (1829–1891). This relationship, between density and refractive index, is now known as the Lorentz–Lorenz formula.

Lorentz's popular Monday morning lectures, which he delivered regularly for many years, became famous. He had great charisma

(Continued)

and skill in dealing with people regardless of whether they were colleagues, politicians or strangers. His exceptional command of languages helped him to preside over international gatherings. He was chairman of the first Solvay Congress, held in the autumn of 1911. This was a meeting organised by the Belgian industrialist Ernst Solvay, to which were invited the most prominent physicists of the time to discuss the transition of classical physics to the revolutionary quantum theory. Lorentz continued as chairman at all further Solvay conferences, in 1913, 1921, 1924 and 1927. Lorentz never really came to terms with quantum mechanics. It is said that once, in discussion with Heisenberg, he was heard to remark: *'It is hard to believe that I am nothing but a matrix!'*

Lorentz was awarded the 1902 Nobel Prize for Physics together with his pupil Pieter Zeeman for their work on the effects of magnetism on the phenomenon of radiation. Characteristically Lorentz began his Nobel lecture by paying tribute to Zeeman, who could not come to Stockholm because of illness. Lorentz had proposed the theory that electromagnetic waves were produced by the oscillations of a tiny electrically charged particle in the atom, later to be called the electron. His calculations showed that if a source which emits light of one frequency were placed between the poles of a strong magnet, the source would emit light of three different frequencies instead of just one. Zeemann had verified this prediction in a series of careful experiments. Lorentz also paid tribute to Clerk Maxwell, whom he described as the creator of the electromagnetic theory of light, and Heinrich Hertz, whose experiments had confirmed the conclusions that Maxwell had drawn from his equations. Lorentz continued: *'There exists in nature a whole range of electromagnetic waves, which, however different their wavelengths may be, are basically all of the same nature.... At the beginning of the range are the waves used in wireless telegraphy, whose propagation was established last summer from the southwest tip of England to as far as the Gulf of Finland.'*

In 1902, Lorentz, in common with most physicists, believed that there exists an *ether*, the 'bearer of light which fills the whole

(Continued)

universe'. Nobody had found any direct evidence of the ether, and the experiments of Michelson and Morley, completed in 1887, had failed to detect an 'ether wind'. Lorentz wondered if in its journey through the ether the earth compressed and dragged it along, resulting in a complicated ether flow similar to the flow of water around the bow of a ship. In his Nobel lecture he developed this topic in descriptive fashion: *'Could we hope to displace the ether with a very large, very fast-moving piston?... in its annual journey around the sun the earth travels through space at a speed many thousand times faster than that of an express train.... We might expect that in these circumstances the earth would push the ether in front of itself, and it would then flow around the planet so as to occupy the space which the earth has just vacated....'* The earth would thus be surrounded by the ether in turbulent motion and the course of light reaching us from the stars would 'be influenced in some way'. Lorentz, however, admitted that 'astronomical observations gave a negative answer to the question of whether the ether moves'.

An earlier possible explanation put forward by Lorentz in 1892 had been that all solid bodies, including Michelson's instruments, are compressed by the ether wind and diminish their dimensions by a tiny bit in the direction of the wind. The correction, by 'pulling space askew to correct for the ether wind', could be applied so as to exactly compensate for the effect of the wind on the velocity of light. Lorentz was unaware that in a short paper published in 1887, the Irish physicist George Fitzgerald had proposed an almost identical contraction. To his credit, Lorentz later took every opportunity to acknowledge that Fitzgerald had published the idea first. It is known as the *Lorentz–Fitzgerald contraction*.

In 1904, Lorentz refined this correction by adjusting not only the parameters of length but also of time, *'pulling both space and time askew'* by what is now known as the *Lorentz transformation* of coordinates and time. Einstein developed the idea into a complete new philosophy of space and time which lay the foundation of the *theory of relativity*.

(Continued)

Lorentz never seemed to accept fully Einstein's theory. He expressed his doubts in a lecture he gave in 1913 as follows: *'As far as this lecturer is concerned he finds a certain satisfaction in the older interpretation according to which the ether possesses at least some substantiality, space and time can be sharply separated, and simultaneity without further specification can be spoken of. Finally it should be noted that the daring assertion that one can never observe velocities larger than the velocity of light contains a hypothetical restriction of what is accessible to us, a restriction which cannot be accepted without some reservation'.*

No account of the history of relativity would be complete without reference to the work of the French mathematician Jules Henri Poincaré. In a paper entitled *'La mesure du temps'*, which appeared in 1898, he wrote: *'...we have no direct intuition about the equality of two time intervals. The simultaneity of two events or the order of their succession, as well as the equality of two time intervals, must be defined in such a way that the statements of the natural laws be as simple as possible'.* In his opening address to the Paris Congress in 1900, Poincaré seriously questioned the existence of an ether. In a communication received on the 5 June 1905 he wrote: *'It seems that the Impossibility of demonstrating absolute motion is a general law of nature.'* Einstein's first paper defining his postulates of relativity was received on 30 June. In subsequent papers and lectures by Poincaré, Einstein is never mentioned. Similarly, we find Poincaré mentioned only once in Einstein's papers. On the other hand, Lorentz acknowledges both Einstein and Poincaré, and often cites them in his works!

There is no doubt of the high esteem in which Lorentz was held by Einstein, who later wrote about him: *'No longer do physicists of the younger generation fully realise, as a rule, the determinant part which H.A. Lorentz played in the formation of the basic principles of theoretical physics. The reason for this curious fact is that they have absorbed Lorentz's fundamental ideas so completely that they are hardly able to realize the full boldness of these ideas and the simplification they have brought into the science of physics....'*

(Continued)

Zuiderzee map from 1658.

In 1920, the genius of Lorentz was enlisted into dealing with a practical problem. Large parts of Holland are below sea level, and the danger of flooding is ever-present. In January 1916, the sea walls around the Zuiderzee failed in two places, leading to widespread flooding, as a consequence of which it was decided to build a dam across the northern part of the Zuiderzee to close it off from the North Sea. The project required complex planning, taking into account the effects of tides, storm surges, and currents between the mainland and the many islands close to the northern coastline.

In 1918, the Dutch parliament appointed a state committee consisting of engineers, oceanographers and meteorologists to take charge of the project and asked Lorentz to preside over the committee. Lorentz was busy, but as a patriotic Dutchman decided to accept the job. He applied his theoretical knowledge of hydrodynamics to calculate the rates of water flow through the intricate network of channels and over sand flats, testing his calculations on models on the simpler cases of the Gulf of Suez and the Bristol Channel. Unexpected effects were discovered, such as the interference of tidal waves and waves reflected from the coastline, which resulted in standing waves and exceptionally strong currents effectively created by the principles of resonance. The dam had to be designed to withstand the worst possible scenario. This work kept Lorentz busy until September 1926, when the report was completed, more than half of it written by Lorentz himself.

(Continued)

Construction work began almost immediately in January 1927 and was completed successfully five years later — but, sadly, not within Lorentz's lifetime.

Lorentz died on 4 February 1928. A contemporary article by O.W. Richardson gave the following description of his funeral: *'The funeral took place at Haarlem on February 10. At the stroke of twelve the slate telegraph and telephone services of Holland were suspended for three minutes as a tribute to the greatest man Holland had produced in our time. It was attended by many colleagues and distinguished physicists from foreign countries. The President, Sir Ernest Rutherford, represented the Royal Society and made an appreciative oration by the graveside.'*

Chapter 16

Relativity

Part 2: Verifiable Predictions

The theory of relativity was developed at a time when there was no means of experimental verification. The concepts are so foreign to the 'everyday world' that it is easy to be sceptical. Perhaps the basic postulates are wrong, or maybe there is a flaw somewhere in the logic of the argument?

Einstein was convinced that his theory was not simply a mathematical fantasy, but a true description of the basic laws of the universe. When asked by someone, 'Professor Einstein, how do you know that you theories are true?', he replied: *'All I can say is, if they are not true, then the Lord has missed a wonderful opportunity.'*

The great achievement of Einstein's work was that it did not end just with a beautiful theory, too good to be overlooked by God. There were concrete predictions, which many years later were to be tested by experiment. In order to experience time dilation space travellers would have to travel near the speed of light. The example of the astronaut's cigar is certainly science fiction but there *are* elementary particles in cosmic radiation which reach us with speeds in the range of about $0.9\,c$ to $0.999\cdots c$, where the time dilation factor becomes significant. Studies of cosmic rays in the 1950s provided evidence that time dilation was a very real phenomenon.

Einstein's best-known prediction is the equivalence of mass and energy according to the equation $E = mc^2$. Confirmation of this equivalence came in 1942 with the production of the first nuclear chain reaction in an experiment by Enrico Fermi's team in a squash court in Chicago. Dramatically and tragically, there followed nuclear bomb explosions in 1945. Mass had been transformed into enormous amounts of energy. We can now produce such energy in controlled fashion in nuclear reactors and we also know that nuclear reactions are the source of the heat and light of the sun.

The relation $E = mc^2$ appears to be completely unrelated to the principles of special relativity. It seems quite mysterious how it could follow from arguments involving symmetries of space and time! In this chapter we try to follow the chain of logic step by step. Once the way has been pointed out, a rudimentary knowledge of mechanics is all that is required to go from the principles of relativity to mass–energy equivalence. The existence of nuclear energy is indeed a very convincing 'proof of the pudding'!

16.1 Time dilation

16.1.1 *Time dilation in action*

The earth is constantly bombarded by *cosmic radiation*, composed mostly of protons of high energy which come to us from other parts of the universe. Luckily, or perhaps by design, the earth's atmosphere shields us from most of this radiation. The protons interact with the air in the upper atmosphere, causing a nuclear reaction in which new particles of matter may be created. A chain of processes is initiated, the end products of which include particles called muons, which are extremely penetrating, and few of them are absorbed by the atmosphere. They are also unstable and exist with a characteristic lifetime of about two millionths of a second.

Introducing the muon:

The muon $\mu^{\pm}$ mass = 207 m_e charge = $\pm e$

Properties: weakly interacting
Mean life = 2×10^{-6} s decays $\mu^{\pm} \rightarrow e^{\pm} + \nu + \nu$
Created at top of atmosphere, observed at ground level
Sometimes called the 'heavy electron'

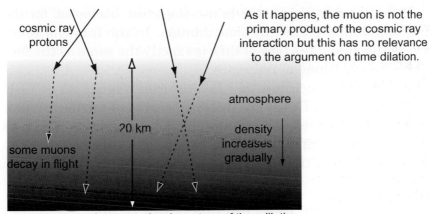

many muons arrive at sea level courtesy of time dilation

Figure 16.1 Cosmic ray interactions.

16.1.2 *Living on borrowed time?*

Most muons are created about 20,000 m above sea level. Their mean life is known to be about 2×10^{-6} s, so that they would be expected to travel about 600 m before they decay, assuming that they travel at almost the speed of light.

How then can so many of them reach sea level, at a distance of 20 km below the point at which they were created? Statistically, the chance of an unstable particle surviving over 30 times its mean life is of the order of $e^{-30} = 10^{-13}$ — a million times smaller than the chance of winning a national lottery!

Experimentally, muons are seen to arrive at sea level in large numbers. Somehow, these travelling muons seem to have discovered the secret of extremely long life!

The secret lies in the fact that, as seen from our frame of reference, the mean lifetime and thus the travel time available before decay have been expanded by a factor of the order of 30: the γ factor. To us the muons appear to live much longer when they are travelling than when they are at rest — a concrete verification of time dilation.

It is probably inaccurate to use the word 'borrowed' for the time apparently gained by time dilation. In the frame of reference of the muon, the flow of time is exactly the same as the flow of time in our frame of reference.

In modern accelerator laboratories such as CERN (Switzerland) and Brookhaven (USA), beams of muons are now produced to order, travelling in a circular storage ring at speeds close to the speed of light. Time dilation keeps them 'alive' many times longer than their average lifetime when observed at rest.

16.2 $E = mc^2$, the most famous result of all

So far our discussion on relativity seems to have brought us no nearer to the ultimate relation, $E = mc^2$, for which Einstein is best known. The reason for this is that we have been dealing exclusively with *kinematics*, the study of motion without reference to force. We will now turn our attention to *dynamics*, the study which encompasses force, energy and momentum. In this context we are interested in how these *dynamical* variables appear in different frames of reference; in particular, frames which travel relative to one another at near the speed of light.

In a collision between two particles the forces of action and reaction are equal and give rise to a rate of change of momentum, but how is this rate of change affected by time dilation when speeds near the speed of light are involved? We start with a fascinating thought experiment of a collision of snooker balls from different relativistic time frames.

16.2.1 *Bringing energy into the picture*

Returning to the relationship $E = mc^2$. The principles of relativity deal with symmetry of inertial frames and with the constancy of the speed of light. It is not at all obvious how these lead to the equivalence of mass and energy. That *mass is a form of energy* becomes particularly relevant in the world of the *atomic nucleus*.

To find the connection between mass and energy we must first look at *momentum* and *energy* from different frames of reference. Let us consider momentum first. The following is an imaginary experiment.

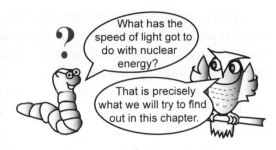

16.2.2 *Conservation of momentum — a thought experiment with snooker balls*

There are two snooker players, one in a railway station, and the other in a passing train. Each strikes a cue ball as in the diagram below. The balls are identical and each player strikes his ball with an identical speed in his own reference frame. The experiment is carefully arranged so that the two balls collide at an open window, just as the train passes the station, then bounce back as in the diagram. It is a head-on collision, so the cue balls bounce straight back along their original paths.

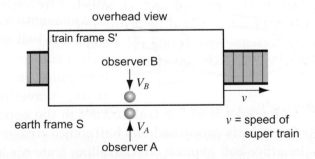

Figure 16.2 An absurd kind of snooker. The diagram combines the two points of view: inside the train as seen in frame S', and outside the train as seen in frame S.

There is complete *symmetry* in the situation. Each player is equally entitled to his point of view. Each considers himself to be stationary in his own frame of reference, and if the cue ball bounces straight back to one player, exactly the same thing happens to the other cue ball for the other player. Incidentally, the train is a 'super-train' travelling at a speed comparable to the speed of light.

To put this discussion into perspective, one of the fastest trains in the world is the TGV (*Train à grande vitesse*); it set a world record speed of 515 km per *hour* in France in 1990. This is still a far cry from our super-train, which, in order to show relativistic effects, would have to travel at somewhere near half the speed of light, or 150,000 km per *second*, about 10,000 times faster than the TGV on its record-breaking run.

This argument is well described by the German word '*Gedankenexperiment*', an experiment performed in one's imagination. The result also comes from the resources of the mind.

The collision is perfectly elastic. The two balls collide 'head-on' and each goes back exactly along its path in the frame of reference of the player who struck it.

To observer A it appears that his own ball returns straight back with reversed velocity and momentum, whereas the other ball has struck only a glancing blow and continues on with the train, travelling from left to right. Observer B in the train comes to the opposite conclusion. As far as he is concerned, *his* ball comes straight back to him, and the other ball glances off, travelling from his left to his right. The situation is symmetrical whether we are using classical or relativistic laws.

Snooker quality control laboratory

Let us assume that at an earlier stage each observer had the opportunity to examine the two snooker balls, side by side in his own laboratory. They were satisfied that the two balls were identical. In particular, their masses were equal.

16.2.3 *Interacting with another time frame*

Our thought experiment involves a super-train travelling near the speed of light. There is a collision between two masses from different time frames. This is something Newton had never envisaged! The laws of dynamics involve time as a basic parameter and we must analyse, in our minds, what new considerations come into play in such a unique situation!

Each observer will notice something peculiar. Each uses his own clock, which measures *proper time*. Time in the other frame appears to run more slowly (remember, the astronaut's cigar lasts longer, and his movements appear slower). Here the other snooker player appears

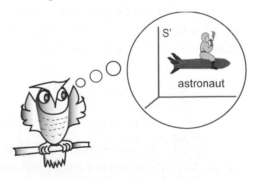

S'

astronaut

Remember the astronaut!

more lethargic, his movements seem slower as viewed from outside his rest frame, and he appears to strike the ball more slowly.

Figure 16.3 illustrates an apparent inconsistency in the dynamics of the collision. For each observer the *perpendicular component of the velocity of the other player's ball is smaller*.

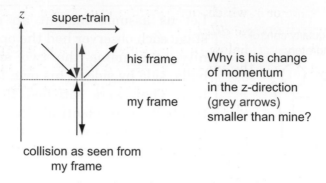

Figure 16.3 Apparent momentum imbalance.

Each observer will infer a momentum imbalance, the change in momentum of his own ball being larger. (Each observer remembers that the two balls have the same mass.)

> We must assume that the law of conservation of momentum holds at every stage and in all frames of reference. If this were not the case, one could identify a 'privileged' frame in which momentum is conserved, as opposed to other frames in which this is not the case. This would contradict the theory of relativity.

An alternative way to express the inconsistency:

According to Newton's third law, action and reaction are equal. But the active and reactive forces are each equal to the rate of change in momentum of the bodies on which they act. This rate of change is slower in the other frame because of time dilation.

16.2.4 *Momentum from another frame of reference*

The two sets of observations can be made consistent by multiplying the classical expression for the momentum of the other ball by the factor γ, which *compensates exactly for the decrease of velocity due to time dilation.*

As will be seen below in Equation (16.2), the juxtaposition of this factor with the mass term can be interpreted as saying that the *mass* of the particle in the other frame *increases* by the relativistic gamma factor,

$$\gamma = \frac{1}{\sqrt{1 - \dfrac{v^2}{c^2}}} \qquad (v \text{ is the speed of the train*})$$

We can define the *rest mass* of an object m_0 as the mass as measured in *its own frame of reference (rest frame)*. Once the object is moving relative to the observer, we of course no longer observe it in its rest frame.

For small velocities, we can use Newton's definition of momentum $p = \text{const} \times v$, where the constant is defined as the *inertial mass.*

In summary, we can define rest mass m_0 by

$$\textit{Classical momentum } p = m_0 v \quad \text{as} \quad v \to 0 \qquad (16.1)$$

If momentum is to be conserved in all frames of reference, then it must be defined as:

$$\boxed{p = \gamma m_0 v = mv} \quad \text{Relativistic momentum formula} \quad (16.2)$$

16.2.5 *A new look at the concept of mass*

Mass is not an invariant quantity. Viewed from another frame of reference, the mass of a particle is always greater than its 'rest mass' m_0 measured in its own frame of reference. *A moving particle has greater mass.*

---

* If we want to be very strict, we can replace v by V_R, the resultant velocity of the ball in the train.

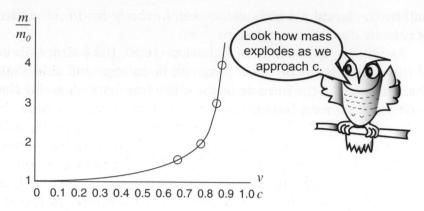

Figure 16.4 Relativistic mass as a function of velocity.

Relativistic mass = γm_0 = mass of an object moving with speed v relative to an observer. m_0 is the mass when the object is at rest relative to the observer.

As the speed approaches c the relativistic mass approaches infinity. If you have mass, you cannot go with the speed of light, not to mention 'faster than light'.

16.2.6 *The relativistic formula for momentum*

The concept of frames of reference gives us an insight into the fundamental symmetry of the laws of nature. Once the concept has been established and the relativistic formula has been derived, we can use it in *any* convenient frame of reference. We can take away the speeding train and simply consider a glancing collision between a snooker ball travelling at high speed from left to right, and another ball travelling in a direction perpendicular to the first ball. (Since we happen to be inhabitants of the earth, the rest frame of the earth is normally the most convenient.)

Even though the high speed component of the first ball is left to right, i.e. *in the x direction*, the observer will find that the *y components* of the velocity of the two balls A and B are *not equal* in magnitude either before or after the collision. If he uses the classical formula for momentum ($p = m_0 v$) he will find an apparent momentum imbalance. The situation is rectified as before, by using the relativistic formula $p = \gamma m_0 v$. The mass of the 'glancing' ball appears bigger because it has *a large component in the x direction* and therefore its *resultant velocity is bigger*. The fact that the main component of v is 'sideways', i.e. in the x direction, is of no consequence. Remember, the gamma factor is independent of direction. *Things which are moving have a greater mass than they had when they were stationary.*

Once we apply the relativistic formula for momentum to *both* of the balls A and B, the momentum will balance exactly in every direction.

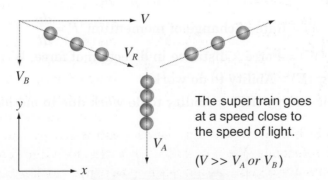

Figure 16.5 The collision as seen from the earth frame of reference.

$$\text{Momentum of A} = \frac{m_0 v_A}{\sqrt{1 - \dfrac{v_A^2}{c^2}}} = m_0 \gamma v_A \qquad (16.3)$$

$$\text{Momentum of B} = \frac{m_0 v_R}{\sqrt{1 - \dfrac{v_R^2}{c^2}}} = m_0 \gamma v_R \qquad (16.4)$$

(m_0 = rest mass of each ball)

16.2.7 *Energy in different frames of reference*

We now turn our attention to the concept of energy, which in classical Newtonian physics is related to work done.

Classical physics

How we calculate kinetic energy

Reminder of classical definitions:

Momentum (p) = mass × velocity $p = mv$

Force (F) = Rate of change of momentum $F = \dfrac{dp}{dt}$

Work (W) = Force × distance in direction of force, $W = Fd$

Energy (E) = Ability to do work

Kinetic energy (KE) = Ability to do work due to motion

Classical expression for kinetic energy

Force = rate of change of momentum

Kinetic energy T = work done in accelerating a mass m from rest to velocity V

= Force × distance in direction of force

$$T = \int F dx = \int \frac{d}{dt}(mv)dx$$

$$= \int mdvv = m\int_0^V vdv$$

$$= m\left[\frac{v^2}{2}\right]_0^V$$

$$\boxed{T = \frac{1}{2}mV^2}$$

16.2.8 *High energy particle accelerators*

At CERN the original accelerator was the *Proton Synchrotron (PS)*, built in 1959. It was at that time the highest energy accelerator in the world, with 200 magnets positioned around a ring of diameter 200 m. It accelerated protons up to a maximum energy of 28 GeV.

The *Super Proton Synchrotron (SPS)*, completed in 1976, had 1000 magnets around a ring 2.2 km in diameter, located in an underground tunnel. At the time the SPS was built it was also the largest accelerator in the world, with a maximum proton energy of 450 GeV.

At the time of writing (2008) the CERN complex consists of a series of machines with increasingly high energies, injecting the beam each time into the next one, which takes over to bring the beam to an even higher energy, and so on. The flagship of the complex, scheduled for completion at the end of the year, will be the Large Hadron Collider (LHC), which will produce colliding beams of protons, each of energy 7 TeV (7000 GeV).

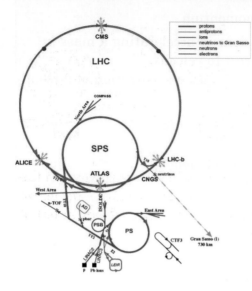

The first step in the process, the original pre-injector of the Proton Synchrotron was a Cockroft–Walton accelerator developed when they first split the atomic nucleus in 1932.

Courtesy of An Post,
Irish Post Office.

The cost of a little extra speed

As far as *speed* is concerned, the work done to accelerate a particle nearer and nearer to the speed of light soon reaches a point of rapidly diminishing return. The speed of 400 GeV protons emerging from the Super Synchrotron is 0.99999753 c, compared to the protons from the original 28 GeV machine, which travel at 0.999362 c. Building the new giant accelerator resulted in an increase in speed of just 0.006%! The γ factor comes into its own in a very practical way in terms of euros or dollars or Swiss francs, in the cost of the accelerators!

However it is energy and not speed which is important, in the quest to investigate the innermost structure of matter or, as we shall see in the next chapter, to make new kinds of matter out of energy.

Calculating the relativistic energy

The classical calculation is of course perfectly valid at 'normal' velocities, when the gamma factor can be taken as equal to 1. But now, as the mass increases, the force required for a given acceleration rapidly becomes greater with increasing velocity, and each increment of the work integral has to be multiplied by the value of γ appropriate to that instant.

In Appendix 16.1 we insert into the integral the relativistic expression for momentum $p = \gamma m_0 v$. Since γ is a function of v, the integration, while somewhat more difficult than in the classical case, can still be done by standard methods, and gives the total work done as

$$W = (m - m_0)\, c^2 = \text{kinetic energy} \qquad (16.5)$$

Work done on a particle = (change in its mass) $\times c^2$

By doing work we create mass

The classical concept that work done is invested into energy, be it kinetic energy, potential energy or heat energy, still holds, but has to be expanded to include mass energy.

For an object in empty space, with no force fields of any kind, there are no potential hills to climb, and only kinetic energy comes into play. Kinetic energy can be defined in a given frame of reference, as the difference between the energy of an object at rest and its energy when in motion.

From (16.5) Kinetic energy $T = mc^2 - m_0c^2$

$$= \text{Total energy} - \text{rest energy}$$

$$(16.6)$$

Einstein's famous relationship $E = mc^2$ gives the total energy, which includes kinetic and rest energy taken together. The rest energy is one of the great new concepts introduced by the theory of relativity. Every particle of matter possesses energy, purely by virtue of its existence.

Mass is a form of energy.

Relativistic and classical kinetic energy

The expression 16.6 appears to bear no resemblance to the classical formula for kinetic energy. There seems to be no reference to the speed of the object. We can show however that the relativistic expression can easily be reduced to the classical form for velocities small compared to the speed of light.

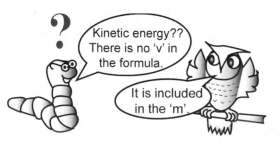

Relativistic formula $KE = mc^2 - m_0c^2$

$$= (\gamma - 1)m_0c^2 = \left[\frac{1}{\sqrt{1 - \dfrac{v^2}{c^2}}} - 1\right]m_0c^2$$

$$= \left(1 + \frac{v^2}{2c^2} + \frac{3v^4}{8c^4} + \cdots - 1\right)m_0c^2$$

$$= \frac{m_0v^2}{2} + \frac{3m_0v^4}{8c^2} + \cdots \qquad (16.7)$$

$$= \frac{m_0v^2}{2} \qquad \text{where } v \ll c$$

The classical formula is an approximation to the relativistic formula when v is small compared with the speed of light.

How many joules of energy in a unit mass?

The relationship $E = mc^2$ also gives the rate of exchange from one form of energy to the other. In SI units the energy contained in one unit of mass (1 kg) equals 9×10^{16} units of energy (joules).
You get a lot of energy for a tiny amount of mass.
Rest mass energy must not be confused with chemical energy of matter, such as the energy obtained from burning coal. Chemical energy is smaller by a large factor. Mass energy is present whenever mass exists, but is generally securely locked into the atomic nucleus and released only under very special circumstances in reactions which transform the atomic nucleus.

Example

What is the energy contained in a grain of sand of mass = 0.001 g (= 10^{-3} kg)? Assuming that it were possible to convert the grain of sand completely into energy, how far, using that energy, could you drive a car against an average resistive force of 1000 newtons?

$$E = mc^2 = 10^{-6} \text{ kg} \times 9 \times 10^{16} = 9 \times 10^{10} \text{ joules}$$

$$E = \text{Force} \times \text{distance} = F \times s$$

$$s = \frac{9 \times 10^{10}}{1000} = 90,000 \text{ km} \quad \text{i.e. twice around the world.}$$

In practice, we can never change mass completely into energy. In nuclear reactions, only a small fraction of mass is turned into energy.

16.2.9 *Nuclear structure*

In the classical world it is self-evident that the mass of any structure is equal to the sum of the masses of its parts. So, for example, *mass of the house = sum of the masses of bricks, mortar, wood, glass and even the coat of paint.* There is no need for mortar to hold the atomic nucleus together. It is composed of

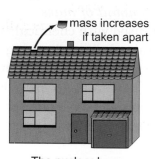

mass increases
if taken apart

The nuclear house

protons and neutrons, which are held together by the strong nuclear force — a strange kind of mortar, which manifests itself in that but the sum of the masses of the protons and neutrons is greater than the mass of the whole nucleus. If we want to take it apart, we must supply energy to create that extra mass. The percentage mass difference is very small, but because of the very unfavourable rate of exchange of mass for energy, it is enough to make it very difficult to take the nucleus apart.

The table below gives a listing of the masses of the constituents, and the total masses of some elements at the bottom of the periodic table:

Nuclear (rest) masses

Name		Symbol	Mass (a.m.u.)†	Sum of parts	Difference
Neutron	O	n	1.008987		
Proton	⊕	p	1.008145		
Deuteron	O⊕	d	2.014741	2.017132	0.002391
Triton	OOO	t	3.016997	3.026119	0.009122
Helium^3	O⊕⊕	He^3	3.016977	3.025277	0.008300
Helium^4	OO⊕⊕	He^4	4.003879	4.034264	0.030385

The accounts in the table above do not balance. The whole is not equal to the sum of its parts.

For example: The sum of the masses of the constituents of stable helium

$$He^4 = 2 \times 1.008987 + 2 \times 1.008145 = 4.034264$$

The helium nucleus has a smaller mass than the sum of the masses of two neutrons and two protons. The deficit is 0.030385 a.m.u. $= 28.3$ MeV/c^2, or 0.76% of the total mass.

The *mass deficit*, Δm, expressed in energy units, is called the *binding energy* of the structure.

$$\text{Binding energy} = (\Delta m)c^2 \qquad (16.8)$$

16.2.10 *Nuclear fusion — nature's way of powering the sun*

The sun is powered by nuclear energy which should last for another 10 billion years. The basic reaction involves making helium out of deuterons. The mass difference is converted into kinetic energy of the reaction products which appears as heat and

---

† 1 atomic mass unit (a.m.u.) $= 1.66 \times 10^{-27}$ kg $= 931.5$ MeV/c^2.

creates a temperature of the order of 10^9 degrees Celsius inside the core of the sun. At this temperature more deuterons are driven together, thus sustaining a *thermonuclear fusion reaction*.

We can calculate the energy released when two neutrons form a helium nucleus, as follows:

$d + d \rightarrow He^3 + n$

Mass 'lost', $\Delta m = 0.003518$ a.m.u.
$\qquad\qquad\quad = 5.84 \times 10^{-30}\,kg$

Energy $= (\Delta m)c^2 = 5.64 \times 10^{-30} \times 9 \times 10^{16}$
$\qquad\qquad\quad = 5.25 \times 10^{-13}$ J

Energy $= 3.3$ MeV

16.2.11 *Nuclear fission*

A fission reaction occurs when the nucleus of a heavy element such as uranium is broken into two or more fragments. An example is a neutron-induced reaction such as

$$neutron + U^{238} \Rightarrow Ba^{145} + Kr^{94}$$

In fission the sum of masses of the parts is *less* than the mass of the whole, and energy is released. Moreover the *fission fragments*, in this case nuclear isotopes of barium and krypton, are neutron-rich and highly unstable, and almost instantaneously eject two or three neutrons. These neutrons in turn can strike other uranium nuclei, causing more fissions. U^{238} is a relatively stable nucleus and fission will occur only by an energetic neutron, so that the chain reaction will not be sustained. However, the isotope U^{235} is much less stable; if a *critical mass* of U^{235} is brought together a stray cosmic ray neutron can initiate an immediate chain reaction, such as the explosion of an atomic bomb.

16.3 The steps from symmetry to nuclear energy

A synopsis

All inertial frames of reference are equivalent.	Symmetry
⇓	
The speed of light is the same for all observers.	Symmetry
⇓	
Time is relative. Things move more slowly in the 'other frame'.	Time dilation
⇓	
To balance momentum their mass must be greater.	Momentum conservation
⇓	
Work must be done to increase mass.	Work makes mass
⇓	
$E = mc^2$	Mass is a form of energy

16.4 Working with relativity

16.4.1 *The 'recoiling gun' revisited*

In elementary textbooks, one of the first examples of conservation of momentum is the 'recoiling gun'.

From the conservation of momentum, the momentum of the recoiling gun is equal and opposite to the momentum of the bullet.

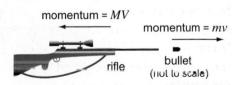

Classical calculation

Mass is a measure of resistance to change of motion, or *inertia*. There is only 'one kind' of mass (It so happens that gravitational attraction is proportional to inertial mass.)

$$MV = mv$$

$$\Rightarrow \qquad \frac{v}{V} = \frac{M}{m} \qquad\qquad (16.9)$$

The speeds of the bullet and the gun recoil are inversely proportional to their masses.

For example, if the mass of the gun is 40 times the mass of the bullet, the initial speed of the bullet will be 40 times that of the recoil.

$$\frac{\text{speed of the bullet}}{\text{recoil speed of the gun}} = \frac{\text{mass of the gun}}{\text{mass of the bullet}}$$

When the classical formula fails

We can see that this formula will not work at very high velocities. For example, if we were to have a 'super-gun' recoiling with a speed of $V = 0.5\ c$, the bullet would, according to the classical calculation, have a speed of $20\ c$ — in breach of all the principles of special relativity!

Relativistic calculation for the imaginary super-gun:

$$\gamma_M M_0 V = \gamma_m m_0 v$$

$$\Rightarrow \frac{v}{V} = \frac{M_0}{m_0}\,\frac{\gamma_M}{\gamma_m} \qquad\qquad (16.10)$$

Example

A super-gun fires a bullet with speed $0.999\,c$ and recoils with speed $0.5\,c$. Calculate the ratio of the mass of the gun to the mass of the bullet, and compare it to the ratio of the speed of the bullet to the recoil speed of the gun.

Going back to Table 15.1 we find

Velocity v	Gamma factor γ
$0.5\,c$	1.15
$0.999\,c$	22.3

Inserting values for the velocities and the γ factor into Equation (16.10),

$$\frac{M_0}{m_0} = \frac{\dfrac{v}{V}}{\dfrac{\gamma_M}{\gamma_m}} = \frac{\dfrac{0.999}{0.5}}{\dfrac{1.15}{22.3}}$$

$$\Rightarrow \frac{M_0}{m_0} = 38.744$$

The ratio of the rest masses gun to bullet turns out to be about 40, and yet the ratio of their speeds is just less than 2.

16.4.2 *Radioactive decay*

A relativistic 'gun and bullet'

Radioactive *alpha decay* presents us with real-life examples of 'gun and bullet' situations in which the decay of one unstable nucleus to a second nucleus and an alpha particle (He^4), which then travel in opposite directions. The recoiling rifle is the second nucleus and the bullet is the alpha particle. (There are just two particles in the final state, after the decay has taken place — see Appendix 16.3.)

A typical example is the decay of radium:

$$_{88}\text{Ra}^{226} \rightarrow {}_{86}\text{Rn}^{222} + \alpha$$

The alpha particle has an energy of 4.8 MeV, which corresponds to a speed of about 0.03 *c*, too small for any noticeable relativistic effects.

However, in *beta decay* the 'bullet' is an electron, which has a mass about 7000 times smaller than that of the alpha particle, and consequently much greater speed. The problem here is that the 'gun' is more like a shotgun, emitting not one 'bullet' but two. The electron is accompanied by a neutrino, which takes up some of the momentum. For example, in the beta decay of bismuth the reaction is

$$_{83}\mathrm{Bi}^{210} \rightarrow {}_{84}\mathrm{Po}^{210} + e^- + \bar{\nu}$$

The energy and momentum of the decay products is shared between the electron and the neutrino in a fashion which is not predetermined. The maximum possible energy of the electron, the so-called *end point energy*, is 1.16 MeV, where the electron

$P_{Po} = 1.591$ MeV/c $P_e = 1.591$ MeV/c

$P_\nu = 0$

is taking all of the available energy, and the neutrino gets none. For that special case, we again have a bullet + recoil gun situation in beta decay.

The kinematics calculation is given in Appendix 16.3, but some of the results are given below:

Answers:

Speed of electron $= 0.952\,c = 2.856 \times 10^5$ km/s
Speed of $_{83}\mathrm{Po}^{210}$ recoil $= 8 \times 10^{-6}\,c = 2.4$ km/s

Ratio of rest masses of 'gun' to 'bullet' $= 3.8 \times 10^5$ } No longer
Ratio of speeds 'bullet' to 'gun' $= 1.19 \times 10^5$ } equal.

A historical interlude: Albert Einstein (1879–1955)

Albert Einstein was born in Ulm, Germany, in 1879. He was slow to talk, but once he started he spoke in complete sentences. Even at the age of four he showed remark-able curiosity about the world and its physical laws. For instance, he was fascinated by the magnetic compass. Here was a needle 'isolated and unreachable, totally enclosed, caught in the grip of an incredible urge to point north'. He later called this his 'first wonder'.

EINSTEIN, Albert
Nobel Laureate PHYSICS 1921
© Nobelstiftelsen

In his autobiography he writes: *'At the age of 12 I experienced a second wonder. It was Euclid's proof that the intersections of the three altitudes of a triangle meet at a point. This, although by no means evident, could be proved with such certainty that any doubt appeared out of the question.'*

As a pupil Einstein was neither particularly good nor bad. He wrote of himself *'My main weakness was a poor memory for both words and texts.'* He sat in class looking bored, dreamy-eyed, and half-smiling, and this attitude was obviously not appreciated by those trying to teach him. He was actually asked to leave school when in fifth grade, his exasperated teacher stating: 'Your mere presence spoils the respect of the class for me.' His dislike of enforced teaching prompted him later to write: *'I believe that it would be possible to rob even a healthy beast of prey of its voraciousness if it were possible, with the aid of a whip, to force it to eat, even when*

(Continued)

550 Let There Be Light

not hungry, especially if the food handed out under such coercion were selected accordingly.'

He was modest, friendly and unpretentious, with an awe of the beauty and logic of physical laws. He wrote: *'The most incomprehensible thing about the world is that it is comprehensible.'* At the same time he recognised the power of his own mind. When judging a scientific theory (his own or someone else's), he asked himself if he would have made the universe the same 'had he been God'!

The examination system for entrance to the prestigious Zurich Institute of Technology failed to uncover Einstein's potential. Mathematics and Physics presented no problems, but he failed Biology and Languages in 1895. He passed at his second attempt, in 1896. Even at the institute, he was not the ideal pupil. The normal curriculum bored him, so instead of going to lectures he spent his time in the Physics laboratory, or else reading advanced research papers. James Clerk Maxwell was one of his heroes; his predictions in electromagnetism were just the kind of logical reasoning which Einstein himself was to employ later in his theory of relativity.

Einstein was fortunate to have a good friend in Marcel Grossmann, (1878–1936) who was conscientious and took good notes. When the time came for examinations Einstein used these for cramming, and duly graduated in 1900 as a teacher of Mathematics and Physics, which gave the qualification for a temporary job in a secondary school. Grossmann remained Einstein's friend for life. They were to publish a number of papers together. It was Grossmann who supplied the mathematical expertise when Einstein was formulating the general theory of relativity in later years.

In 1902, Einstein obtained more stable employment at the Bern patent office, where he worked until 1909. He must have made a reasonable impression, as he was promoted in 1906 to 'Technical Expert, Second Class'. The job at the patent office could not have been too demanding, since he was able to do most of his original work on relativity in his spare time — and maybe also during office hours!

(Continued)

This was an exciting time in the history of physics. Max Planck had proposed the light quantum hypothesis in 1900. It took a couple of years to gain general acceptance, but when it did it was just the kind of thing to appeal to Einstein's sense of adventure. What Planck had just done was 'to wipe the slate clean of preconceived prejudices', to quote the words to be used by Einstein some years later concerning his ideas on space and time.

Further milestones:

1903: Married Mileva Maric; 2 sons, born 1904 and 1910
1909: Associate Professor of Theoretical Physics, Zurich
1914: Professor at Kaiser Wilhelm Institute, Berlin
1919: Divorced from Mileva
1919: Married Elsa
1921: Nobel Prize
1932: Professorship at Princeton

Einstein's famous paper on the special theory of relativity was published in 1905 and was entitled 'On the Electrodynamics of Moving Bodies'. The title reflects his deep insight into Maxwell's theory of electromagnetism. Maxwell's work involved the motion of electric charges and the propagation of electromagnetic waves. Einstein realised that motion was relative, and that it was necessary to consider the description of the phenomena from different frames of reference.

Einstein received the 1921 Nobel Prize for Physics for 'the discovery of the law of the photoelectric effect'. It is remarkable that the great man wrote a number of papers during the first years of the century, any one of which could have been deserving of the prize. In his paper published in 1905, he extended Planck's concept of the quantum of energy to a concept of 'atoms of light'. It took many years for this concept to be accepted by the physics community. One may surmise that the Nobel committee were nervous about awarding him the prize for the theory of relativity since it was at that time not well established, and there was some controversy as to the originators of the principles of the special theory.

(Continued)

Einstein had a great sense of humour. Writing about the time when his visa application to the US was opposed by a group of women who claimed he was 'a communist at heart', he said: *'Never before had I experienced from the fair sex such energetic rejection, or if I had then certainly not from so many at once.'*

In a letter from an unknown correspondent, Einstein was asked: 'Because of gravity a person is sometimes upright on the spherical earth, sometimes upside down, sometimes at left angles, sometimes at right angles. Would it be that while upside down, people do foolish things, like falling in love?' Einstein replied: *'Falling in love is not at all the most stupid thing that people do, and gravitation cannot be held responsible for it.'*

We also find in Einstein's writings serious thoughts about the universe and religion. In response to a question from a Sunday school in New York City, 'Do scientists pray?' he said: *'Everyone who is seriously involved in the pursuit of science becomes convinced that a Spirit is manifest in the Laws of the Universe, vastly superior to that of man, in the face of which we, with our modest prowess, must feel humble. In this way the pursuit of science leads to a religious feeling of a special sort, which is indeed quite different from the religiosity of someone more naive.'* In another letter to the National Conference of Christians and Jews in the US, he wrote: *'If the believers of the present day religions would earnestly try to think and act in the spirit of the* founders *of these religions, then no hostility on the basis of religion would exist among the followers of the different faiths'.*

In 1932, Einstein accepted an arrangement at the University of Princeton by which he was to spend five months a year there and the remaining seven months in Berlin. Shortly afterwards the Nazis came to power, and what was intended only as a visit became a permanent post. He became a US citizen in 1940.

In a letter dated 2 August 1939, Einstein told President Franklin Roosevelt that experimental work by Frédéne and Irene Joliot in France and Enrico Fermi and Leó Szilard in America had shown that

(Continued)

it might be possible to set up a nuclear chain reaction in uranium. It was conceivable that the vast amounts of power released could be used for *'extremely powerful bombs of a new type'*. He also mentioned that Germany had stopped the sale of uranium from Czechoslovakian mines which had just been taken over. Roosevelt replied in October that he would set up a board consisting of the head of the Bureau of Standards and chosen representatives from the Army and Navy to investigate the possibilities of Einstein's suggestion — a prompt reply but hardly one to indicate the extreme urgency of the situation.

Einstein wrote again, both in March and in April 1940, to stress that interest in uranium had intensified in Germany and that the Kaiser Wilhelm Institute had been taken over by the government to carry out highly secret research on uranium.

He was not to know that by then the US government had realised the potential of nuclear power. The Manhattan Project to develop an atomic bomb was about to be launched. Einstein, however, was considered a 'left-leaning political activist' and a 'security risk' and those who had been engaged on the project were forbidden to consult with him.

Einstein wrote a fourth and final letter, dated 25 March 1945. However, it failed to reach President Roosevelt before his death on 12 April 1945. Obviously, Einstein was not aware that the atomic bomb project was already well under way. The first test explosion was in fact carried out in New Mexico in July 1945.

When Einstein heard of the bombing of Nagasaki in August 1945, he was shattered. His equation $E = mc^2$ had explained where nuclear energy came from, but not how to make a bomb. With great sadness he now saw that a terrible weapon of war had been constructed based on his discovery of the mind.

In the last decade of his life, Einstein dedicated himself to the cause of nuclear disarmament. *'The war is won,'* he said in December 1945, *'but the peace is not.'* One of the last acts of his life was to sign a manifesto urging all nations to give up nuclear weapons.

Albert Einstein died on April 18, 1955 at Princeton.

Appendix 16.1 Deriving the relativistic formula for kinetic energy $T = mc^2 - m_0c^2$

Relativistic expression for momentum, $p = \gamma m_0 v$*

Force = Rate of change of momentum

$$F = \frac{d}{dt}(\gamma m_0 v)$$

$$= m_0\gamma\frac{dv}{dt} + m_0 v\frac{d\gamma}{dt}$$

It's all happening in this frame.

$$\frac{d\gamma}{dt} = \frac{d\gamma}{dv}\frac{dv}{dt}$$

$$\frac{d\gamma}{dv} = \frac{d}{dv}\left(1-\frac{v^2}{c^2}\right)^{-\frac{1}{2}} = \frac{v}{c^2}\left(1-\frac{v^2}{c^2}\right)^{-\frac{3}{2}} = \frac{\gamma^3 v}{c^2} \tag{A.1}$$

$$\frac{d\gamma}{dt} = \frac{\gamma^3 v}{c^2}\frac{dv}{dt}$$

$$F = m_0\gamma\frac{dv}{dt} + m_0 v\frac{\gamma^3 v}{c^2}\frac{dv}{dt} = m_0\gamma\frac{dv}{dt}\left(1+\frac{v^2}{c^2}\gamma^2\right)$$

$$\gamma^2 = \frac{1}{1-\frac{v^2}{c^2}} \Rightarrow \frac{v^2}{c^2} = 1 - \frac{1}{\gamma^2}$$

$$\Rightarrow F = m_0\gamma\frac{dv}{dt}\left(1+\gamma^2\left(1-\frac{1}{\gamma^2}\right)\right) = m_0\gamma^3\frac{dv}{dt}$$

$$= m_0\frac{c^2}{v}\frac{d\gamma}{dt} \text{ (using Equation (A.1))}$$

---

* Note that the relativistic diagram is identical to that for the classical calculation.

$$T = \int_0^x F dx = m_0 \int_0^t F v dt = m_0 c^2 \int_0^\gamma d\gamma = m_0 c^2 (\gamma - 1)$$

(where the upper limit of the integral, $\gamma = \dfrac{1}{\sqrt{1 - \dfrac{V^2}{c^2}}}$)

$$T = mc^2 - m_0 c^2$$

The final step:

Total energy, mc^2 = kinetic energy + rest energy

The expression for 'invariant mass' — a useful relationship:

Total Energy $\quad E = mc^2 = \gamma m_0 c^2$

Momentum $\quad\quad p = \gamma m_0 v$

$$E^2 - p^2 c^2 = \gamma^2 m_0^2 c^2 (c^2 - v^2)$$

$$= \gamma^2 m_0^2 c^4 \left(1 - \frac{v^2}{c^2} \right)$$

$$= (m_0 c^2)^2 = (\text{Rest mass})^2$$

Since the rest mass is a constant, the expression $E^2 - p^2 c^2$ is invariant, i.e. the same for all observers.

The above can be written as

Reminds me of Pythagoras

$$E^2 = p_x^2 c^2 + p_y^2 c^2 + p_z^2 c^2 + m_0^2 c^4 \quad\quad (A.2)$$

This is another example of an invariant quantity such as the invariant interval in space time (p. 506).

Appendix 16.2 Dimensions and units of energy

Dimensions of energy

Dimensions of *force* = Mass × acceleration
$$= [M][L][T]^{-2}$$

Dimensions of *work* = Dimensions of energy
$$= \text{force} \times \text{distance}$$
$$= [M][L]^2[T]^{-2}$$

Dimensions of *rest energy* = m_0c^2
$$= [M][L][T]^{-1}[L][T]^{-1}$$
$$= [M][L]^2[T]^{-2}$$

As one would expect, *the dimensions of classical energy and of mass–energy are identical.*

SI units (based on the mks system)

In using *SI units*, relativistic energy will come out in joules and the units and dimensions of momentum, speed, kinetic energy and mass will be consistent.

Example:

Mass of proton $m_p = 1.6726 \times 10^{-27}$ kg
Rest energy of proton $E = m_0c^2 = 1.5053$ J

The electron-volt system

It is more usual and numerically less cumbersome to use units based on the *electron-volt*.

An electron-volt (eV) is the kinetic energy gained or lost by an electron (or any other particle with the same size charge) when it moves through a potential difference of 1 volt.

Since moving a charge of 1 coulomb through a potential of 1 volt takes 1 joule of work and the magnitude of the charge of electron is $e = 1.602$ coulombs,

$$\Rightarrow \quad 1 \text{ eV} = 1.602 \times 10^{-19} \text{ joules}$$

(which will give it a speed of 593 km/s, which is 0.2% of the speed of light).

$$10^3 \text{ eV} = 1 \text{ keV (kilo electron-volt)}$$
$$10^3 \text{ keV} = 1 \text{ MeV (mega electron-volt)}$$
$$10^3 \text{ McV} = 1 \text{ GeV (giga electron-volt)}$$

We can now use electron-volts/c^2 as a unit of mass. (In practice bigger units such as MeV/c^2 and GeV/c^2 are more commonly used in high energy particle physics.)

Expressed in these units the mass and rest mass energy of a particle are numerically equal.

> *Example:*
>
> Mass of proton $m_p = 938.3$ MeV/c^2
> Rest energy of proton $E = m_0 c^2 = 938.3$ MeV

Similarly, since the dimensions of momentum are mass × velocity, it is convenient to express momentum in units of eV/c [in this way the units are self-consistent in Equation (A.2)].

A summary of units of mass, energy and momentum in the electron-volt system:

Physical quantity	Units
Mass, m_0	MeV/c^2
Rest energy, $E = m_0 c^2$	MeV
Momentum, p	MeV/c

The advantage of using these units is that the factor c effectively disappears in relativistic equations such as (A.2), as is shown in the example in Appendix 16.3.

Useful relationships using electron-volt energy units:

Relativistic mass $m = \gamma m_0$

Total energy $E^2 = m_0{}^2 + p^2$

$$\gamma = \frac{E}{m} = \frac{1}{\sqrt{1 - \beta^2}} \qquad \left(\beta = \frac{v}{c} \right)$$

Momentum $p = m\beta = \gamma m_0 \beta$

Appendix 16.3 Relativistic analysis of the beta decay of bismuth 210

$$_{83}\mathrm{Bi}^{210} \rightarrow {}_{84}\mathrm{Po}^{210} + \mathrm{e}^- + \nu \qquad \mathrm{e}^- \text{ end-point energy} = 1.16 \text{ MeV}$$

A feature of beta decay is that there are three particles present after the decay: the emitted electron, the recoil nucleus and a *neutrino* (a 'three-body final state'). As a result, the energy of any of these decay products is not unique. We consider here the special case in which the neutrino takes a negligible part of the energy available. The process can then be considered as a simple projectile-recoil 'two-body final state', with the electron at its maximum possible energy (the 'end-point energy' of the beta ray spectrum).

The electron and recoil nucleus go off with equal and opposite momenta:

$P_{Po} = 1.591 \text{ MeV/c} \qquad\qquad P_e = 1.591 \text{ MeV/c}$

$P_\nu = 0$

Rest energy of electron $= m_0 = 0.511$ MeV/c^2

Total energy of electron $=$ rest energy $+$ KE $= 0.5511 + 1.16$
$$= 1.671 \text{ MeV}/c^2$$

Gamma factor for electron $= \dfrac{\text{total energy}}{\text{rest energy}} = \dfrac{E}{m_0}$

$$= \dfrac{1.671}{0.511} = 3.27$$

(Momentum of electron)$^2 = E^2 - m_0^2 = 1.671^2 - 0.511^2$
$$= 2.792 - 0.261 = 2.531$$

Momentum of electron $= 1.591$ MeV/c

Speed of electron $= c\sqrt{1 - \dfrac{1}{\gamma^2}}$ $c = (1 - 0.3058^2)\, c = 0.952\, c$

$$= 2.856 \times 10^5 \text{ km/s}$$

Momentum $= \gamma m_0 v = 3.27 \times 0.511 \times 0.952$
$$= 1.591 \text{ MeV}/c \text{ (checks out!)}$$
$$= \text{Momentum of } _{84}\text{Po}^{210} \text{ recoil}$$

Rest energy of $_{84}\text{Po}^{210}$ recoil $= m_0 = 209.9829$ u
$$= 1.956 \times 10^5 \text{ MeV}/c^2$$

Total energy of $_{84}\text{Po}^{210}$ $= \sqrt{m_0^2 + p^2}$ $= 1.956 \times 10^5$ MeV/c^2

Gamma factor for $_{84}\text{Po}^{210} = 1.0000$

Speed of $_{84}\text{Po}^{210}$ recoil $= \dfrac{P}{m_0} = \dfrac{1.591}{1.956 \times 10^5} = 8.1 \times 10^{-6}\, c$

$$= 2.4 \text{ km/s}$$

Ratio of rest masses of 'gun' to 'bullet' $= 3.8 \times 10^5$

Ratio of speeds of 'bullet' to 'gun' $= 1.19 \times 10^5$

Chapter 17

Epilogue

Then there was light …

Many of the most fundamental phenomena in physics are related in some way or other to light and its properties. This has been the theme of the book. We concluded with Einstein's theory of special relativity, which on the basis of the universal constancy of the speed of light, leads to the equivalence of matter and energy. When Einstein derived his famous equation in his 'laboratory of the mind', he could hardly have imagined that within half a century it would be possible to build machines which could turn energy into matter at the precise rate of exchange which he had predicted.

When it became possible to create matter at particle accelerators, a window opened into a new world. Myriads of new fundamental particles were discovered. The exploration of the physical laws governing this microscopic, sub-nuclear world became a major and exciting goal. Experimental physicists from many countries pooled their skills and funding to build experiments of unprecedented size and complexity, driving themselves to the very edge of technological feasibility.

Einstein's words 'The most incomprehensible thing about the world is that it is comprehensible' applied equally to the new world. As theoretical physicists untangled the properties and the interrelations between particles, they found that by logical argument they could predict the existence of hitherto-unobserved particles.

The final chapter of our story of light and its place in the universe concludes with the prediction of the existence of a family of three particles — with masses about 100 times the mass of the proton — which act as the carriers of the weak nuclear force in a similar manner to the function of the light quantum which carries the electromagnetic force. It was a testimony to the high esteem in which the theorists were held when a colliding beam accelerator of unprecedented size and complexity was constructed to look for this *'heavy light'*.

17.1 Making matter out of energy

17.1.1 *Collisions make particles*

The equation $E = mc^2$ gives the rate of exchange when matter is converted into energy. It is well known that such conversion can occur, and the term 'nuclear energy' is now part of everyday language. What is less well known is that the opposite also occurs. Basically, if enough energy is concentrated at one point, matter may be created out of this energy. In practice this only happens at the nuclear or sub-nuclear level — when, for example, a proton of high energy collides with another proton or neutron.

In the everyday 'household' world, we know that if an object collides with another object, its energy must go somewhere. For example, if two cars travelling at 100 km/h collide, both cars will be broken into pieces, and fragments may be thrown in all directions. The kinetic energy of the cars is converted into the kinetic energy of flying debris, work is done in distorting metal, heat is generated, and the cars may even go on fire. For a plane which crashes at 1000 km/h the devastation is even greater. But what happens if a proton travelling at almost 300,000 km/s crashes into a stationary proton? Protons cannot be distorted or broken into fragments, so where does the energy go?

In this case matter appears, as if 'out of nothing', in the form of a jet of particles, mostly in the forward direction. These newly

born particles may
be the familiar con-
stituents of matter
such as protons and
neutrons (accompa-
nied by their anti-
particles), or 'new'
particles which are
unstable and decay
within a tiny frac-
tion of a second.

17.1.2 *Prediction and discovery of the π meson*

The repulsive electrostatic force between touching protons is more than enough to cause them to fly apart, yet protons together with neutrons form a tightly bound nuclear core. An even stronger force must hold them together. In 1935 Hideki Yukawa (1907–1981) published a paper entitled 'On the Interaction of Elementary Particles', in which he proposed the existence of a new particle which would provide that force. He proposed that it would play the role of an *exchange particle* analogous to the role of the photon in the electromagnetic interaction between charges. Yukawa's calculations predicted that this new particle should have a mass of the order of 200 times the mass of the electron (unlike the photon, which has zero mass). Yukawa was working in Japan at that time, virtually isolated from physicists in Europe and the US. The advent of World War II made this isolation even greater and there was little time or opportunity to build experiments to look for the particle.

A candidate for Yukawa's particle had been observed by Carl Anderson (1905–1991) and Seth Neddermeyer (1907–1988) in cosmic ray interactions in a cloud chamber as early as 1937, but it turned out later to be the wrong particle. Its mass was about right, but it lacked the most crucial property — namely, it did not interact strongly with protons and neutrons. How could the carrier of the nuclear force, exchanged between protons and neutrons, not interact strongly with them? (What they had in fact observed was the muon, which we already met in the previous chapter.)

Once the war was over, it was realised that Yukawa's particle was a key player in the mechanism of nuclear interactions, and the hunt began in earnest. In 1947 Cecil Powell (1903–1969) from the H.H. Wills Laboratory at the University of

Launching a balloon. *Courtesy of Physics Department, University of Bristol.*

Bristol launched a number of balloon flights carrying nuclear photographic emulsion detectors to the top of the atmosphere. Nuclear emulsion is basically a stack of sheets of photographic film pressed together to form a solid block of material which is highly sensitive to the passage of charged particles. As the particles pass through the material they create a latent image. When developed, the silver grains in the emulsion look, to quote Powell's words, 'like beads on an invisible string'

Hideki Yukawa and Cecil Powell. *Courtesy of Physics Department, University of Bristol.*

and form a three-dimensional picture which can be examined under a microscope at high magnification.

Success came quickly, with the observation of an event in which one of the outgoing tracks was identified as being due to a particle with characteristics matching those predicted by Yukawa. The track terminated as the particle came to rest in the emulsion, and a new track appeared as Yukawa's particle decayed into another new particle, later identified as a muon. As the mass of the Yukawa particle was intermediate between that of the electron and that of the proton, it was given the name π *meson* (or *pion*), derived from the Greek word μεσων ('middle').

Hideki Yukawa received the Nobel Prize for Physics in 1949 for the prediction of the π meson, and Cecil Powell in 1950 for its observation. Their discoveries marked the beginning of a new chapter in the history of physics. In the words of Powell, 'There was revealed a whole new world. It was as if suddenly we had broken into a walled orchard, where protected trees flourished and all kinds of exotic fruit had ripened in great profusion.'

Exposing particle detectors to cosmic rays was initially the only method of studying high energy interactions. However, within the next five years, accelerators built in the US, Europe and the Soviet Union were capable of producing proton beams of higher and higher energy. By aiming these protons into a target, matter could be created out of energy in a controlled manner. The gates to Powell's orchard of exotic fruit were open!

In addition to the known particles, large numbers of new particles, with different masses and properties, were found. The trail of exploration uncovered a wide variety of species in this *'fundamental particle zoo'*. The Big Bang of creation was being reproduced on a mini-scale.

17.1.3 *The forces between the particles*

Strong and weak nuclear force

Fundamental particles interact with each other via the *strong* and *weak* nuclear forces and the *electromagnetic* force (*gravitational* attraction is too weak to play a significant role). The strong force binds nucleons together in the atomic nucleus and is also responsible for the interaction mechanism through which new particles are created. The force has a typical reaction time of 10^{-22} s, which is approximately the time it takes a particle travelling near the speed of light to cross the diameter of a nucleus. Particles which are subject to the strong nuclear force are classified under the generic name of *hadrons* (from the Greek 'αδροσ, meaning 'robust'). The weak force which is responsible for radioactive beta decay is also responsible for the decay of relatively long-lived particles, such as the decay $\pi \to \mu + \nu$ with mean life 2.6×10^{-8} s. Both the strong and weak forces have a short range and do not extend beyond the nuclear diameter. The magnitude of the strong force exceeds that of the weak by a factor of about 10^{12}. To put this into perspective, let us assume that an ant can push with a force equivalent to a weight corresponding to 0.1 g. Compare that with a scrum of eight rugby players pushing with a total force equivalent to a weight of about 1000 kg. The ratio of these two forces is a factor of about 10^{7}, still far short of the ratio of the strong to the weak nuclear force. In fact the equivalent ratio is the force of about 100,000 scrums divided by the force of an ant!

Worker ant. *Courtesy of USDA APHISPPQ Archives.*

Wales vs Australia, Cardiff, November 2006.

Comparing forces exerted by ants and rugby players:

$$\frac{100,000 \text{ scrums}}{1 \text{ ant}} = \frac{\text{Strong nuclear force}}{\text{Weak nuclear force}}$$

There are some particles which are completely insensitive to the strong nuclear force. These are called *leptons*. The *photon* is also a member of this 'club of total immunity'.

Club of Immunity from the Strong Nuclear Force (members' list, 1964)				
Name	Photon	Neutrino	Electron	Muon
Symbol	γ	$\nu_e \nu_\mu$	e	μ
Mass	0	~0	m_e	$207\, m_e$

Electromagnetic force

Particles which are electrically charged also experience the *electromagnetic force*. Just as leptons are immune to the strong nuclear force, electrically neutral particles do not feel electric or magnetic forces. Electromagnetic forces are about 100 times weaker than the strong nuclear force, and are subject to the laws already well documented by Clerk Maxwell in the 19th century. Electric and magnetic fields are used to accelerate and guide the protons in accelerator rings. They are also used to separate and control secondary beams of new charged particles produced at the target.

17.1.4 *The laws of the world of fundamental particles*

The relevance of Einstein's theory of relativity is very obvious in the world of fundamental particles. The relation $E = mc^2$ governs the creation processes of all particles, the now as well as the old. When these new particles subsequently decay into other particles the energy released is governed by the same equation. The relationship is now so well established that units of mass and units of energy are interchangeable. Things move fast in this world; speeds close to the speed of light are the norm. Time dilation is a significant factor for nuclear and sub-nuclear particles in motion.

Einstein was not the only one who could look with satisfaction at phenomena in the sub-nuclear world. Max Planck's 'act of despair' and 'revolutionary hypothesis' was that light energy comes in quantised units. Niels Bohr extended the concept to quantised angular momentum of electron orbits. We now find that quantisation also is the norm rather than the exception in the particle world. Particles themselves have intrinsic angular momentum, which comes in quantised units. Its value is independent of where they are or what they are doing. Finally, the philosophy of quantum mechanics, developed at Bohr's Institute at Copenhagen, rules throughout. Processes are based on probability and not determinism.

The Heisenberg uncertainty principle comes to life in a very real way. When written in the form $\Delta E \Delta t = h/2\pi$ it can be interpreted as saying that Nature allows energy conservation to be violated by an amount ΔE provided that it takes place over a short enough time interval, Δt. The principle takes on very real significance in the case of particles which decay via the strong interaction and have very short lifetimes. Such particles, also known as *resonances,* have a lifespan of the order of 10^{-22} s, which means that they have an intrinsic uncertainty ΔE in mass energy. This uncertainty, called the *resonance width,* can be as much as 20%.

Dirac's prediction of the positron as the antiparticle to the electron extends to all members of the particle zoo. There is a rule which applies very strictly to some classes of particles, in that they cannot be created singly; they must be accompanied by an antiparticle. We cannot add an extra proton to the universe without balancing it with an antiproton, or make a neutron without an anti-neutron. This rule does not apply to mesons, any number of which can be created (provided that the usual conditions apply).

Neutrinos

The law of conservation of momentum is strictly observed in the sub-nuclear world. It had been invoked as far back as 1931 by

Wolfgang Pauli (1900–1958), when he was developing his theory of radioactive beta decay. He proposed a 'desperate remedy' to explain the experimental results of apparently missing momentum by postulating that it was carried away by a mysterious particle. He called it the 'little neutral one', or *neutrino*. The mysterious particle seemed to have neither mass nor charge and did not seem to interact at all. In fact it appeared to have no properties other than the ability to carry energy and momentum. Like the emperor's new clothes, the existence of the neutrino had to be taken on trust. Thirty years later interactions of neutrinos were seen at Los Alamos and subsequently at Brookhaven and CERN. In fact there was evidence for not just one, but two kinds of neutrino, which were labelled v_e and v_μ. Pauli was vindicated.

17.1.5 *Quarks*

Enjoying the fruit in the orchard of elementary particles was not without its problems. In 1934, there were protons, neutrons and electrons. These were the fundamental building blocks of matter. We had a seemingly simple model of what the universe was made of. The picture suddenly changed and became anything but simple. It seemed incomprehensible that Nature should have myriads of fundamental entities, with a whole spectrum of masses from zero to larger than the proton, which had no visible structure.

It was up to the theoretical physicists to find order in the midst of apparent chaos. Murray Gell-Mann (1929–), Yuval Ne'eman (1925–2006) and George Zweig (1937–) of the California Institute of Technology were convinced that the principles of symmetry must apply to the world of the very small just as they do in classical physics. They used an algebra originally developed in the 19th century by a Norwegian mathematician, Marius Sophus Lie (1842–1899), to group the particles into mathematical families. Technically the symmetry was called

SU(3) (a special unitary symmetry group of arrays of size 3×3). In 1963, Gell-Mann suggested a conceptually easier description in which the three abstract mathematical entities were identified as real objects, which he named '*quarks*'.

The quarks were given the somewhat peculiar names of u (up), d (down) and s (strange). Protons, neutrons and heavier particles are made of three quarks. So, for example, the formula for a proton is uud, and that for a neutron is udd. The antiproton and the antineutron consist of three corresponding antiquarks. Mesons are made up of one quark and one antiquark; for example, the constituents of the π^- meson are ūd.

> The quark model refers only to *hadrons*, the particles which are subject to the strong nuclear force. *Leptons*, such as the electron, the muon and the neutrino, have no sub-structure.

All normal matter is made up exclusively of u and d quarks. The s quark is a constituent of particles that were discovered in the early 1950s and called '*strange*' because their behaviour was not well understood at the time.

A total of 27 states can be constructed out of the 3 quarks. They are arranged as one decuplet, two octets and one singlet. Their properties corresponded well with properties of families of known particles, with one notable exception. The heaviest member of the decuplet, made from three s quarks, had not been observed. As one can imagine, a huge effort was made immediately to find the missing particle. Within a few months a new particle, named Ω^-, was found at Brookhaven and its properties slotted it very neatly into the vacant place in the SU(3) decuplet. It seemed that the problem was solved, and in 1969 Gell-Mann received the Nobel Prize for 'classifying elementary particles'.

Gell-Mann had the sound in his mind before the spelling of 'quark', which rhymes with 'stork'. In his occasional perusals of

James Joyce's Finnegans Wake he had come across the following passage:

> *'Three quarks for Muster Mark'!*
> *Sure he hasn't got much of a bark,*
> *And sure any he has*
> *It's all beside the mark*

Whatever about the pronunciation, the number 3 fitted nicely into his scheme, despite Joyce's obvious intent that it should rhyme with 'bark!'

The story of quarks and their literary connection was the product of the imagination of a physicist with wide interests and a sharp sense of humour. Whether quarks are real physical objects or abstract mathematical entities is not important; the essential thing is that the representation gives results and makes further predictions which agree with experiment.

17.1.6 *Charm*

In August 1964, a paper was published by James Bjorken (1956–) and Sheldon Lee Glashow (1932–), two American theorists working at the University of Copenhagen. It suggested the existence of a wider symmetry named SU(4), of which SU(3) was a sub-group. Despite the fact that it pointed towards a basic unification of weak and electromagnetic interactions, it attracted little attention at the time. This new symmetry implied that there should be a fourth quark. The authors even suggested the name '*charmed*' for the new quark, to quote their own words, 'as an expression of pleasure at a delightful idea'. They were, however, careful to point out that 'this model is vulnerable to rapid destruction by experimentalists'.

The rapid destruction of charm did not take place; far from it. Ten years later, there was a dramatic development when, on 11 November 1974, two experimental groups, one under

Samuel Ting (1936–) at Brookhaven and the other led by Burton Richter (1931–) at Stanford, working on completely different experiments, simultaneously announced the observation of a new resonant state, which decayed approximately 10^{-19} s after creation. It had a mass more than three times the mass of the proton. The discovery of a new resonance in itself was nothing special, but in this case there was something very anomalous in the way it behaved. Resonant states decay in typically 10^{-22} s, but this particle lived about 1000 times longer than expected. Richter's paper concluded with the statement: 'It is difficult to understand how, without involving new quantum numbers or selection rules, a resonance which decays to hadrons could be so narrow.' In plain English: 'We have discovered some kind of matter which we have never seen before.' In 1976, the Nobel Prize was awarded jointly to Richter and Ting *for their pioneering work in the discovery of a heavy elementary particle of a new kind*.

Ting called the new particle J, while Richter chose the Greek letter ψ. The physics community diplomatically adopted the name J/ψ. More important than the name of the particle was its structure, which seemed most likely to be a combination of a charmed quark and a charmed antiquark.

 Just like their charge, the 'charm' of quark and anti-quark 'cancels', so J/ψ is said to possess *'hidden charm'*.

To make the existence of the charmed quark credible, it was necessary to find particles containing one charmed quark with one or more of the known quarks u, d and s. Throughout 1975 and 1976, experiments at accelerator laboratories on both sides of the Atlantic searched for what was called 'naked' charm. As thousands of bubble chamber pictures and billions of electronic counts were analysed at Stanford, CERN, Brookhaven and Fermilab, more and more evidence of charmed particles began to accumulate. This evidence, although convincing, was based on the characteristics of decay products and was of necessity indirect. The lifetime of charmed particles was expected to be about 10^{-12} s, which meant that even at the speed of light they would travel no further than a fraction of a millimetre before decay. This distance was far too short to resolve the points of creation and decay using the above techniques.

17.1.7 *The return of photographic emulsion*

In the 1970s, nuclear emulsions were very much 'old' technology. There were much more efficient ways of detecting subnuclear particles. The reaction products from every pulse of an accelerator could be monitored by electronic counters; particle tracks were made visible as they passed through spark chambers; bubble chamber photographs provided beautiful pictures of interactions *as they occurred.*

However, nuclear emulsions had one very important advantage over all other techniques at that time — their large resolving power. A length of 1 mm looks 'like a mile' under a microscope, while it could barely be resolved in any other form of detector at that time.

Eric Burhop (1911–1980) of University College London realised that it might be possible to re-construct the creation and subsequent decay of a charmed particle by combining the advantages of different experimental techniques. In emulsion it is difficult to find the interesting events of special interest, unless one knows exactly where to look. The complete scanning under microscope of a 10-litre stack might well take one observer 100 years to complete. There might only be one or two charm-producing interactions in a whole stack — trying to find them would be worse than looking for a needle in a haystack!

The task of finding the events would be much easier, if it were somehow possible to pinpoint them to within a small volume of the order of 1 cm^3. Burhop organised a collaboration of research groups from universities in Brussels, Dublin, London, Strasbourg and Rome in a combined effort to look for charm, by incorporating nuclear emulsions and spark chambers in a hybrid detector. Strasbourg had special expertise in 'wide-gap' spark chambers. These had much better directional resolution than conventional spark chambers, but were more difficult to construct and operate. They could be used to map the trajectories of product particles and project them back to their origin to within a volume of a few mm^3. The Rome group had a long tradition in both emulsion and electronics. Electronic circuits would trigger the spark chambers at the instant of an interesting interaction. L'Université Libre de Bruxelles, University College London and University College Dublin had participated in joint emulsion experiments over the previous 20 years.

Interactions involving high-energy neutrinos were theoretically the most efficient way of creating charmed particles. Neutrinos of course are uncharged, and weakly interacting. Nearly all neutrinos reaching us from the sun go right through the earth and continue onwards across the universe! In fact only about one neutrino in 100,000 will interact on its way from one side of the earth to the other. On the rare occasions when a neutrino interacts with a nucleus, there are practically no restrictions on the kind of matter which will be created out of energy. (When hadrons interact two bundles of quarks crash together, and 'ordinary' quarks are more likely to turn their energy into other 'ordinary' quarks.)

> Neutrinos they are very small.
> They have no charge and have no mass.
> And do not interact at all.
> The earth is just a silly ball?
> To them, through which they simply pass.
>
> *John Updike*

In January 1976, the equipment was brought across the Atlantic to the Fermilab accelerator in Batavia, near Chicago, and exposed to a neutrino beam of high energy and intensity over a period of three months. During that time the counters clicked 250 times to register that debris from what was presumed to be a neutrino interaction had come from the emulsion

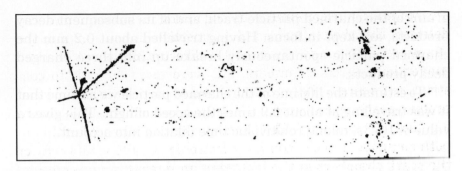

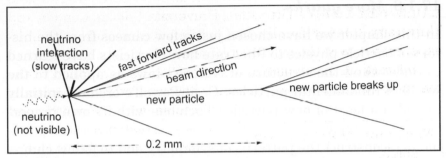

Figure 17.1 First picture of the track of a charmed particle. *Courtesy of European Emulsion Collaboration.*

stack. The emulsions were brought back, developed and shared out among the participating laboratories, and the search began in the areas reconstructed from tracks in the spark chambers. By November 1976, a total of 29 neutrino interactions had been located. One of these, found in Brussels, was of great interest and is reproduced in Figure 17.1.

The incoming neutrino interacts with the nucleus of an atom in nuclear emulsion (probably silver or bromine). Fragments of the nucleus scatter in all directions (short, dark tracks). Particles created out of energy shoot forwards towards the right at a speed close to the speed of light, leaving light tracks of minimum ionisation.

Nuclear emulsion is a three-dimensional medium, and in the photograph most of the tracks go out of focus almost immediately. Figure 17.1 is a mosaic of photographs in which the grains of the charmed particle track, and of its subsequent decay products, are kept in focus. Having travelled about 0.2 mm the charmed particle spontaneously breaks up into three charged decay products.

To estimate the lifetime of the charmed particle we assume that it was travelling at about 0.9 times the speed of light. This gives a value of 10^{-12} s, taking relativistic time dilation into account.

17.1.8 *More quarks*

In this chapter we have chosen just a few cameos from the history of particle physics to illustrate how physicists have learned to understand the structure of Nature's laws as applied to the world of fundamental particles. Cecil Powell's orchard initially provided a feast of new particles, each one with its own properties and characteristics. As these were studied, it became possible to reconstruct the patterns of symmetry and to follow chains of reasoning, leading to the prediction of as-yet-undiscovered species. The charmed quark was just one example.

From charm to beauty

The charmed quark was an important piece in the jigsaw puzzle of the world of the very small. Its place was in the section for *hadrons*, particles which respond to the strong nuclear force. The fit was perfect, but the jigsaw was not yet complete. There were two more vacant places. In 1977, Leon Lederman (1922–) and his team at Fermilab discovered a resonant state called Υ, with a mass about three times larger than that of J/ψ. This state behaved in the same sort of anomalous way as J/ψ; its lifetime was much too long. The most likely explanation was that Υ was a combination of a new quark, rather picturesquely called the '*beauty*' (or b) quark, and its antiquark. Υ has 'hidden beauty' in the same sort of way that J/ψ has 'hidden charm'.

The one remaining vacant place was not filled for some time. The particle designated to fill the space was given a name before it was seen. Eventually, in 1995, the existence of the t (or *'truth'*) quark was confirmed by two huge experimental groups working independently at Fermilab.

17.1.9 *The innermost shell of the nuclear onion*

While the above experiments were searching for quarks, others were looking into the genealogy of the lepton family. In 1975, Martin L. Perl (1921–) and his team at the Stanford Linear Accelerator published the discovery of a sibling to the electron and the muon — a super-heavy electron. This particle, almost 3500 times heavier than the electron, was given the name *'Tau'*. It was expected that each of the three leptons would have an accompanying neutrino. As of the end of the 20th century, this completed the list of the fundamental, indivisible building blocks of matter:

6 quarks $\begin{pmatrix} u \\ d \end{pmatrix} \begin{pmatrix} c \\ s \end{pmatrix} \begin{pmatrix} t \\ b \end{pmatrix}$	6 leptons $\begin{pmatrix} e \\ \nu_e \end{pmatrix} \begin{pmatrix} \mu \\ \nu_\mu \end{pmatrix} \begin{pmatrix} \tau \\ \nu_\tau \end{pmatrix}$
Building blocks of the strongly interacting particles	Present members of the Club of Immunity from the Strong Force

17.2 A unified theory of weak and electromagnetic forces

17.2.1 *The role of light as the carrier of the electromagnetic force*

Maxwell's famous paper, published in 1864, linked electricity and magnetism in one unified theory. This great unifying step forwards brought with it a bonus, in that it identified light as an electromagnetic wave which could carry energy and information.

It also presented light as a communicator, a carrier of force between electrically charged particles.

In the 20th century the picture changed, but not in essence. Light is sometimes represented by an electromagnetic wave and at other times it behaves like a particle. The electric force is transmitted in quantum 'lumps' as electrons emit photons and absorb photons. According to the theory of quantum electrodynamics, all of chemistry can be reduced to these fundamental processes. The theory is basic and elegant and makes the most accurate predictions known to science — which is as much as one can demand from any theory! Quantum electrodynamics, however, has nothing to say about nuclear forces.

17.2.2 *Unification — the long hard road*

In 1979, Sheldon Glashow (1932–), Abdus Salaam (1926–1996) and Steven Weinberg (1933–) shared the Nobel Prize for 'contributions to the theory of the unified weak and electromagnetic interaction between elementary particles...'. The theory had taken a long time to mature. From the late 1950s, the Nobel laureates had been working sometimes together and sometimes independently, seeking a single theory which would encompass both weak and electromagnetic forces. The goal was to describe the weak and electromagnetic interactions in terms of a mathematical symmetry. As Salaam later pointed out, 'The electroweak picture was being pieced together with almost exasperating slowness. At first we were trying to gauge the wrong symmetry.'

As one technical problem was overcome, a new one appeared. These excerpts from the lectures at the Nobel ceremony give an indication of the difficulties in formulating a coherent new theory:

Weinberg: *'At that point we had reached the right solution, but to the wrong problem.'*

Glashow: *'I had "solved" the problem of strangeness changing neutral currents by suppressing all neutral currents; the baby was lost with the bath water.'*

There were other problems, one of them being that the formalism was 'not renormalizable,' which meant that the calculations encountered infinities. This latter problem was solved in 1971 by the Dutch physicist Gerard 't Hooft (1946–). In the words of theorist Sidney Coleman (1937–2007): 't Hooft's work turned the Weinberg–Salaam frog into an enchanted prince.'

The end of the road

Despite the difficulties in formulating the electroweak theory, experimental evidence was emerging that the theory was on the right track. The theory had predicted that there should exist *'weak neutral currents'* — weak interactions in which the particles involved do not exchange electric charge. No such interactions had ever been seen until 1973, when a neutral current neutrino interaction was observed at CERN. In 1978, an experiment at the Stanford linear accelerator showed some delicate interference effects between electromagnetic and weak forces, as predicted by the theory. The time was right to look for the jewel in the crown — *the carrier of the weak force.*

17.2.3 *The heavy photon*

If the weak and electromagnetic interactions are different manifestations of the same basic law of nature, there must be a particle which acts as a carrier of the weak force, just as the photon carries the electromagnetic force. The electroweak theory made specific predictions about this particle. First of all there should be not one but three particles: one with positive charge, one with negative charge and one neutral. Secondly, unlike the 'ordinary' photon, they would have mass. Not only that, but they would be heavier than any other fundamental particle — almost 100 times heavier then the proton. In fact, the electroweak force carriers had accurately predicted masses and even names before they had ever been observed.

Two examples of interactions mediated by a heavy photon are given below:

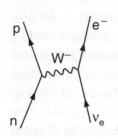

Neutrino–neutron interaction

A neutrino interacts with a neutron by exchanging a virtual W^- particle. The neutron changes into a proton while the neutrino changes into an electron.

$$v_e + n \rightarrow p + e^-$$

To make a *real* W particle an amount of energy equal to its mass energy has to be supplied. When acting as the communicator of the weak nuclear force the W^- is said to be in a *virtual* state, because it has to 'borrow' this energy courtesy of the Heisenberg uncertainty principle $\Delta E \Delta t = h/2\pi$. In the case of the W particle, the amount of borrowed energy (ΔE) is very large, so the 'borrowed' time (Δt) is very short. As a consequence the range of the weak interaction is short even on a nuclear scale.

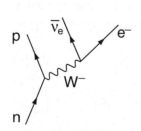

Neutron beta decay

A free neutron decays by emitting a W^-, which in turn decays into an electron and an antineutrino.

$$n \rightarrow p + e^- + \overline{v}_e$$

The Nobel committee showed great confidence in the electroweak theory by awarding the 1979 Nobel Prize to Glashow, Salaam and Weinberg, without waiting for proof that the W and Z particles really existed. Perhaps they felt that it would not be possible to create such heavy particles in the foreseeable future since no accelerators at that time could produce enough energy. There was one possibility, but it appeared technically almost impossible to realise. If, instead of accelerated protons colliding with stationary protons in a fixed target, they could be made to move in opposite directions and crash into one another, the available energy would be much larger.

The difficulty of crashing nuclear size particles into one another head-on can be illustrated by the following analogy. Let us represent the target proton by a disc of diameter $\sim 10^{-15}$ m and make the reasonable assumption that we can focus protons from an accelerator into an area of 1 mm^2. Then the chance of hitting the target proton is only about 1 in 10^{11}. This is roughly the same as the chance of hitting a grain of sand with a second grain of sand in an area equal to the size of France!

Despite such apparent difficulties the technology for building colliding electron and positron beams had already been developed in Frascati in Italy and at Stanford in the US. At CERN, intersecting storage rings had been built to give proton–proton collisions. However, the energy of the beams in all these machines was much too low to produce either a W or a Z particle.

In 1978, the scientific policy committee at CERN showed the same confidence as the Nobel committee by deciding to adapt the giant 7 km ring at the 400 GeV Super Proton Synchrotron (SPS) to accommodate antiprotons circulating in the opposite direction to the protons. The project was quite overwhelming in its complexity.

Anti-protons had first to be created at the target of the 28 GeV Proton Synchrotron (PS). This was the original accelerator around which the CERN laboratory had grown from the 1950s. Now CERN's original prima ballerina became simply a member of the supporting cast. The antiprotons were then sent to another ring, the Anti-proton Accumulator (AA), in which they circled repeatedly to build up energy.

Forty hours and 30,000 pulses later, when about 6×10^{11} protons had been accumulated, they were sent the 'wrong way' into the one-way traffic of protons in the SPS. The two beams were very delicately focused to ensure that the particles collided in designated cross-over areas where complex detector systems were set up to detect and measure the products of the very high energy proton/antiproton collisions. The analysis of the enormous amount of debris created by the collision of 3 quarks with 3 antiquarks was in itself a monumental task faced by two collaborations, composed of over 20 research teams from Europe and the US.

There is no room in this epilogue to do justice to the full story of the experiment. Suffice it to say that in 1982 the first W particle was observed, followed by the Z^0 in 1983. Carlo Rubbia (1934–), the driving force behind the experiment, and Simon van der Meer (1925–), the architect of the beam focusing which made it possible, received the Nobel Prize in 1984. Glashow, Salaam and Weinberg, from 'the class of 1979', were special guests at the ceremony as the King of Sweden presented the prize *'for the discovery of the massive short-lived W particle and Z particle'*.

17.2.4 *The full circle*

The story of light has turned full circle. We started with light as a communicator, enabling us to see, to get information from the rest of the universe, to investigate the distant past. We discussed the central part played by light in the basic theories of quantum mechanics and relativity, and finally, its role as the communicator of the electric force which is responsible for the

Table 17.1 The heavy photons predicted by the Weinberg–Glashow–Salam model.

Name	Predicted mass (GeV/c^2)	Observed mass (GeV/c^2)
$W^{\pm}$	82 ± 2.4	80.9 ± 1.5
Z^0	94 ± 2.5	95.6 ± 1.4

Mass of proton in energy units = 0.938 GeV/c^2.

chemistry of atoms and molecules. Ingenious application of these electromagnetic forces has made it possible to build accelerators and detectors which probe even further into the nuclear forces and constituents. Here we have seen another form of light, acting as the communicator of the weak nuclear force.

We can hardly do better than to conclude by presenting a portion of Salaam's acceptance speech at the Nobel Prize ceremony in 1979. He famously quoted the following verses from *Al-Quran* (*Sura Al-Mulk*, 3&4):

Thou seest not, in the creation of the
    All-merciful any imperfection,
Return thy gaze, seest thou any fis-
    sure. Then Return thy gaze, again
    and again.
Thy gaze, Comes back to thee daz-
    zled, aweary.

He then proceeded to say: 'This in effect is the faith of all physicists; the deeper we seek, the more is our wonder excited, the more is the dazzlement for our gaze.'

Abdus Salaam (1926–1996).
*Courtesy of the Pakistan
Academy of Science.*

Index

585